SURFACES AND SUPERPOSITION

ERNEST W. ADAMS

Foreword by Patrick Suppes

Center for the Study of
Language and Information
Stanford, California

Center for the Study of Language and Information
Leland Stanford Junior University
Printed in the United States
05 04 03 02 01   5 4 3 2 1

Library of Congress Cataloging-in-Publication Data

Adams, Ernest W. (Ernest Wilcox), 1926–
Surfaces and superposition / Ernest W. Adams.
p. cm.
Includes bibliographical references.

ISBN 1-57586-279-4 (acid-free paper)
ISBN 1-57586-280-8 (pbk. : acid-free paper)

1. Geometry. I. Title.
QA445.A25 2001
516—dc21 2001025290
CIP

∞ The acid-free paper used in this book meets the minimum requirements of the American National Standard for Information Sciences—Permanence of Paper for Printed Library Materials, ANSI Z39.48-1984.

CSLI was founded early in 1983 by researchers from Stanford University, SRI International, and Xerox PARC to further research and development of integrated theories of language, information, and computation. CSLI headquarters and CSLI Publications are located on the campus of Stanford University.

CSLI Publications reports new developments in the study of language, information, and computation. In addition to lecture notes, our publications include monographs, working papers, revised dissertations, and conference proceedings. Our aim is to make new results, ideas, and approaches available as quickly as possible. Please visit our web site at
http://cslipublications.stanford.edu/
for comments on this and other titles, as well as for changes and corrections by the author and publisher.

To my brother,
an archaeologist for all seasons,
William Y. Adams

Contents

Foreword

In the enormous literature on the foundations of geometry, Ernest Adams' *Surfaces and Superposition* occupies a unique position. Using standard mathematical concepts and theorems from point set topology, as well as psychological theories and results from experiments on perception, Adams presents an extended philosophical analysis of applications of topology to our ordinary experience of surfaces in this insightful work. It is the use of results from mathematics and psychology, while remaining strictly philosophical in method and style, that accounts for this book's remarkable depth.

From another standpoint, the present work constitutes an extended commentary on certain essential features of Euclidean geometry, especially as systematized in Euclid's *Elements*. Using Euclid's controversial concept of the superposition of two figures, Adams—with his natural empirical bent—rightly points out that, although superposition can be avoided in axiomatic developments of geometry, it is an essential aspect of the standard use of measuring instruments in geometry. What he has to say here is just as important and original as his application of topological concepts.

There is much else to be remarked on, but I will restrict myself to another major thread of the book. This is the detailed analysis in Chapter 4 of the concept of physical abstraction and the many interwoven remarks in later chapters on the problems of giving a proper philosophical account of the nature of abstraction. I am not sure I fully understand all aspects of his viewpoint, but I certainly agree with Adams' criticisms of what Russell, Whitehead and some other prominent philosophers have had to say on this difficult subject.

Many readers, not just philosophers, will find much of interest to think about and reflect on in *Surfaces and Superposition*. The unsolved problems that are carefully delineated in numerous passages constitute a worthy challenge to those interested in deepening the foundations of geometry.

I have known Ernest Adams for more than forty years, in fact since we were young men together at Stanford in the 1950s. We are separated by only a few years in age. Almost from the beginning, when he was writing his dissertation and I was his thesis advisor, I really thought of him as a younger colleague rather than a student. He listened intently to comments and suggestions, always holding on to his own independent and original ways of dealing with philosophical ideas.

Over the years he has contributed to a surprisingly wide range of topics in philosophy. His doctoral thesis, a contribution to the philosophy of physics, was on the foundations of rigid body mechanics, remnants of which appear in the chapter on rigid frames and in other parts of the book. Beginning in 1956, a year after he completed his thesis, and over the next decade or so, he wrote a number of papers on utility theory and game theory. Utility theory, especially, is close to the general theory of measurement, and already by the mid 1960s he was publishing in this area as well. Some of his skepticism about overly formal approaches to conceptual problems came out early in his criticisms of representational theories of measurement, and echoes are to be found in the present work. This interest in measurement theory has persisted through the decades and may be found in his 1992 book, *Archaeological Typology and Practical Reality*, on the problems of classification in archeology, written with his brother William.

Still another important strand of Adams' philosophical work began in the 1960s. This is his probabilistic approach to the analysis of conditional sentences in ordinary language, summarized in his 1975 book on the logic of conditionals. It is the body of work, to which he is still contributing, that is probably best known among philosophers.

During the same decade, in 1961, Adams published his first article on geometry whose title, "The Empirical Foundations of Elementary Geometry," already announces the overall theme of the present work. The many long and substantive articles he has published since on the foundations of geometry attest to the permanence of his interest. What is special about the current book is that it has the feeling of a work on which the author has been reflecting for much of his career. The many details and leisurely asides, often as historical footnotes, provide the signs to those of us who know him well that surfaces and superposition engages him at the deepest level.

Patrick Suppes

Preface

The originally planned subtitle of this work, "Field notes on some geometrical excavations," now seems too flippant for so august a production as a book. Nevertheless, an archaeological metaphor is in some ways more fitting for this book's origins, aims, and methods than the 'foundational' or 'architectural' metaphor that is common for studies in the 'conceptual underpinnings' of scientific disciplines. Let us contrast these metaphors, beginning with the architectural one.

There has been a tendency in the past half century to picture theories, and geometrical theories in particular, as formal structures whose superstructures of 'theorems' are based on—deduced from—'primitive' postulates and concepts that are not themselves deduced from or defined in terms of anything else within the structure. The mathematician concerned with deductive structure generally ignores the non-deductive, 'primitive' side of his subject, but the epistemologist seeking justifications for its postulates and the meaning of the concepts in terms of which they are formulated looks for an 'interpretation', total or otherwise, that as it were 'bolts them to a foundation in the world of facts'.

There are important disagreements as to the nature of the foundations, as to the facts and the bolts that tie theories to them, but in most cases they seem to have been pictured somewhat like the flat cement slabs that commonly serve as foundations for small structures. Still metaphorically, they are formed of undifferentiated 'matter', or perhaps Tractarian 'objects' all having the same properties and among which the same relations apply. Moreover, the discovery that Geometry has an empirical element makes it natural to suppose that the facts underlying it have sensory components—sense data, appearances, or the like. Thus, Russell's wonderful "The relation of sense data to physics," in which the 'bolts' that connect the sense data, first to Geometry and then to Physics, are set-theoretical constructions. Reichenbach's 'coordinating definitions' are less radical but nevertheless related efforts to connect geometrical concepts to concrete things, which in his case were rigid rods. In each case there was just one kind of concrete 'fact', part of an 'experiential foundation' secure enough to bear the weight of the theory built upon it, and entities of the theory are bolted to it by connections of specified kinds—constructions in one case and coordinative definitions in the other.

This 'concrete slab picture' was the way I conceived things when I first began these studies, now some 40 years ago. This guided my first efforts to arrive at an understanding of the rôle of Geometry in our dealings with the world, but through repeated efforts to work out the details I came to regard this as misguided, and to adopt a rather different attitude, one that is associated with a different metaphor, namely an archaeological one. This has something in common with the architectural metaphor.

An architectural foundation supports the building that is built on it, but it is *hidden* below or perhaps around the superstructure, which is naked to the eye. And, the fact that the substructure, the 'underpinning', is hidden makes it difficult to map or 'reconstruct'. It must be 'unearthed', as it were, and doing that requires an effort quite unlike that involved in building its superstructure. And here is the important point: one should not begin by assuming that the underpinning is like one kind of common architectural foundation, much less that it has a single level with a well defined outline. Conceivably it could be like the roots of a mountain, that stretch down to an indeterminate depth and outward an indeterminate distance. Or it could be like an archaeological site that has been occupied and built and rebuilt over for centuries or millennia. In any case, the archaeologist must excavate, perhaps layer by layer, perhaps by 'trenching', and he cannot predict with any certainty in advance what he will find. Moreover, what he will find is not likely to have the order and completeness that is found in the buildings that are above ground, since even if the buildings that were there in earlier times had this order, their crumbled remains may only consist of fragments piled up helter-skelter. Then the archaeologist will begin by compiling field notes, noting the locations of the fragments, their forms, probable uses, etc., which may provide the raw materials on which will be based the report that he eventually hopes to write.

My 'conceptual excavations' have not been unlike archaeological ones. I dug and trenched in ways that initially seemed plausible, and found myself sifting through 'conceptual debris' that I thought might be relevant to my inquiry. I might even have proceeded à la Foucault or Derrida, and dug into purely linguistic materials, were it not that a hint from Euclid pointed me in another direction. That is the relation of geometrical concepts and propositions to practical procedures. Thus, Proposition 1 of Book I, which is the first demonstrated proposition in *The Elements*, is "On a given finite straight line to construct an equilateral triangle" (Heath, 1925, p. 241), which sounds and is like a culinary recipe. Moreover, the very concepts in terms of which geometrical propositions are stated are characterized in constructive terms, e.g., parallel lines are defined as " ... lines ... that do not meet *when they are produced* ..." (Definition 23 of Book I). Of course, following Plato, modern mathematicians reject constructibility in the concrete realm and replace it by existence in an abstract or 'ideal' one, but when we seek 'Geometry's application to the world' we may not be ill-advised to examine its technological origins. This seems to me to have enormous implications.

One is that if 'the facts underlying Geometry' have to do with how to do things, then they transcend the static realm of 'eternal existents' to which it is generally thought that its truths should correspond. The other has to do with the things that Geometry instructs us how to construct. What are those? Take the equilateral triangle of Proposition 1. It is something that is drawn on a plane surface. And what can be drawn on a surface but a drawing? It is true that this brings us face to face with the traditional question concerning the function of geometrical drawings. Are they as Plato had it, only 'reminders' of the things that Geometry is really about, which are in a timeless, ideal realm which has no necessary connection to the realm of the senses (and is this really so different from current popular mathematical philosophies)? I think that if there is any truth in this, it is only in a special sense.

I am going to suppose that while geometrical theory can be regarded as being about *idealized* drawings on *idealized* surfaces, its propositions inform us about real drawings on real surfaces in somewhat the same way that the proposition that adding 1 gallon to 1 gallon of water yields 2 gallons of water informs us concerning the result of pouring the contents of a 1 gallon can of water into a container already containing 1 gallon of it, even though neither container contains exactly 1 gallon, and what they contain isn't pure H_2O. Of course, even granting that Geometry can be used to inform us about practical drawings (e.g., Euclid's directions do tell us how to draw real triangles on fairly flat surfaces that are fair approximations of ideal equilateral triangles), the obvious question is: how is it that this discipline plays such an important rôle in modern science? Well, at present that lies close to the horizon in my 'excavations'. What lies more immediately to hand is at a lower and possibly more primitive level: what are real diagrams, which idealized geometrical theory supposedly informs us of? This question is especially acute because our findings suggest that real diagrams are not the 'stuff of physical theory'. But I am going to suggest that being clearer about the 'stuff of diagrams' helps to clarify their relation not only to geometrical theory but to physical theory as well. And, we also find more order than one might expect at this primitive level, that links it first to applied topology, and via that to Geometry. A brief word about that is suitable to this preface, if only as a word of warning.

Take a triangle drawn on a flattish surface, perhaps following the directions given in the demonstration of Proposition 1. A similar triangle appears below, though, significantly, what

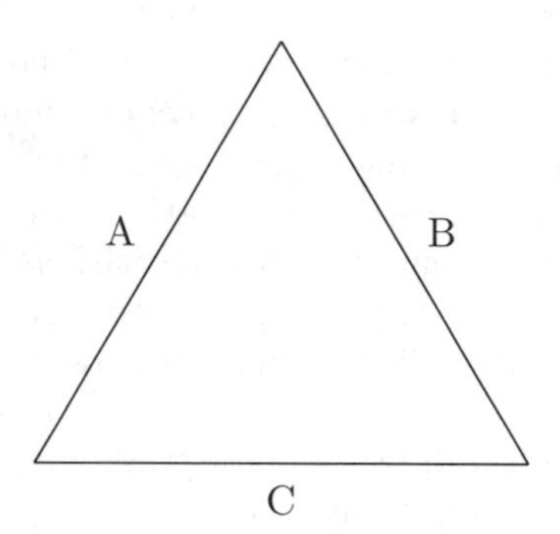

actually lies before the reader was printed and not drawn. In any case it is a figure that itself has features: three solid and fairly straight sides, and an inside and an outside. Moreover the sides have *inner* and *outer* edges that are themselves fairly straight, and the edges meet at points, or 'vertices'. Even putting aside the geometrical idea of straightness that this brief inventory includes, note the essentially 'proto-geometrical' ideas it involves: of edges, of insides and outsides, and of edges meeting at points. If the ideas of sides and edges enter seriously into geometrical theory, as they did in *The Elements* (cf. Proposition 7 of Book I), we must examine them in the same way Euclid did the crucial concept of parallelism. And that leads us to Topology, which at least in its origins as *Analysis Situs* (Leibniz, 1956, pp. 254–258) aimed to systematize and examine the logical interconnections among intuitions that underlie our 'ordinary usages' of terms like 'inside', 'outside', 'boundary', and so on. Two points of fundamental importance are connected with this.

One is that an intermediate objective of the theory to be developed here is to 'bolt' topological theory to a foundation in empirical facts and phenomena, beginning with the most fundamental concepts of Topology, that of an *open space*, or an open set. But the distinction between open and closed spaces, or at least open and non-open spaces, is problematic. The most obvious fact is that the distinction is not a metrical one, still less a *geo*metrical one. For example, the distance between the space inside the above triangle, which is naturally regarded as an open space, and the boundary of the triangle, which must be added to it in order to create a closed space, cannot be measured, nor can it be discerned by the eye or any other organ of sense. And yet the distinction is crucial to applied Topology.

This brings us to a rather radical speculation. To make the distinction in the case of the triangle, we imagine how it was drawn or otherwise came into being. For instance, if it was drawn or printed on white paper, then we conceive it to include its boundary points or edges, while its interior is 'background', whose boundary belongs properly to the triangle and not to the interior. On the other hand, if the paper had originally been black, and the white 'interior' had been printed on *that*, it would have included its boundary, and it would have individuated a closed set. Nor is the distinction between foreground and field entirely *ad hoc*, which brings us to the deepest level so far reached in our 'excavation'.

Take the sides of the triangle that are labelled 'A' and 'B'. How do we know that they come together at the top of the triangle; in fact, how do we 'identify' them at all? How do we know that the 'A' and the 'B' label distinct 'things', and are not placed adjacent to just one thing: the whole triangle? Our tentative answer is that we conceive the triangle as having been formed in a particular way, namely by drawing what we have been calling its 'sides' separately. Moreover, these imagined 'acts of construction' not only identify the things constructed, they also determine their relations of incidence, of touching or being separate. The details, some of which are discussed in section 2.3, are messy, but they bring out something of great importance. That is that not only are the fundamental concepts of applied Topology defined 'genetically', but so are the identities of the very objects to which they apply.[1]

Summing up what was said above, three levels have so far been encountered in our excavations, aspects of which will be discussed in what follows. These are: (1) that of *diagrams*, which Geometry provides 'recipes' for creating and about which it provides idealized information; (2) the analysis of topological aspects of the diagrams such as interiors and edges; and (3) 'genetic' characterizations of fundamental topological concepts, which are linked to the very idea of concrete 'thing'. Now, in what follows the discussions of (1) and (3), the diagram and genetic characterization levels, will remain largely at the 'field note' stage. The discussion of level (2), the analysis of topological aspects of diagrams,

[1]The fundamental link between Topology and physical identity should come as no surprise, given the connection of the latter to *continuity*, which is a topological concept.

will be more systematic. It will set forth a deductive theory whereby, starting from intersection relations among parts of diagrams like the sides of the triangle above, one arrives first at *points* of intersection, then at open spaces, and then at boundary concepts, dimension, and linearity, in each case describing 'operational tests' for determining, e.g., that the surfaces on which diagrams are produced are two-dimensional. This is the part of the discussion already mentioned, that aims to 'bolt abstract topological theory to a world of diagrammatic facts'. However, the reader is warned that this 'operational' theory turns out to be anything but 'elementary', and, like topological theory in its early days, it is still only partially worked out. Our excavation is anything but complete, even at its most carefully studied level.

And what about levels as yet hardly touched, including connections with *physical* geometry—i.e., with the geometrical concepts that enter into current physical theories? Obviously these theories are not concerned with diagrams; at best diagrams appear as 'representations' in physical papers and treatises. Well, a first, tentative step in that direction is taken by extending the account of diagrams and figures on the surfaces of bodies to *superpositions* of these bodies that result when the they are fitted together. This in turn leads in two directions: (1) to an account of *spatial measurement* carried out with the use of measuring rods that must be superimposed on objects in order to measure them, and (2) to an account of the *spaces* associated with actual and possible *frameworks* of fitted together bodies. However, simple as the idea of superposition may seem, the fact that superimposing one body on another may conceal the very diagrams that were 'unearthed' at a lower level gives rise to new difficulties not unlike ones that result when one level in an archaeological site has to be disturbed in order to get at a another one. Given this and other difficulties, our discussion of superposition and things that one might hope to account for in terms of it, such as spatial measurement and spaces, remains at the field note level. Still more so possible connections with geometrical concepts of physical theories, discussions of which, except for very cursory comments on the relation of 'superficial signs' to *matter*, are dismissed here as being peculiar to the theories in question.[2]

[2]In any event, given the state of modern Physics, it would seem unthinkable to attempt to account for the geometrical concepts that enter into its theories independently of the conceptions of *light* that they presuppose. And light is a very difficult concept, as one recognizes in the too often forgotten fact that we do not see images on the retinas of our eyes, much less the light that falls on them, and in Einstein's insight that we cannot follow a light signal. Given this and other difficulties, light and its properties are left entirely out of consideration here.

Concluding, I wish to acknowledge the invaluable help and encouragement in carrying out the studies reported here that I have received from individuals and institutions. My greatest debt is to Professor David Shwayder, who, in the early stages of this work, conducted with me seminars on Space and Time, during which views that evolved in one form or another into the *leitmotifs* that inform this study began to take shape. Moreover, Professor Shwayder has over the years rendered me the most valuable help that one writer can offer another, namely that of detailed, careful reading and criticism of the latter's efforts. Chapters 16–21 of Professor Shwayder's *Statement and Referent*, 1992, sketch his own views on topics treated here. More detailed expositions have still to appear, which, together with increasing divergence from my own views, is why these ideas are not discussed in this book. Professor Vann McGee is another with whom I have had profitable discussions ranging over a very wide range of topics, including ones touching on space and time. And, I cannot fail to mention my debt to Professor Patrick Suppes who has helped and encouraged me in almost all my efforts ever since graduate school. I should also like to express my deepest appreciation to Dikran Karagueuzian, the Director of CSLI Publications, as well as members of his staff who helped in the preparation of this book, Christine Sosa, Lauri Kanerva, and Max Etchemendy, who prepared the book's many diagrams.

As to institutional assistance, I am grateful to the Guggenheim Foundation, the Institute for Advanced Study in the Behavioral Sciences, and the National Science Foundation for grants and fellowships in support of the present studies. The present volume is the too long delayed and all too partial recompense for this assistance.

Part I

Preliminaries

1

Characteristics of the Approach: The Case of Points

1.1 Introduction

This study takes some first steps towards the development of a theory of applied Geometry. The goal of such a theory should be to give 'empirical definitions' of the fundamental concepts of Geometry such as "point", "straight line". and so on. However, we will make only limited and probably disappointingly little progress towards that goal, namely to account for applications of the concepts of point set topology to the surfaces of material bodies, and the extension of that to superpositions of these surfaces.

It may seem strange that in order to reach Geometry we should detour by way of point set topology, given that Geometry is generally regarded as an elementary mathematical subject and point set topology uses the full resources of higher-order set theory. But there are compelling reasons for taking this course. In some ways Geometry stands to its foundation as Arithmetic stands to *its* foundation, involving as that does concepts of cardinal numbers and one-one correspondences. The superstructure, 'the mechanics of Arithmetic', is among the first things learned in school while its 'underpinning' in one-one correspondences and cardinal numbers is learned later, if at all (and it is significant that the latter science was developed much later than elementary Arithmetic). So too, 'probing' or 'excavating' the foundation of elementary Geometry uncovers ideas that are less elementary than those of the subject for which they serve as foundation. These include the topological concepts of a boundary, an interior, a line, of dimension, and so on. But it has proved to be an arduous enterprise even to relate these ideas to concrete application, and that, together with ideas relating to superposition, will be our enterprise in this essay.

As the reader may already imagine, the way we undertake this enterprise will be quite unorthodox, at least in relation to recent trends in the foundations of Geometry, and this chapter will comment on differences between it and other works on the foundations of Geometry with which the reader may be familiar. The following section will note certain themes that are characteristic of our approach, and contrast them with more orthodox ideas, and succeeding sections will illustrate them by giving a brief sketch of their application to the case of *points*.

1.2 Characteristics of the Approach

Beyond the topological focus, the most basic difference between our approach to the foundations Geometry and others is that even though our approach is not specifically Euclidean, it follows Euclid in placing the theory of the plane (more exactly, the *surface*) before that of Space and solids.[1] In the present case, however, Space with a capital 'S' never enters the picture, and while bodies and their changing relations do, it is by way of their *possible superpositions* and fittings together. This allows us to treat superposition in a Euclidean way, which is more realistic than Helmholtz's, and which marks our approach's most radical departure from orthodoxy. Let us comment on that at more length.

Euclid has been criticized for giving superposition a rôle in his theory e.g., in the proof of the all-important Proposition 4 of Book I of *The Elements*, that triangles with two sides and included angle are congruent. Thus, Russell wrote:

> The fourth proposition is the first in which Euclid employs the method of superposition—a method which, since he will make any detour to avoid it, he evidently dislikes, and rightly, since it has no logical validity, and strikes every intelligent child as a juggle. (*The Principles of Mathematics*, Second Edition, p. 405.)

and Hilbert and other modern writers on the foundations of Geometry prove essentially the same proposition without reference to superposition (Chapter 11 will return to these points). The author would agree with Russell on this matter, but at the same time insist on the following point made by Euclid's great translator:

[1] It is notable that concepts of space only enter the picture in Books XI–XIII of Euclid's *Elements* (Heath, 1956, Vol. III). As will be noted below, the theory developed in this work is restricted mainly to topological aspects of the foundations of geometry, and its two-dimensional part is compatible with all of the standard two-dimensional geometries.

> In the note on *Common Notion 4* I have already mentioned that Euclid obviously used the method of superposition with reluctance, and I have given, after Veronese for the most part, the reason for holding that the method is not admissible as a *theoretical* means of proving equality, although it may be of use as a *practical test*, and thus may furnish an empirical basis on which to found a postulate. (Heath, 1956, Vol. I, p. 249).

Following Heath, the author would suggest that superposition can be regarded as a fundamental *physical test* of geometrical equality, i.e., of congruence. However, and this is the cardinal point, it cannot serve as the basis of an *empirical* theory if it is interpreted in the way Helmholtz does in his celebrated memoir "On the fact underlying Geometry" (1868), where it is a relation between three dimensional physical bodies and regions of space. That is both because regions of space are not empirical observables,[2], and because things like material bodies that can be observed cannot be superimposed on one another is such a way as to be *totally congruent*, i.e., fill exactly the same space.[3] But no such objection applies to coincidences that arise when *surfaces* are superimposed and concrete things on them are brought together face-to-face, so long as those things are not themselves three dimensional physical bodies. However, the crucial point is that if this is to be possible it requires us to recognize the existence of certain concrete observables in the external world that are not physical bodies—certain 'quiddities' that transcend the traditional empiricist mind-body dualism.

Concreta that can fill this rôle are what we will call *surface features*, examples of which include bumps, dents, scratches, sticky spots, and, preeminently, visible marks and figures like the letter 'S' and the hollow triangle, '⧍'. These are undeniably publicly observable things in the external world; in fact, they are on the page's surface, although they are not parts of it. That they might be *material*, or collections of material particles, and therefore be three dimensional, will be considered and rejected in section 2.2. But the point to make here is that the properties

[2]That is why we cannot accept Whitehead's theory of points as arrived at by processes of extensive abstraction, starting with spatial regions. Analogous objections apply to mereological theories of spatial 'structure', at least so long as they take the 'parts' that they deal with to be parts *of* something like three-dimensional continua.

[3]The same point was made by Rudolf Hertz in commenting on Helmholtz's paper, cf. Helmholtz, 1868, Lowe translation, in Cohen and Elkana, 1977, pp. 43 and 62. But as stated, Chapter 4 will characterize the *physical abstraction relations* that subsist between certain physical observables and abstract spacelike regions with which they coincide.

and relations of surface features like these will be the foundation on which the theory of surface topology to be developed in Parts II and III of this work rests. As such they will be of fundamental importance for us, and all of Chapter 2 will be devoted to this metaphysically unfamiliar kind of 'thing', which, because it has no place in the world outlook of modern science, the reader may be uninclined to take seriously. But it is hoped that she or he will at least accord them provisional acceptance, if only to see what this may explain.

Let us then note certain immediate consequences of taking them to be the fundamental objects of geometrical application.

One is that Geometry is not treated as the science of space á la Newton, Kant, and innumerable others, and we are led to a relativism that is even more radical than that of Leibniz. Moreover, the fact that we treat Geometry as a science of concrete, two-dimensional, and immaterial objects implies that we do not treat it as a branch of Physics. In fact, the problem of characterizing the geometrical concepts that are presupposed in the formulations of this or that physical theory, Newtonian, Einsteinian, or whatever, is regarded as peculiar to that theory.[4]

More generally, not only is our theory non-physical, but we do not regard Geometry as empirical in the ordinary sense at all. Although we hold that the primary concrete observables that Geometry *applies to* are two dimensional, non-material features of bodies' surfaces,[5] we will also hold that those are not geometrical objects 'proper', i.e., they are not what the variables of geometrical theory range over. Rather, the relation between geometrical entities and these concrete observables will be held to be one of *abstraction*, which is more like the relation between *universals* and the particulars that fall under them.[6] In a way the relation of Geometry to application is better modeled on that of Arithmetic to the concrete, as analyzed in Cantor's work, where natural numbers are arrived at by a process of abstraction, starting with concrete objects and moving first to classes that have those objects as members,

---

[4]The reader may reasonably ask what relevance this study might have to the philosopher of Physics who seeks to illuminate the geometrical presuppositions of the theories that concern her or him. This will be returned to briefly and inconclusively in Chapter 16, but here it may be said that the present study seeks to make explicit and analyze things that are part of our common heritage, which are presumably assumed, possibly unconsciously, in all geometrical thinking, including that of physicists.

[5]To be sure, it will be held that geometrical entities like points stand in well defined relations to objects that can be discriminated by the senses, but it will not be held that they *are* such objects, or that they are *physical* in the ordinary sense.

[6]It follows that on this view, while Geometry can be applied, and accounting for its applications is one of our principle objectives, applied Geometry is not an 'interpreted formal system' after the fashion of Nagel, 1961, Chapter 8, and others.

but which are not themselves objects in the ordinary sense.

The 'Cantorian model' of the relation between abstract mathematical entities—classes in particular—and concrete particulars is significant both for the similarity of this relation to the relation between abstract geometrical entities and concrete *'sensibilia'*, and for important dissimilarities. Perhaps the most important similarity resides in the existential presuppositions of theories of classes and theories of geometrical entities. That entities having the properties affirmed of classes in the theory of classes derives from the use in everyday speech of expressions like "the class of persons born in the United States between 1880 and 1890". Usually users of these expressions do not ask whether there exist *entities* answering to such descriptions, but it is plausible that the ready acceptance of Cantor's Postulate of Abstraction depends on the ubiquitousness of this usage. And, while there is a sense in which this makes the postulate *a priori*, its grounds lie in the nature of our linguistic, rather than in that of our perceptual apparatus.

As to geometrical theory, we shall maintain a similar thesis with respect to certain postulates of existence that are presupposed in it. For example, we are accustomed to saying things like "He returned the book to its place on the shelf" without questioning the existence of an entity denoted by "its place on the shelf," but it can be argued that the ubiquitousness in ordinary speech of usages like this underpins the sort of abstraction that is fundamental to Geometry. That is what gives certain principles of Geometry an *a priori* character, which we will argue is also synthetic—although not Kantian.

Another similarity between the set-theoretical and the Geometrical cases is related to the fact that while the ready acceptance of the postulates of Cantor's theory derives from their underpinning in ordinary usage, Cantor *regimented* that usage for his purposes, specifically for applications to mathematics. For example, he assumed the atemporality and extensionality of classes, though neither is presupposed in everyday usage. Thus, modern set theory does not envisage the possibility that a thing might belong to a class at one time but not at another, in spite of the fact that it makes good sense in everyday usage to say things like "Jones joined the class of millionaires in 1990." And, in spite of the fact that the classes of unicorns and of golden mountains are both empty, ordinary usage does not recognize them as identical. As to Geometry, we will suggest that its existential presuppositions also regiment ordinary usage, but in a way that differs from that of the theory of classes.

As said, in our theory geometrical entities will be treated as being like abstract *universals*, whose extensions at any one time are Cantorian classes, more than they are like the classes that are their extensions.

That points can be regarded as universals arises from the fact that they are what concrete objects that *meet* in them have in common. In this respect they are like abstract *colors*, which are what objects having those colors have in common. Furthermore, universals like colors have both temporal and modal aspects. An object can change color over time, and an abstract color can exist even if nothing has it, if something *could* have it (think of Hume's imagined shade of blue). Similarly, something can be at a point at one time but not at another, and there may be points in space where nothing is; i.e., our spaces are *container spaces.*

However, our theory of geometrical entities like points will introduce modality in a special way, namely by way of *postulates of constructability* that are closely akin to Euclidean constructability postulates, e.g., that it is possible to draw a straight line from any point to any point (Postulate 1 of Book I of *The Elements*, Heath, 1956, Vol. I, p. 195). Note that this is possibility in *this* world, and our theory of it will be modeled more closely on Euclid's usages than it is on widely known possible worlds characterizations.

Another important aspect of constructive modality brings us back to our primary reason for approaching the foundation of Geometry in a way that allows us to deal realistically with superposition. The most important use of constructions in Euclid's theory arises in giving *operational tests* for geometrical properties, perhaps the most famous of which is the test for the parallelism of straight lines in the plane, i.e., they are defined as lines that do not meet if prolonged (Definition 23 of Euclid's *Elements*, Heath, *op. cit*, p. 154). Thus, we do not tell simply by looking at them that the lines in the figure '−|' are not parallel; we *demonstrate* this by prolonging them, say to form the figure '−+', in which the extended lines do intersect. Similarly, in the proof of the 'infamous' Proposition 4 of Book I (Heath, 1956, Vol. I, p. 247), two triangles are not proved to be congruent by simply inspecting them, but rather by superimposing one on the other to see whether they coincide. And it is important that this operation has an empirical side: something has to be seen, but an active test must be carried out to make that possible.

Finally, returning to the topological theme, we will be concerned with topological aspects of Euclid's theory, e.g., as in Definition 3 of Book I of *The Elements*, "The extremities of a line are points" (Heath, 1956, Vol. I, p. 165). With caution, extremities may be equated to *boundaries* in the topological sense, but which Euclid left unanalyzed because he lacked the set-theoretical machinery needed to characterize them in a precise way. Combined with set-theoretical machinery, which we will use freely, our constructive approach allows us to do this in the case of the topologies of surfaces, which also leads to characterizations of other topological concepts such as those of an *interior*, of *connectedness* (or continuity), of *dimension*, and of *linearity*— though not of *straightness*, which is a metric concept. While metrical concepts are discussed at some length in Chapters 14 and 15, they are not fully dealt with in this work.[7]

Below we will illustrate the foregoing themes in their application to *points*, but let us summarize them first.

First and foremost, because the two-dimensionality of bodies' surfaces makes it possible to characterize their superpositions in a realistic way, we follow Euclid in taking the surface, rather than the three dimensional *space*, as the point of departure; in fact we never arrive at a single Space of the kind that philosophies of Physics often seek to characterize. It follows that the concrete objects of our theory are not three-dimensional, hence they are not physical in an ordinary sense, and therefore Geometry, construed as a theory of such objects, cannot be a branch of Physics. Second, even though applied Geometry may have empirical content it is not an ordinary empirical theory, and it is not even an interpreted formal system. Rather, geometrical entities like points are related to concrete observables by *abstraction*, somewhat as Cantor's classes are related to the objects that are their members. However, they are even more like *universals* of which classes are extensions than they are like classes, in that they are temporally variable and non-extensional. This non-extensionality is related to *constructability* of the kind that enters into Euclid's theory, rather than being a possible-worlds concept, and an empiricist-operationalist program involving it will be returned to in Chapter 3.

Now we will illustrate the above themes by sketching their application to the case of *points on surfaces*.

1.3 Illustrations in the Case of Points on Surfaces

We see that the dot "•" is fairly small, but it is not as small as a single point; in fact if it were that small it would be too small to be seen. We also see that the three segments composing the figure ⧖ do not meet in any point while those composing the figure ⚹ do. But how do we account for these things, e.g., for seeing that the segments composing the ⚹ have in common something that is too small to be seen?[8] For answer, we take

[7]It will be argued in Chapter 15 that while topologies of bodies' surfaces and even of the matter that forms them can be characterized along lines developed in this work, topologies of *spaces* cannot be characterized in a similar way.

[8]The fruitlessness of Euclid's and others' attempts to define *point*, e.g., as "that which has no part" and as the extremity of a line (Definitions 1 and 3 of Book I of *The Elements*, Heath, 1956, Vol. I, p. 153) is suggestive of the difficulties involved in this concept. Aristotle seems closest to the mark in identifying points with locations

note of two things: (1) the distinction, emphasized by certain ordinary language theorists, between *seeing* and *seeing that*, and (2) the analogy between seeing that two segments like those in the figure −| are not parallel and seeing that the segments in the figure X̲ do not have a point in common.

In the case of the figure −|, visual observation of the segments forming it leads to an intellectual conclusion: that carrying out the Euclidean test for parallelism would establish that the segments are not parallel. If the horizontal segment were extended it would meet the vertical one, as in −+. This suggests that we should look for analogous sensory and intellectual components involved in 'seeing that' the segments forming X̲ do not meet in a common point. But what *test* might establish this: that the three segments do not meet in a common point? Euclid did not describe a test that would be analogous to his test for non-parallelism,[9] but we will now describe one, whose properties will be examined at length in Part II.

Imagine the figure X̲ enlarged as in Figure 1A below:

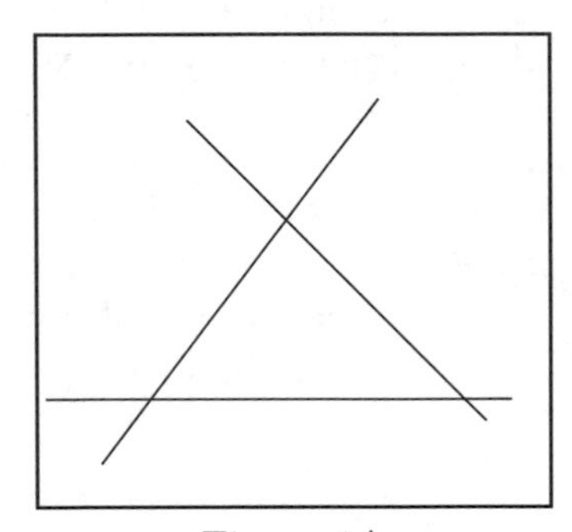

Figure 1A

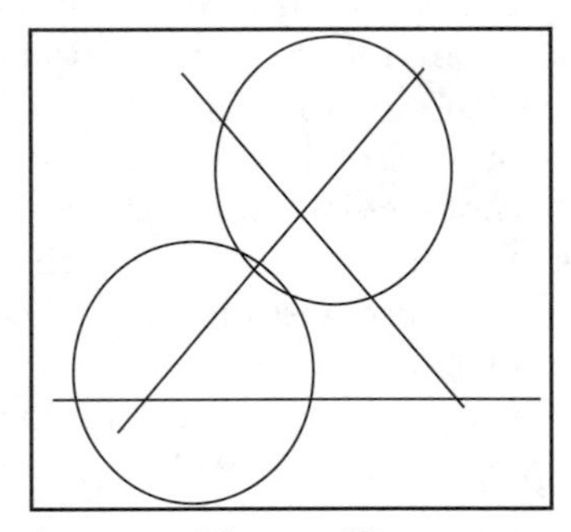

Figure 1B

We still see that the segments in Figure 1A are not coincident, but a test like the one whose result is pictured in Figure 1B would *demonstrate* this. It pictures two 'auxiliary ovals' drawn over the left-hand segment, which *cover* it in the sense that anything touching the segment would

("we can make no distinction between a point and the *place* (τόποσ) where it is" (*Physics*, IV. I, 299 a 30, quoted from Heath, *op. cit*, p. 156), but of course *location* must still be analyzed. Footnote 9 comments further on this matter.

[9]It must not be thought that devising such tests is easy. Some writers have regarded Euclid's test for parallelism (Definition 23 of Book I, Heath, *op. cit*, pp. 191–4), and the associated Postulate 5 (Heath, *op. cit.*, pp. 202–20), as being among his most important original contributions to geometrical theory. Heath's discussion makes it clear that it is by no means self-evident what 'the right test' for parallelism is.

have to touch one of them,[10] while *neither oval touches both of the other segments.* Given this, there could not be a point that is common to all three segments. If such a point existed it would be a point of the left-hand segment, and therefore it would be a point of one of the ovals that cover it. But then it couldn't be a point of both of the other segments, since neither oval touches both of them. On the other hand, if such a 'separation test' were applied to an enlargement of the figure, at least one of the ovals covering the left slanting segment would have to touch both of the other segments. This would prove the coincidence of the segments in the same way that failure of the two segments to meet when extended would demonstrate their parallelism.

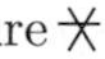

There is another important point to note about the separation test. Because it is complicated it is not self-evident that it really does provide necessary and sufficient conditions for coincidence. Therefore it is an important part of the theory of coincidence to *justify* this test and ones like it, in the present case by proving that two or more segments meet in a common point if and only if none of them can be covered with figures like the ovals in Figure 1B, in such a way that no one of the covering figures touches all of the given segments. The proofs of many of Euclid's Propositions have the same function, i.e., of showing that the constructive methods described in these propositions have the properties required of them, e.g., showing that a triangle constructed by such and such a method *is* equilateral, which is what the proof of Proposition I of Book I does (Heath, *op.cit.*, p. 241).

Another task of the theory is to explore interconnections between tests for different things. For example, one way of showing that the dot "•" isn't as small as a single point is to draw two lines, each of which touches it, but which don't touch each other, as in the figure. If the dot were as small as a point and both of the vertical segments touched it, they would have a point in common and therefore they would have to touch each other. But here we have a second test involving points, and we would like to know what its relation is to the first one, which seems to be utterly different. The theory of coincidence seeks to demonstrate interconnections between such tests, e.g., that any mark that touched all of the segments forming the figure could also be touched by lines that didn't touch each other, as in the figure.

Another thing that the theory of coincidence seeks to do is explain various intuitions, e.g., that points might be said to have no parts, or

[10] Section 2.4 points out that the *interiors* of regions outlined by visible ovals are in a sense 'secondary features', and 'observing' that they stand in topological relations, like covering, to other visible things requires special analysis. But this will be put off for now.

that anything large enough to be seen is too large to be at a single point—although small dots can 'approximate' and 'represent' them for this reason, which is connected to the fact that it is *hard* to draw separate lines that both touch a dot like "•".

Various other aspects of a theory of points can be noted briefly. One is that it involves *postulates of constructability* that are analogous in many respects to Postulates 1–3 of Book I of Euclid's *Elements*, e.g. that it is possible to draw a line from any point to any point. This postulate guarantees the possibility of carrying out an operation that is involved in many of the proofs in Geometrical theory, e.g., the proof of Propostion 5 of Book 1 (that the base angles of isosceles triangles are equal, and so are their complementary angles; Heath, *op. cit.*, p. 251). An analogous postulate in the theory of coincidence is needed to guarantee the possibility of drawing the kinds of ovals or similar figures that are involved in the separation test described above, which demonstrates the non-coincidence of lines or other figures. In fact, a general *Separation Postulate* will be stated as basic principle 6.3.1 in Chapter 6, where the theory of coincidence is developed systematically.[11]

Another aspect of the theory has to do with the *justification* for affirming the existence of abstract points and other geometrical entities, especially when they are too small or too 'thin' to be seen. What justifies our saying that the segments forming either of the figures + or ⚹ have points in common? It is here that we follow Cantor's example and simply *postulate* the existence of these entities, thus making explicit what is presupposed in the ordinary use of expressions like "the point of intersection of the segments."[12] The postulate is a principle of 'classical physical abstraction' (Adams, 1993), which is *a priori* in the sense that it derives from pre-existing usage, and it is synthetic because it depends in part on the Separation Postulate.[13]

As said, the following chapter will discuss these matters in more

[11]It is to be noted that this postulate only requires the possibility of 'freehand' drawings, in contrast to 'ideal' geometrical figures, which is in keeping with our topological, 'pre-geometrical' approach.

[12]The most direct geometrical analogue to Cantor's axiom would be to postulate the existence of a *place* at which any thing is at any time, where the *exact place* of the thing is appropriately taken to be coextensive with it. Including points in the category of places (or at least of *locations*) is a further step, since they are too small to be the exact places of any concrete things. Footnote 8 suggests that Aristotle may have done this, and section 4.6 will comment on extending the range of the abstract *beyond* what is instantiated in the concrete.

[13]By contrast, so far as I am aware the synthetic aspect of Kant's *a priori* had to do with the qualities of 'phenomena', and nothing to do with what can or cannot be constructed in the real, 'noumenal' world. But perhaps Kant would have held that what we are calling surface features *are* in the phenomenal world.

detail, but let us turn to other matters first.

1.4 Relevance to Geometry

The reader may wonder what the foregoing considerations and the theory to be built on them have to do with the science of Geometry. Since Plato we have been taught that diagrams like Figures 1a and 1b are mere 'visual representations' of real geometrical objects. Thus, in the *Meno* Plato held that Geometry is not concerned with visual diagrams, or as Wedberg put it, he held that "There are no truly Euclidean objects in the sensible world" (Wedberg, 1955, p. 49). More generally, Plato held that

> ... no one who has even a slight acquaintance with geometry will deny that the nature of this science is in flat contradiction with the absurd language used by mathematicians ... They constantly talk of 'operations' like 'squaring', 'applying', 'adding', and so on, as though the object were to *do* something, whereas the true purpose of the whole subject is knowledge—knowledge, moreover, of what eternally exists, not of anything that comes to be this or that at some time ... (Chapter XXVI, §2, of *The Republic*, p. 244 of the Cornford translation).

Of course, mathematicians no longer use 'absurd words' like 'apply' in the way that Euclid used them, e.g., in the proof of Proposition 4 of Book I of *The Elements*, which states that triangles are congruent if two sides and their included angles are equal (Heath, *op. cit.*, p. 247).[14] But the theory developed here will be non-Platonic, and it will be supposed to apply to visible figures, which are not supposed to be mere representations of something *else*.[15]

But even supposing that *our* theory applies literally to Figures 1a and 1b and others like them, it isn't clear that they are objects of *geometrical theory*. We have already set aside the geometrical aspects of theories of Physics (e.g., Newtonian or relativistic mechanics) as being special to those theories, but are these not *the* empirical applications of Geometry?

[14]Note that Hilbert took substantially this proposition without proof, as an axiom, namely Axiom IV.6, on p. 15 of the Townsend translation of *The Foundations of Geometry*, 1902. But as will be pointed out in section 11.1, even as late as the middle of the last century, geometry texts like Legendre's celebrated *Éléments de Géométrie* continued to speak of 'applying' figures.

[15]In fact, in some ways our approach fits in better with Berkeley's characterization of abstract ideas as particulars that are 'rendered universal' by the fact that we suppose that certain things that are true of them are true of all figures like them (cf. section 15 of the Introduction to *On the Principles of Human Knowledge*).

It might seem that whatever is left over after those aspects have been set aside are things of naive, everyday speech such as what might be referred to as 'the edge of the field', to which the rules of thumb of carpenters, surveyors, and others unversed in modern physical theory (and perhaps classical geometry as well) apply. But that would be too simple.

Even in ancient Greece persons applying Geometry had little use for theory. As Plato said "...for such purposes a small amount of geometry and arithmetic will be enough" (p. 243 of Cornford translation of *The Republic*; the purposes that Plato referred to were applications in 'warlike operations'). But he did hint that theory is useful in attaining exactitude. For instance, commenting on the place of astronomy in the education of the Guardians, he wrote:

> ...we must use the ... heavens as a *model* to illustrate our study of those realities, just as one might use diagrams exquisitely drawn by some artist like Daedalus. An expert in geometry, meeting with such designs, would admire their finished workmanship, but he would think it absurd to study them in all earnest with the expectation of finding in their proportions the *exact ratio* of any one number to another. (p. 248 of Cornford translation, my italics).[16]

Exactitude is suggestive. Actual diagrams are not drawn by ideal artists and they do not exactly conform to Euclidean specifications. Nor did Euclid attempt to explain how the theory developed in *The Elements* applies to things that don't exactly conform to these specifications. Applying the theory was largely a matter of following rules of thumb, and to some extent this is still the case. But the theory developed in the present essay aims to explain these applications in part, as the example of geometrical points illustrates. Thus, it explains why small dots are not geometrical points, but the smaller they are the closer they come to instantiating points because they come closer to satisfying the requirement that any two things that touch them must touch each other.

In some ways *lines* furnish a better example. Again, no observable thing, even a two-dimensional feature on a surface, is thin enough to be a geometrical line, but it follows from the theory developed in Part II, especially in Chapter 10, that the *edges* of observable things like the long rectangle below are one-dimensional:

[16]The 'realities' referred to were the *general* laws of motion, both celestial and terrestrial, which were little understood in Plato's time. The author's article "Idealization in applied first-order logic" (Adams, 1999) discusses a 'Platonic model' of the relation between inexact empirical theories and the idealized models in terms of which they are often formalized in first-order logic.

Chapter 10 proves this by developing a Platonic conception of an 'ideal line' that a concrete thing may 'partake of', which we characterize here as being topologically *homeomorphic* to the ideal line. The theory developed in Part III even goes some way beyond this, in the direction of characterizing *straightedges and straightness* (cf. section 14.2), which is a part of the theory of *superposition*. This is fundamental to *Euclidean metrology*, and, concluding this section, we will make two comments concerning its relations both to Geometry and to a branch of modern science.

To superimpose one figure on another is essentially to 'apply' one to the other in the sense intended, for instance, in the proof mentioned above of Proposition 4 of Book I of *The Elements*. Analyzing this operation is a first step towards a full-scale analysis of the metrical concepts of Geometry, and while this essay only goes a short way towards doing this, it does solve certain 'philosophical' problems that exercised classical thinkers, which are beginning to receive attention once again. These concern the idea of bodies having *common boundaries* or *faces*. Ancient Greek philosophers, e.g., Sextus Empiricus (cf. Mates, 199., p.) as well as modern writers (e.g., Stroll, 1988, Zimmerman, 1996) have puzzled over the question: when two physical bodies are superimposed face-to-face, do the faces that were separate before they were superimposed become fused into one? If they do then the whole bodies 'become one' by fusing into a single body, while if they do not fuse then they must still be separate, and therefore they haven't really been superimposed. The solutions that are offered to this and related problems in Part III are given in terms of *composite surface topologies* which are formed from the surface topologies of their components when they are superimposed. This is too complicated to describe at this point, but the essential point is that the faces that were separate originally, and the features on them, *retain their identities* when they are superimposed, but the topological *spaces* that are generated as a result of their being superimposed are somewhat complicated *extensions* of the component spaces, which are described in detail in Chapter 12.

Superposition is important for another reason. Theories of visual perception, like those of Locke or of Marr, 1982, tend to suppose that visual images 'resemble' surface features such as the long rectangle pictured above, and these resemblances provide 'our knowledge of the external world'. Typical of the problems that arise in this approach is that of explaining how, if all we are 'given' are two-dimensional visual images, we arrive at the idea that 'the world' is three-dimensional. The fact is that what we perceive by sight *are* features on surfaces, like the long rectangle, but 'transferring' them to mental images leaves out something crucial. Surfaces and their features can be superimposed, and the facts of superposition are what actually inform us both of spatial dimensionality and of metrical facts, but 'superpositional facts' are not transferred to mental images. This is discussed in Chapter 17, which is the penultimate chapter of this essay, and it suggests that visual perception theorists should pay much closer attention to superposition than they have heretofore.

A final point about our overall program is as follows.

1.5 An Empiricist-Operationalist Program

As said earlier, *The Elements* develops a largely constructive-operational program that defines and analyzes interconnections between concepts of elementary geometry. For instance, Definition 23 of Book I characterizes parallelism as the property of straight lines of not meeting when they are extended or 'produced' (Heath, 1956, Vol. I, p. 165), and Proposition 1 of Book I describes a method for constructing an equilateral triangle on a given base (Heath, *op. cit.*, p. 241). The present work applies the operational approach to the topological concepts that underpin geometry, e.g., that of an *extremity*, which enters in Definition 3 of Book I of *The Elements*, namely "The extremities of a line are points" (Heath, *op. cit.*, p. 153).

The present work seeks to give operational definitions that link topological concepts directly to empirical observables, and specifically to the surface features that were discussed in section 1.2. Hence our program may be said to be 'operational', and it is akin in spirit to P.C. Bridgman's old program for analyzing concepts of Physics.[17] This has already been illustrated in the case of the concept of *coincidence*, as applied to the figures ⚹ and ⨯, in the first of which the segments meet in a point, and in the second of which they do not. Figure 1B pictured the result of carrying out an operation in which one of the segments of ⨯ is covered by ovals, the upshot of which is to prove that the three segments are not coincident. This procedure is operational and the ovals that are produced are surface features that are observed by sight, hence it fits into an empirico-operationalist program. This work will extend the method to more properly topological concepts, including ones mentioned earlier, of a boundary, continuity, dimension, and linearity, as well as the most

[17]Cf., "...the concept is synonymous with the corresponding set of operations. If the concept is physical, as of length, the operations are actual physical operations, namely those by which length is measured ..." (*The Logic of Modern Physics*, 1927, p. 5).

fundamental topological concept, beyond that of a point, namely that of an *open set*. Both the operational definitions and their justifications and interrelations are complicated, and details will be set forth at length in Part II, which is inspired by if not modeled on the pattern of Euclid's *Elements*.

But we must note a serious limitation.

1.6 The Problem of Appearance and Reality

The triangle in Figure 1a appears to divide the page on which it appears into an 'inside' and an 'outside', and the constructive *proof* of this is to verify that any continuous line drawn from one side to the other must cross at least one of the triangle's sides.[18] Now, a doubtful reader *could* try to verify the impossibility of doing this by actually trying to draw such a line, but would failure to do so after repeated efforts really prove that that would be impossible? Failure might only show that the reader was too unskillful, or too clumsy, or be using drawing instruments that are too crude. Can we ever *really* prove a physical impossibility—that something can't be done? For answer, we will proceed as follows.

What we will take for possibility is the *appearance of possibility*. Our 'datum' will be the fact that it *appears* to be impossible to draw a continuous line from the inside to the outside of the triangle in Figure 1a, and our theory will concern interconnections between these appearances. That is what will give it, and we would argue give classical geometry as well, its empirical aspect.

But the appearance is itself an idealization that may not correspond to reality, and this is closely connected with another kind of idealization. Although it looks as though it would be impossible to draw a continuous line from the inside to the outside of the triangle, closer inspection—say under a magnifying glass—might reveal that the lines forming the triangle's sides were nothing but closely packed dots—'pixels'. Still closer inspection would certainly reveal that the dots themselves were assemblages of smaller particles whose light-reflecting properties created the appearance of continuity. In any case, though, it *should* be possible to 'thread ones' way' among the particles from inside to outside in such a way as to avoid touching any of them. The 'reality' would conflict with the appearance, and how are we to deal with that? The answer is to stick resolutely with the appearance, while admitting that to speak of 'the' appearance is a gross oversimplification. And, we would suggest

[18]This topological version of Pasch's Axiom can be regarded as a special case of the Jordan Curve Theorem (Courant and Robbins, 1941, pp. 244–246). It is also closely related to the test for a one surface feature to cover the boundary of another that is described in Theorem 8.2.2 of section 8.2 of the present essay.

that the practical application of classical geometrical theory also deals with simplified appearances, e.g., that it should be possible to draw a straight line from any point to any point. These and similar claims are idealizations, but the very fact that they are about appearances is part of what gives geometrical theory an empirical aspect. However, we can always ask: what is their relation to *reality*?

This question has plagued philosophers and philosophically inclined physicists from ancient times to the present. Plato held that the objects of science, and especially of geometry, are accessible only to thought, and not accessible to sensory observation. Descartes held that the senses are only guaranteed to yield veridical information about the external world when they consist of clear and distinct ideas, whose validity is guaranteed by a benevolent deity whose own existence must be inferred by pure reason. Leibniz held that geometry is not the science of absolute Space, but rather of relations between material entities (¶ 62 of the Fifth Letter to Clarke), but he neglected their surfaces. Berkeley and others following him (e.g., Reid, Mach) hold that the proper entities of science are 'ideas' that are directly accessible to observation, but are not material. And, Kant held that the geometrical content of our ideas conforms to laws of our own perceptual apparatus that have no counterpart in the external world. None of these views accords a place to the 'appearances' that concern us here, of bodies' surfaces and their features, so it appears that learned opinion is against us. On the other hand, the diversity of these opinions suggests that the door may be open to other approaches, and in particular to one that is inspired by the status of *diagrams* like Figure 1a, or those that appear in the planar books of Euclid's *Elements.*

But of course, treating diagrams as appearances does not resolve the old appearance and reality problems that Plato *et. al.* attempted to solve. Appearances *are* appearances, and how can we claim that surface features like Figure 1a are publicly observable if they only appear to certain people under certain circumstances? How can we claim that they are 'out in the external world', in the same places as bodies are, if they are mere appearances and bodies are 'really there'? To hold this seems to start down Berkeley's slippery slope towards holding that appearances are all that we have knowledge of, and bodies are mere congeries of them.

Well, any epistemologist knows that these are tremendous questions, and the author does not pretend to be an epistemologist. He can only forthrightly admit that these difficulties lie on the road ahead. But that is far ahead, and they will only be reverted to in the last two chapters of this essay. In the meantime we will simply ignore them, and pretend that we can all 'see' Figure 1, and that it lies on a particular page of this essay.

The following chapters turn to details, but let us first briefly summarize the themes to be dealt with in them.

1.7 Summary of Themes of Following Chapters

The remaining chapters of Part I will develop the ideas just discussed in more detail, but still informally. Part II presents a deductive theory of the topologies of the surfaces of bodies, which defines abstract *points* and *open sets* in these spaces in such a way as to 'coordinate' them with concrete observables on these surfaces, and in terms of this it argues for the surfaces' two-dimensionality, the one-dimensionality of the observables' boundaries, and other intuitive surface properties.

Part III extends the discussion to consider properties of the 'composite topologies' that result when one surface is superimposed on another, or several surfaces are 'fitted together'. Classical paradoxes of superposition like the 'fusion problem' described above are resolved, and new concepts like *orientability* are brought into the picture. A static and relational concept of rigidity is introduced, that holds between bodies if they can only be fitted together in 'minimum' numbers of ways. Although the development is informal and not worked out in detail, this suggests a way of building on the theory developed in Part I, to define topologies of 'frameworks' consisting of bodies that are rigidly fitted together.

Part IV consists of unsystematic remarks on two further ramifications of the present investigation. Chapter 16 is concerned with relations between the present theory and modern physical theory, which are difficult to define because of the fact that even Newtonian physics deals with material bodies in 'absolute space', whereas the present investigation is primarily concerned with non-material features of bodies' surfaces, and certain relative spaces that are defined by rigid frameworks that these features define. Nevertheless, it is suggested that there are at least two important connections between these realms. One has to do with *measurement*, which is essential to the application of physical theory, but which at its simplest must be carried out by 'applying'—superimposing—measuring rods to the bodies with which physical theory is concerned. The other has to do with *matter*, which is the 'substrate' of which bodies are formed, but which can be 'recognized within the bodies' by the kinds of marks that appear on their surfaces.

Chapter 17 discusses very briefly connections between the present theory and certain *theories of perception*, both in philosophy and psychology, that derive from the fact that the visual images that figure in these theories closely resemble surface features like the star '⚹', which we argue are 'in the external world'. In fact, we very tentatively suggest that adding *non-material* but external and publicly observable surface features to traditional empiricism's mind-body dualism may contribute to solving skeptical difficulties associated with this dualism. Chapter 18, the final chapter, gives a still briefer résumé of the aims and claims of this work, followed by inconclusive comments on six important objections to its principal themes.

Concluding, it must be emphasized that the only systematic, in fact deductively organized, part of this work is Part II, which describes an 'empirico-operational foundation' for an application of present day mathematical topology (especially point-set topology) to the surfaces of material bodies. Part I deals informally and unsystematically with various '*leitmotifs*' that guide the deductive theory in Part II, especially its foundation in an unfamiliar ontological category of 'objects', namely two-dimensional and non-material 'features' of the surfaces of bodies, and principles of abstraction analogous to those of Cantor's theory of sets, which are presupposed in moving from the level of concrete surface features to abstract entities, especially points, that are instantiated in them. Part III is a largely informal sketch of how the theory developed in Part II might be extended to describe the 'composite surface topologies' that are formed when the surfaces of two or more bodies are superimposed, which, arguably, makes it possible to account for metrical concepts of 'geometry proper', such as length and distance. Part IV consists of still more informal reflections on 'applications and implications' of the ideas developed in Parts I–III for modern science, specifically Physics and theories of perception in Psychology. These remarks can at most provide 'food for thought' to students whose primary interests are in these subjects, and the author can only say in his own defense for including them is that these are the topics in which he was originally interested, and it was by prolonged reflection on them that he was led to the views expounded here.

2

The Concrete Superficial

2.1 Introduction

The figures spoken of in the previous chapter, like X , ✳ , • , and | are not assumed to have the familiar physical properties of bodies or collections of bodies, and this chapter will discuss those properties that may strike the reader as puzzling. Among these are: (1) their two-dimensionality, (2) their non-materiality, which follows from their two-dimensionality, (3) their distinctness from the surfaces on which they lie, (4) their *locations on* these surfaces, as well as 'in space', and (5) their concreteness and accessibility to public sensory observation. Of these, (4) and (5) may be obvious, since to have arrived at this point readers must have seen, say, the X , and also to have seen where it lies on the page—on its top surface. But its two-dimensionality and non-materiality as well as its distinctness from the surface and even from the parts of it that it occupies may be less obvious. We will begin with these, and then turn to serious problems that arise in analyzing relations between concrete surface features that are fundamental to the theory that will be developed in Parts II and III.

2.2 Immateriality and Two-Dimensionality[1]

It might be admitted that if the X lay exactly, entirely, *on* the surface of the page then it would be two dimensional and therefore immaterial. But it could be objected that it is really composed of ink particles, and these are material and therefore they do not lie entirely on the surface. This is true in one signification of "the X ", because it could be taken to refer to the collection of ink particles. But it could also be taken to refer to the black-colored parts of the particles' surfaces that are visible to the eye of an observer looking at the surface, which only approximates

[1] The ideas put forth in this section have been developed elsewhere, especially in Adams, 1984 and Adams, 1994.

the collection of particles if the particles themselves are so small that all parts of their surfaces are close to their visible parts. However, parts of surfaces are not plausibly supposed to be material; they may have length and breadth, but not depth,[2] and it would make no sense to try to measure their physical properties like weight.

Non-materiality is perhaps more evident in the case of other features that bodies' surfaces may be exhibit. Bumps, dents, and scratches are often discernible on surfaces, but it is implausible that they should be thought to be material. Further afield, hot, sticky, and shiny spots are 'things' that are concrete and discernable to the senses, even if not necessarily by sight, and they would hardly be said to be material.

But, even though these 'sensible features' may be coextensive with parts of bodies' surfaces, we cannot assume that they are parts of their surfaces.[3]

---

[2]Chapter 9 will present a more detailed account of the dimensionality of surfaces and their features, but this is topological and requires a much more complicated 'underpinning'.

[3]Note the discussion and rejection of what Price calls the 'naive realist' view that sense data are literally parts of bodies surfaces, in Chapter 1 of *Perception*. Thus, bodies' surfaces may have parts where there are *no* discernable features, and even more to the point, two distinct features may be *coextensive* and coincide in the part of the surface on which they lie, but be different from each other and therefore different from the part of it on which they lie. A bump may coincide with a visible red spot, but it may only be discernible by feel and not be the same as the spot, since it may remain while the spot disappears. How we ascertain that the bump and the spot coincide, when they do coincide, will be returned to in section 2.5, but the fact that either of them may appear or disappear suggests a better way of picturing the relation between them and the surfaces on which they lie. That relation is more closely akin to the relation between *bodies* and the *spaces* in which they may move.

The key fact is that a surface may exhibit a feature at one time and not at another, and even while it is exhibited it may move or change relative to the surface. A red spot may appear, grow larger, and then diminish and ultimately disappear, while all the while being referred to as "the red spot." True, such features are typically less mobile than bodies moving freely in space, but they *can* move, and the 'spaces' in which they move are bodies' surfaces. Of course, we may ask how we ascertain that they *do* move, but this is the analogue of the problem of distinguishing between relative and absolute motion 'in Space', which we already know to be very difficult. This will be returned to in section 2.6, where we will distinguish a class of 'placemarking features' that do not move in relation to one another, which, incidentally, allows us to distinguish *solid* bodies, or at least bodies with solid surfaces, that have such features from fluid ones that lack them. But there are several other things to discuss first, beginning with the topics of incidence and identity, which are interconnected.

2.3 Incidence and Identity[4]

The complementary *incidence relations* of *separateness* and *binary coincidence* (touching) are illustrated in the two diagrams below:

Diagram 1. Separate cross and ring

Diagram 2. Touching cross and ring

These are the primitive relations of the theory of surface topology that will be developed in Parts II and III, but they cannot be taken for granted in a foundational study such as this purports to be. We do take it for granted that the *relata* of the relations, the cross and the ring, can be publicly observed, but it no more follows from this that the relation between them is empirical than it follows from the fact that two persons can be observed that their relative age can be directly observed. This section is concerned with how we observe incidence relations between concrete surface features, but this turns out to be a surprisingly complicated matter, which is, moreover, inseparable from the problem of how we *individuate* the things that stand in these relations.

Superficially, the determination of separation and coincidence seems unproblematic. We determine that the cross and ring in Diagram 1 are separate by tracing over the extent of one of them, say the cross, and finding that we never come in contact with the other.[5] But our theory requires us to consider incidence relations between *parts* of figures, e.g.,

[4]This section deals with the most 'foundational' problem discussed in this work, namely that of individuating the surface features that are central to the theory to be developed in detail in Parts II and III. This problem and its relation to the more general problem of physical identity is discussed in Adams, 1986, but the 'solution' proposed here is new. However, the solution is tentative and speculative, and it deserves much more careful scrutiny than the author has yet given it.

[5]This 'continuous tracing' method is closely related to criteria of identity-over-time discussed in Wiggins, 1971, and elsewhere, which is applied here to establishing the separateness of surface features. Here as elsewhere we ignore 'the problem of microstructure', in which things (surface features in our case) that *look* continuous are seen to be composed of discrete dots when they are viewed from close up or under a magnifying glass.

between the segments forming the 'tepee', X, and we run into difficulty in trying to apply the tracing test to them. This difficulty even arises in applying the test to the 'cross-plus-ring' in Diagram 2, of which the cross and the ring are parts. That is because the cross and the ring are 'embedded' in the figure, hence their extents are not well-defined, and therefore we cannot trace over them in a precise way to determine whether there is path from one to the other that never goes outside of them. In fact, the fundamental difficulty is to say precisely what the cross and the ring *are*. The problem may more easily be grasped in reference to parts of the single cross in Diagram 3 below:

Diagram 3. Single cross

Suppose that we attempted to determine whether the left arm of the cross touches the right arm by trying to trace a path from one arm to the other while remaining continuously in contact with one or other of them. The problem is that because the arms are fused into the whole cross, we can never be sure that we haven't slid off of them onto another part of the cross. But speaking of fusion does suggest something: namely that the problem of incidence that arises from fusion is very close to well known problems of *identity* that arise from this. If we point to the arms with right and left hands and ask "Is this separate from this?" are we sure we aren't really pointing at the same thing? This is a familiar problem (cf. works of Strawson, 1959, Wiggins, 1971, Hirsch, 1982, and others), and all that we are doing here raising it in a new context.

A partial solution to the identity problem that is available in the case of physical bodies but not in that of their surface features is instructive. If the arms of a wooden cross do not completely fuse with it when they are nailed to it, they can be detached later and identified as parts among those that were originally used to form it. Of course this does not work with the arms of the cross that is drawn in Diagram 3: once they have been drawn they cannot be detached later in a way that allows us to be sure that what is detached is the same as what was originally drawn. However, the fact that both the wooden cross and the drawn cross were originally formed from 'components', wooden arms in one case and 'drawn' ones in the other, suggests that these components may be identified 'genetically', by reference to the processes of their formation. And, this in turn suggests that their incidence relations may also be characterized in these terms. Of course *defining* parts or components by reference how they are produced has the serious drawback of making determinations of incidence relations between them very difficult practically—akin to the difficulty of determining the relative ages of persons just from their appearances—and this difficulty will be returned to later. But the genetic characterization has several advantages, and we will note two of them before returning to incidence.

One advantage is that it allows us to explain why it is natural to describe the cross in Diagram 3 as being formed of crossing vertical and horizontal segments, even though it is logically possible that it could have been formed by superimposing two 'corners' like those depicted below:

Diagram 4. Corner components

There is no way of telling by direct visual inspection how the cross in Diagram 3 was formed, and we might even say with the mereologists that any region covered by the cross was a 'component' of it. But according to our genetic characterization of componentiality, the cross in Diagram 3 would be formed of linear components if it was *drawn* in that way, e.g., by first drawing a vertical stroke and then crossing it with a horizontal stroke.[6] This would be much more natural than to produce the cross by superimposing corners like those in Diagram 4, even though *post hoc* visual inspection cannot conclusively determine how it was drawn.[7] The next point is related.

If components are individuated genetically they are not fundamentally individuated in terms of their *forms*. Therefore, it is not an *essential* property of a component to be, say, a straight vertical segment. The advantage of recognizing this is that it allows us to avoid begging questions that could arise if, for instance, the component were fundamentally de-

[6]This is not meant to argue that the mereologists are 'wrong', since terms like 'part', 'component', 'constituent', and so on can be variously interpreted. All that can be argued here is that the genetic interpretation has important advantages for our purposes.

[7]Of course the cross that the reader sees in the diagram was *printed* and not formed by 'strokes' in the way described; hence it doesn't have these components. At best it 'represents' a figure that has them. If it is objected that Geometry applies just as much to printed figures as to ones that are drawn, this can be met if superposition is used to establish that printed figures are equal, geometrically, to figures that are drawn in the way we are considering. But it will be important that the *topological equality* that we will be concerned with in Parts II and III is more fine-grained than geometrical equality; in fact a fundamental concept of these topologies is defined in terms of genetic feature-components. This is commented on briefly in section 2.4.

fined as 'the vertical segment', which would presuppose the concept of being parallel to an edge of the page. We do not want to presuppose geometrical concepts in a study of the foundations of geometry.[8]

The third and perhaps most important advantage of the genetic characterization of feature components is that it permits us to give least a partly empirical characterization of incidence relations involving them. Go back to the cross in Diagram 3, and suppose that it was in fact formed by first drawing a vertical stroke and then crossing it with a horizontal stroke. It is not implausible that a person witnessing this process might observe the 'encounter' that takes place when the horizontal stroke *meets* the vertical one—even if at a later date no one could verify conclusively that the encounter took place. In a sense, this would be like verifying that one person was older than another, which could only be done conclusively by someone who observed both parties at birth, although in practice all kinds of practical cues to age might be used. Analogous cues in the case of the cross might derive from its appearance of having been drawn with a straightedge, e.g., in a way in which the horizontal line would have had to encounter the vertical one, given that that had already been drawn.

Clearly the foregoing is highly speculative, but at least it suggests a way of interpreting incidence relations among components empirically, albeit the interpretation is more complicated than one might wish.[9] However, let us end this section by noting one more advantage of the genetic interpretation of componentiality, which leads into the topic of the following section, which will also be very important as far as concerns the

[8]Of course the component can be *described* as "the vertical segment", but being vertical is no more an essential property of it than *standing* is an essential property of a person who is described as 'the person who is standing up'.

[9]And there is still a problem. Consider a cross with a plaque over its center

and suppose that the vertical segment was drawn first, the plaque second, and the horizontal segment last. Applying the 'encounter' test to determine whether the vertical and horizontal components were coincident would require showing that the horizontal segment encountered the vertical one when it was drawn. But even supposing that we could see it being drawn we might not be able to see its encounter with the vertical one if the plaque concealed the part of the vertical one that the horizontal one encountered when it was being drawn. We might say that the horizontal segment coincided with the vertical one if, when it was drawn, it *would have encountered* the vertical one, if the plaque had not been there. But even if this should be right, its counterfactuality would make it non-empirical.

'foundation' of our theory of surface topology. That is that this mode of individuation allows us to make a 'figure-ground' distinction. The figure below illustrates this:

Diagram 5. Hollow Square

This is a hollow square, but is the hollow a component of it? According to the genetic interpretation this cannot be determined without knowing how the figure was formed. If the white square had itself been drawn, as it were 'over' the center of the black one, and this counted as part of the figure, then it would be a component of it. But if it were a region left over after the rim of the square was drawn, then it would not be a component. This may seem to be a distinction without a difference, but it will be crucial to the topological theory because, as will be seen in Chapter 7, the all-important distinction between closed and open point sets will be based on it.[10] Of course that lies far in the future, but the following section makes a more immediate use of the distinction.

2.4 Asides on Dependent Surface Features

Recall that the hollow in the hollow square in Diagram 5 is only regarded as a *component* of it if it was drawn in the process of drawing the figure. If it was not produced in that way, i.e., if it was only left over after the rest of the figure was drawn, it is part of the figure's ground but not a component of it.[11] But there is a sense in which the hollow can be seen even if it is not a figure or component of one, and this introduces a category of *dependent surface features* that are so called because observing these features depends on observing independent figures like the rim of the square. They also stand in dependent incidence relations to other things and to one another. Consider the following.

The tiny circle inside the hollow in the figure below would no doubt be said to be *incident in the hollow*:

[10]For this reason it would be hard to overemphasize the importance of our figure-ground distinction, and the genetic characterization of surface features that it rests on, which is all too cursorily discussed here. Note that superposition doesn't establish a difference between a figure and its ground—between an open point set and its closure, and this shows that a *non-geometrical* distinction is fundamental to topology. If the distinction is 'merely conceptual', it is one of the most important conceptual discoveries in mathematics of our time. Recent philosophical discussions of the distinction will be commented on in section 7.1.

[11]We will see in Chapter 7 that the hollow is a *component*, in the *topological sense*, of the complement of the set of points of the square itself (cf. Kelley, 1955, p. 54).

Diagram 6. Hollow with circle in it

However, this kind of incidence is not determined genetically, since the circle could not have been 'encountered' when something else was drawn. In fact, the very idea of being *inside* the hollow is a topological one, and analyzing incidence relations involving such 'topological constructs' must wait on an exact characterization of basic surface topologies, which is developed in Part II.[12]

For similar reasons, although we can 'see' the outer boundary of the solid square in Diagram 6, we will not assume that its incidence relations to other things are determined in the way that the fundamental incidence relation between the horizontal and the vertical segments forming the cross in Figure 1 is determined. And, since boundaries are themselves topological entities, defining their incidence relations to other things waits on the exact topological characterization of them that will be given in Chapter 8.[13]

Hollows, insides and outsides, and boundaries are not the only dependent features that can be seen or otherwise sensed on surfaces, but we will not try to provide an exhaustive list of such things here. We will only say that most entities whose incidence relations cannot be accounted for by the methods outlined in the previous section will not be included in the category of independent surface features, and they will included, if at all, among the dependent features like hollows and boundaries. But questions arise about multi-modal incidence judgments, which will be commented on very sketchily in the next section.

[12]The reader may be reminded of Figure 1b in Chapter 1, in which a side of a triangle is pictured as covered by two hollow ovals. In view of the present remarks, the judgment that hollows 'cover' things like the sides of triangles requires careful analysis.

[13]Note the contrast between our approach to boundary concepts and approaches to surfaces and other 'extremities' that have been put forward in recent philosophical work. For example, although Stroll (1988) focuses on *surfaces* rather than on the boundaries of things *on* them, he treats them essentially as independent entities that differ only in 'shape' from the things they are extremities of—i.e., of physical bodies. This would imply that incidence relations, e.g., touching, between surfaces and other entities are determined by the same means as they are between 'primary' physical bodies. But in our account, incidence relations between dependent topological constructs and other things are not determined in the same way as they are between 'primary' surface concreta.

2.5 Multi-Modal Incidence Judgments

Return to the bump and red spot noted in section 2.2, and the question of how their coincidence might be determined. The bump and spot would probably simply be 'found', and if so they would not be construed as having been generated by any particular process. Therefore it is questionable that they should be regarded as components of a complex 'bumpy spot'. Given this, their coincidence doesn't seem to be determinable in the way described in the previous section.[14] Nor does the tracing test for determining the separateness of non-componential features seem to apply, since it is hard to see how one might 'trace a continuous path' from the visible spot to the tangible bump. Although we would argue *contra* Berkeley that the bump and spot are in the same place in the external world, nevertheless there seems to be some sense in which they 'belong to different worlds'.[15]

But how do we tell that the bump and the spot are in the same place? This seems to involve eye-hand coordination, which infants learn very early. But beyond noting that this appears to be an acquired skill that involves temporal coincidences,[16] we have nothing to say about its details.

2.6 Standard Surface Features[17]

We know that features like colored spots can appear and grow on surfaces; in fact, our genetic characterization of their components tacitly assumes this, since the 'encountering' test for incidence presupposes that they encounter one another in the process of being drawn. On the other hand, we have tacitly assumed that these components remain unchanged

[14]Perhaps only artificially created surface features like drawings can have genetic components to which tracing tests for incidence relations can be applied. If that seems overly restrictive, it should be recalled that the primary application of the surface topology theory of that will be developed later is to *constructions* drawn on these surfaces.

[15]See especially paragraph 115 of the *New Theory of Vision*, "The two distinct provinces of sight and touch should be considered apart, and their objects have no intercourse, no manner of relation one to another, in point of distance or position." However, this seems to conflict with Berkeley's view that visible ideas stand to tangible things much as words stand to the things they stand for—'visible ideas' constitute the "Universal language of nature," which will be returned to in section 2.8.

[16]Note that Kant held that the temporal order is multi-modal. cf. Allison, 1983, p. 102 "Kant does add that time, as the form of inner sense, is a formal, *a priori*, condition of *all* appearances, since all appearances, as modifications of the mind, belong to inner sense."

[17]The matters discussed in this section will be returned to in section 4.7, where they will be linked with physical abstraction.

once they have been drawn.[18] But what is it for a surface feature or a component of it to remain unchanged? This is analogous to the question of what it is for a body to remain unchanged in size, shape, or location in space over a period of time, and it invites similar answers. In this section we will outline the first of two conceptions of 'topological constancy' and change, an 'empirical' one that is modeled on Leibniz's conceptions of relative rest and motion (par. 47 of his 5th letter to Clarke), and the following section will speculate on a less empirical and more 'absolutist' conception. The Leibnizian idea is that once a component has been produced it remains unchanged if it remains unchanged *in its relations to other things*. It only remains to make clear what relations remain unchanged, and to what things they remain unchanged. Specifying these things adapts Leibniz's views to the 'spaces' of features on bodies' surfaces.

As concerns the relations that should remain unchanged, we suggest that the feature or component in question should not change in its relations of incidence to other features. For instance, once the horizontal and vertical components of the cross in Diagram 3 have been produced, they should not change in their relation of separation from or touching one another. Thus, the topological analogue of Leibniz's essentially metrical concepts of rest and motion substitutes incidence relations for distance relations.[19]

As to what the feature or component in question should maintain unchanging incidence relations to, we suggest that these should be *topologically standard features, that have the property of maintaining constant incidence relations to one another*. Not all surface features are standard in this sense: for instance, the red spot was imagined as gradually growing, and in the process possibly coming to touch features that it hadn't originally touched. It is important, however, that the existence of these 'standard place-marking features' is a matter of empirical fact. Only certain kinds of surfaces can bear marks that do not move, grow, shrink, or alter their shapes, and that therefore approximate topological standardness. These include pieces of paper on which characters can be written or printed, boxes that can be labeled, and solid parts of the earth's surface on which surveyor's reference marks can be placed. These marks act in many ways like the fixed stars in space, which move only imperceptibly relative to one another, and with respect to which the positions and motions of other astronomical objects are described. And, as the fixed stars

[18] But we should not say that incidence relations between genetic components only obtain at the time they are drawn. That would be like saying that relations of older and younger between persons obtain only at the times they are born.

[19] Note that this makes size and shape as well as motion relative.

aren't necessarily assumed to have fixed positions in absolute space, so the 'Leibnizian conception' of topological constancy and change does not assume that location references like surveyor's marks have fixed positions in 'absolute' surface spaces. Four other points related to this are the following.

It is useful for astronomical and geographical reference that the fixed stars are nearly everywhere in the night sky, and that location marks can be 'created' at will, nearly anywhere on many surfaces. In the astronomical case, we don't have to say that a comet is so and so many degrees distant from, say, the Pole Star, if it is very close to this star. Similarly, in the geographical case we don't have to say that a tower is located at such and such latitude and longitude if it is located near the top of a known mountain.[20]

The second point is that the fact that fixed stars and surface location markers have two important properties, (1) of maintaining constant positions relative to one another and (2) of existing or being positionable nearly everywhere, explains the choice of these things as location references. If we chose, say, the sun or the comets as standard astronomical references we would have to make constant reference to the times at which things were located at or near them. Similarly, if we chose *people* as standard geographical location references then we would have to specify the times at which other things or people were near them.[21]

The third point explains why a reflection in a mirror does not even have a location in or on the mirror. That is because even at a given moment, the standard surface features that it is 'incident in' depend on the position of the observer. The reflection of a door that is behind someone looking in the mirror may appear to that person to be coincident with a smudge on the mirror's surface, but appear to be very far from the smudge to someone standing elsewhere.

The last point is that the surfaces of bodies of water, like oceans and lakes,[22] that do not support place-locating standards with the properties

[20]However, an important difference between the fixed stars and surface reference marks is that the fixed stars only locate astronomical objects in the Star Map, and not 'in space', whereas surface reference marks locate things everywhere on the surface. The lack of marks that mark the positions of things like satellites, clouds, or birds moving about 'in empty space' makes that space more like the surface of a liquid, as noted below.

[21]The importance of practical usefulness is related to the information-theoretic account of standards developed in the Appendix of Adams, 1966. This will be returned to in Chapter 4.

[22]That the surface of a lake is generally regarded as including only the lake's top surface, and not all of the 'extremity of the lake', suggests that ordinarily when we speak of the surface of something we mean the part of its surface that is *accessible*.

noted above, nevertheless present features, e.g., water waves. But these tend to be ephemeral, and even when they exist they change constantly in their relations of incidence to one another. Even an artificially placed buoy is of only limited usefulness, since, while it can help to determine locations relative to the land (if it is anchored to the bottom of the ocean), it doesn't help to determine locations relative to floating objects. But liquid surfaces bring us to a less empirical and more absolutist conception of surfaces and the 'spaces' they support.

2.7 The Substantiality of Surfaces

The Leibnizian conception of surface constancy and change suggests that surfaces stand to the features on them as Space stands to the bodies that move about in it. But the surfaces that we are concerned with seem to require something more, and one indication of this is Euclid's definition of a surface as the extremity of a *solid* (Book IX, Def. 2), which seems not to apply to the surface of a liquid like a lake. We shall not assume that the bodies whose surfaces we are concerned are perfectly solid or rigid, but we must at least assume that they are resistant to the touch, because that is presupposed in the 'tracing test' for feature coincidence described in section 2.3.[23] Of course, the fact that our surfaces are those of 'topologically solid' bodies does not imply that the *the surfaces* are material. As Euclid also says, the surfaces of solids have length and breadth only (Book I, Def. 5), which is inconsistent with their being material. Hence it is implausible that surfaces should be especially thin material bodies, since that would imply that they should have such material properties as weight and thickness.[24]

But that our surfaces should be surfaces of material bodies does suggest an alternative characterization of constancy and change relative to them. That is that a feature like a location mark on a body's surface should remain in a constant relation to the matter of which the body is formed. What these constant relations might be is difficult to spell out exactly, but a water-wave is a counterexample. The wave might not change in its relations of incidence to other waves, but its 'material substrate', i.e., the water in the wave, is constantly changing. However, the

[23]Given that our surfaces must be resistant to the touch, it is implausible that they should include the ones referred to in Aristotle's characterization of *the place of a body* as 'the surface of the *surrounding* body' (*Physics*, Book IV, Chapter 4; McKeon, 1941, p. 276).

[24]But note that Avrum Stroll says "...to be at a physical place entails having some thickness" (Stroll, 1988, p. 42), which appears to be inconsistent with the kind of surfaces that we are concerned with being in physical places. Of course they are not in places on *surfaces*, but we would argue that they are in places in most physical spaces.

substrate cannot be observed *apart* from its sensory manifestations, and so the wave's 'absolute motion' relative to the water can only be inferred from changes in observable features of the wave and their relations to other observables. There are two points to make about this.

One is that the fact that the surface of a lake doesn't support observable features that stand in constant incidence relations to one another is explained by the liquidity of the water in the lake, and the fact that observable features of the liquid, e.g., ripples on the lake's surface, do not stand in constant relations to one another.[25]

The other point harks back to the Leibnizian motif, that surface features are to surface spaces as bodies are to physical spaces. In the light of what was just said, we might now say that surface spaces are more like an 'aether' that has physical properties, e.g., curvature, than they are like 'the void'. However, the fact that it is harder to find bodies in the aether that maintain constant incidence relations to each other than it is to find features on 'topologically solid' surfaces that maintain constant incidence relations to *them*selves suggests that there is less evidence for the existence of an 'aetherial substrate' to Space than there is for the existence of material substrates to bodies' surfaces.

But this leads to concluding comments on some general philosophical issues.

2.8 Ontological and Epistemological Remarks

The evidence just noted, to the effect that that bodies' surfaces are supported by material substrates, is inferential or 'abductive' in Peirce's sense, and obviously such reasoning was rejected by idealists like Berkeley, while Russell, Whitehead, and others attempted to 'construct' the substrates. The present comments do not pretend to resolve the epistemological issues, but they do suggest taking another 'dimension' into consideration.

In "The relation of sense data to physics" Russell constructs a penny from its 'appearances in perspective space' (Russell, 1914, section vii).[26] These appearances are elliptical in most perspectives, but Russell said nothing about the penny's surface, the lettering and other features on its faces, or the milling or smoothness of its edge. Unlike the appearance in a perspective of being elliptical, the lettering and milling are not

[25]Weyl's comment on the similarity between material substance and Space is worth quoting in this connection: "But in a completely homogeneous substance without any quality the recognition of the same place is as impossible as that of the same point in homogeneous space" (Weyl, 1949, p. 165).

[26]Russell says that a 'sensibilium', like appearing elliptical from a perspective, is physical and not subjective, but calling this an *appearance* gives it away.

appearances, and they are not 'in' perspectives. They are where the penny is, they 'inhere' in it, and if inferences there be, *they* give the penny its appearance, and its look and feel of substantiality. But how came it to be that they were overlooked in the 'epistemology of the penny'? Perhaps it was only traditional mind-matter dualism, but why this dualism? Plausibly, something deeper lies behind that.

Body and mind are generally conceived of as causal agents, and as Physics describes the causal laws in accord with which bodies act on bodies, so traditional philosophies attempted to describe the laws in accord with which minds act either on themselves or on bodies. Bodies influence bodies by physical forces, and minds influence bodies by acts of will. But only the *agents*, the force-producing bodies or the will-producing minds, are conceived to be real, self-subsistent things. However each of these things has *qualities*, and these are not self-subsistent. The preeminent qualities of bodies are their physical properties, size, shape, mass, etc., while the preeminent qualities of minds are properties of 'appearances', 'sense impressions', 'presentations', or 'ideas'.[27] But surfaces, and more especially surface *features*, have no place in this scheme, and the primary reason, one suspects, is that they are not conceived to be self-subsistent causal agents. Without the body there is no surface, and without the surface there is no surface feature.

But we are now suggesting that the existence of a body's surface is inferred from its surface features, and therefore these features play an important role in our knowledge of the external world.[28] In fact, we find things in the traditional picture of the mind that suspiciously resemble them, particularly in dimensionality.[29] Berkeley's 'visible moon', which is a 'visible idea', is described as a "round luminous plane about 30 visible points in diameter" (*New Theory of Vision*, par. 44), and Nelson Goodman says that "A visual presentation of a baseball, for example, is spatially bidimensional, whereas the baseball, for example, is spatially tridimensional" (*The Structure of Appearance*, p. 95).[30] H.H. Price even

[27]We ignore here the traditional philosophical distinction between primary and secondary physical qualities, the primary ones of which correspond to qualities of ideas, and the secondary ones of which, like color, really only inhere in ideas that do not correspond to physical properties.

[28]Note how close this is to Berkeley's language analogy, expressed in the famous dictum "...the proper objects of Vision constitute the Universal Language of Nature," (*New Theory of Vision*, 1st edition, par. 147).

[29]Many of the points to be touched on in this section will be returned to in Chapter 17.

[30]But note that the *roundness* that Berkeley ascribes to the visible moon is an attribute of something that might appear in a two dimensional *picture* or 'representation' of the moon, but not an attribute of the moon itself, or of any feature of it.

says "Naive Realism holds that in the case of a visual or tactual sense datum, belonging to means the same as *being a part of the surface of*: in that literal sense in which the surface of one side of this page is part of the whole surface of this page," (*Perception*, p. 26), and in a footnote he adds that *belonging to* is sometimes called 'being an appearance of'. Later he even makes the telling remark that "'Surface', it is true, is a substantive in grammar; but it is not the name of a particular existent ..." (*Perception*, p. 106), which, if 'self-subsistent' is substituted for 'existent', agrees fairly closely with what we have said above.[31]

But there is one crucial way in which appearances as they have traditionally been conceived differ from surfaces and their features. Surfaces can be superimposed face to face but appearances cannot be, and in fact superposition plays almost no part in traditional theories of perception.[32] Moreover, the neglect of superposition is what leads to all of the difficulties that sense-datum empiricism has in accounting for geometrical ideas, especially distance and most especially distance from the observer. Certain philosophical problems of superposition will be returned to in Chapter 11, but here we will simply remark that perhaps the most important philosophical advantage of our approach to the basis of geometry is that it highlights the role of superposition as the route to distance and externality.[33]

[31]But we emphatically do not agree with the immediately following remark that surfaces are *attributes*.

[32]This must be qualified. For instance, Berkeley and Price both have very interesting 'deconstructive' comments on it. Berkeley's remarks in entry #528 of Notebook A of the *Philosophical Commentaries* (p. 301 of the Ayers edition) are too long to summarize, but they are akin to Price's comment on p. 237 of *Perception*, "...it is obvious that one sense-datum A cannot literally be superimposed on another B, for that would require that B as a whole, or some part of it, shall exist when not actually sensed." Of course, this only proves that surfaces and surface features are not sense data. This is related to the special problems of superposition discussed in Chapter 11.

[33]Chapter 17 comments on what seem to the author to be serious problems in current psychological theories of visual perception that arise from neglecting the 'tactile operation' of superposition.

3

The Logic of Constructability

3.1 Introduction

Key postulates and propositions of Euclid's *Elements* are formulated as propositions concerning the possibility of constructing or producing things having particular properties. The instance, Postulate 1 of Book I is stated as "Let the following be postulated: to draw a straight line from any point to any point" (Heath, p. 195), and Proposition 1 of that book is "On a given finite straight line to construct an equilateral triangle" (Heath, p. 241). Paraphrasing, the postulate asks the reader to grant the possibility of drawing a straight line from any point to any point, and the proposition both describes a method, using just the postulated construction 'tools', for constructing an equilateral triangle on a given line, and it proves that what is constructed by this method is an equilateral triangle. Putting aside the illocutionary force of asking the reader to grant something, which is not the same as asserting it, the thing to stress here is that both the postulate and the proposition concern not the *existence* of things of different kinds, but rather ways of *producing* things of these kinds.[1] This is in contrast to modern expositions even of 'Euclidean' Geometry, which typically reformulate propositions concerning methods for constructing things as ones concerning the existence of the things constructed. What is the significance of this change, and is it important from the point of view of theory? It is important to say something about this, because our operational formulation of the 'topological underpinning' of geometrical theory is constructive in much the same way as Euclid's formulation of the theory itself is.

[1]But recall Plato's animadversions against the 'absurd language of the mathematicians' who regard mathematics as being concerned with *doing* things. Whether or not Euclid, who was allegedly a platonist, was following a preexisting constructivist tradition in formulating geometrical propositions, more or less against his better judgment, that tradition had some validity in our judgment, and that is part of our reason for following it here.

Here are two observations. First, plausibly Euclid was concerned with *applications* in a way that modern pure geometry tends not to be. Many of the propositions of *The Elements* resemble *recipes* for doing various things, e.g., for constructing equilateral triangles.[2] Of course there is much more than that to *The Elements*, since many if not most of its propositions are theorems that state *facts*, such as the Pythagorean Theorem (Proposition 47 of Book I, Heath, pp. 349–369). Moreover even its recipes, or *problems*, contain arguments proving that what is produced by the given method meets the specified requirements, e.g., of being an equilateral triangle. In the course of time this 'factual-deductive' side of Geometry has come to dominate the subject to the point of excluding its technological side,[3] but persons seeking to understand the applications of Geometry are not ill-advised to look to its older and more technologically oriented formulations.

The second point, which may seem more important to modern methodologists of science, has to do with the epistemic status of geometrical propositions. Clearly these cannot be assumed to be straightforwardly empirical. E.g., if straight lines were assumed to be concrete empirical observables, the proposition that there *exists* a straight line connecting any two points would be blatantly inconsistent with the empirical facts. This is related to Protagoras's criticism of the science of Geometry, which Plato responded to by putting forth the theory that Geometry is concerned with entities that transcend the senses and can only be grasped by reason, and which can be viewed as a direct progenitor of Kant's doctrines. Absent an empirical side we may even be led to a view like Poincaré's that the primary test of geometrical theory is absence of contradiction (cf. Poincaré's paper "Non-euclidean geometries," in *Science and Hypothesis*, esp. p. 50). But technological constructivism is an intermediate position between simple empiricism and idealism. There may not *exist* a 'concrete' straight line between two points, but it may be possible to *draw* one. At any rate, Postulate 1 asks the reader to grant this possibility, which may not be so difficult when the field of operation is

[2]Note how similar the 'labels' that Euclid uses for his constructive propositions are to labels of recipes. Thus, the label of Proposition 1 of Book I of *The Elements* is "On a given straight line to construct an equilateral triangle," which looks like the label or heading of a recipe for making equilateral triangles.

[3]This is not the place to enter in detail into the history of the evolution of Geometry from technological-constructive to factual-deductive formulations, but it is interesting to note that while the 2nd edition of Legendre's great textbook *Éléments de Géométrie* maintains the distinction between problems and theorems (e.g., Problem V of Book II is stated as "Diviser un angle ou un arc donné en deux parties égales"), in the 15th edition the 'Problems' are relegated to a special section (pp. 49–59).

restricted to a plane surface.[4] Of course, even in that application what is to be granted is idealized, since it isn't plausible that perfectly straight lines can be drawn or that perfectly flat planes are available to draw them on. But the topological theory to be developed in this work will show how certain of these idealizations can be dealt with. E.g., Chapter 10 deals with the seeming problem that arises because the things that are drawn to 'represent' straight lines aren't perfectly one-dimensional. In any event, the constructive formulation of the present theory hews as closely as possible to 'realistic' possibilities, and its plausibility depends on this—though admittedly it still involves idealization.

Let us now briefly consider the logic of constructability, as it is exemplified in Euclid's theory and as it will be exemplified in the topological theory to be developed in Parts II and III.

3.2 The Logic of Constructability

The claim that it is possible to produce something that doesn't exist is clearly a modal one, and it should be formalizable within the framework of modal logic. Unfortunately the only branch of modal logic with which author is at all familiar is that which deals with propositions like "It is possible that there is water on Mars," which can be analyzed in terms of possible worlds. But taking the expression 'possible world' literally, it seems inappropriate to formulate a statement like "It is possible to draw a line from point A to point B" as being about any world other than our own. So, putting aside the possibility of a full-scale possible worlds formulation, let us sketch a more modest alternative.

There is *a* sense in which constructability in Euclid's theory can be treated logically as existence: namely as existence or membership in a *class of constructabilia*. Then, that a line can be drawn between any two points can be interpreted as saying that, given any two points, there exists something in the class of constructibilia that is a line containing the two points. This suggests that constructability can be formalized in a two-sorted logic, one of whose domains is the class of things in this world,[5] and the other, partially overlapping one, is the class of things that can be constructed. If points were assumed to be concrete existents they would belong to the first class,[6] while the straight line that could

[4]It must be admitted that Books XI–XIII apply the postulate to points in space, e.g., in the proof of Proposition XI.11 (Heath, Vol. III, p. 292). In the author's view the assumptions of the three books on solid geometry are much less plausibly grounded than the ones of the books on plane geometry are. This is commented on in section 13.2.

[5]The temporal side of existence, i.e., the possibility that something may exist at one time but not at another, is left aside here.

[6]But, as section 1.3 suggests and as will be returned to below, we will not assume that.

be drawn between them would belong to the second.[7]

Geometrical *constructabilia* are assumed to have all the standard geometrical properties, and to be subject to all of the geometrical operations that actually existing objects are subject to. Thus, constructible straight lines have the same kinds of extremities, i.e., points, as existing lines have (Definition 3 of Book I),[8] and, paraphrasing Proposition 4 of Book 1, two constructible triangles must be congruent if they have two sides and the included angles equal. These things follow because the general propositions of geometrical theory are understood to apply to all figures, existing or constructible. This includes generalizations about constructablility itself, e.g., that an equilateral triangle can be constructed on any constructible line.[9]

The foregoing leads to two remarks on the *inductive logic* of constructability, i.e., about how generalizations like "it is possible to draw a line between any two points" are confirmed. Now, it seems to the author that natural laws of all kinds apply not only to all actually existing things but to all 'naturally possible' ones, and these include geometrical constructions. But how we confirm general *technological* propositions, about what can always be *done*, e.g., how we confirm a proposition to the effect that straight lines can always be drawn between two points, is less easy to specify. Presumably practice in drawing lines gives us confidence in the possibility of doing this, at least approximately, while no such practice induces us to believe that infinite lines can be drawn, would explain why we postulate the first but not the second. Nor, interestingly, is the possibility of creating a concrete feature as small as a single point postulated, though propositions involving them are ubiquitous throughout *The Elements*. This leads to two final sets of comments.

[7]Our formulation is obviously only a first approximation, since it ignores the fact that Euclid distinguishes between constructions of different sorts. Straight lines are *drawn* (Postulate 1), or *extended* (Postulate 2), circles are *described* (Postulate 3), equilateral triangles are *constructed* (Proposition 1), and so on. In any event, the theory developed in Parts II and III only requires two domains—the actual existents and the constructibles.

[8]Note that this implicitly assumes that straight lines are what we would call straight line *segments*. Euclid occasionally speaks of infinite straight lines, as in Proposition 12 of Book I, but these are explicitly characterized as such. I find nothing that seems to imply their constructability, and if they are not then those that exist are non-constructible existents.

[9]A point made earlier should be reiterated here, that it is the fact that constructions are in the *plane* that makes it plausible that they can always be carried out, e.g., that a straight line can always be drawn between two points. Moreover, this seems to assume that the things that are drawn are conceived 'genetically', as in section 2.3. Otherwise the presence of the vertical segment between the two dots in the figure '• | •' would preclude drawing a straight segment connecting them.

One is that while there is no explicit reference in *The Elements* to a class of constructible figures, nevertheless such a class could be regarded as a primitive concept of the theory. The class is closed under the explicitly postulated construction operations, and it seems to be limited to things whose constructability is *practicable*, even if idealized. But there is a large range of variability between ruler and compass constructions and *possible* geometrical constructions. What can be constructed with ruler and compass is an interesting and profound question,[10] but classical geometers studied other means of construction, e.g., the Spiral of Archimedes that was used in solving the problems of trisecting the angle and duplicating the cube. This shows that these mathematicians were concerned not just with the 'Euclidean' constructions but with other 'practicable constructions' as well (cf. Knorr, *The Ancient Tradition of Geometric Problems)*.

But even beyond possible constructibles like Archimedean Spirals, more things between heaven and earth are dreamt of in the 'philosophy' of *The Elements*. These include infinite lines, which by no stretch of the imagination can be regarded as constructible.[11]

As suggested in section 2.3, the most ubiquitous of all 'elements of *The Elements*', namely *points*, are abstract things like universals, that *concreta* have in common. What concreta that have points in common have is *location*, and infinite lines are *loci*. The following chapter will consider the 'logic of locus abstraction' in more detail.

[10]See Chapter VIII, by L.E. Dickson, in Young, 1955.

[11]Note Gauss's remark "I must protest against the use of infinite magnitude as something completed, which is never permissible in mathematics. Infinite is merely a way of speaking.." (quoted from Bell, *Men of Mathematics*, p. 556).

4

Remarks on Physical Abstraction

4.1 Introduction

This chapter extends the discussion of geometrical abstraction that was begun in sections 1.1 and 3.3, and tries to make it plausible that concepts of geometrical abstraction derive from 'proto-geometrical usage'. For example, "its place on the shelf" can be regarded as a proto-geometrical part of the ordinary language statement "He returned the book to its place on the shelf," since it seems to refer to a 'place', and that is related in turn the concept of a geometrical point. We will also try to make it plausible that proto-geometrical abstractions are closely related, ideologically, to other physical abstractions[1], including colors, times, directions and shapes, as well as to classes. These matters will be discussed unsystematically, but we will approach all of them by way of certain expressions of definite description form that refer to the abstract entities in question, and the existence and uniqueness assumptions that Russellian analyses of these expressions seem to presuppose.

Here are examples of statements involving definite descriptions of the kind that we are concerned with, which are underlined:

S_{class} Marco Polo belongs to *the class of persons born in Venice in the 13th century.*

S_{place} Marco Polo came back to *the place from which he started.*

S_{color} That turquoise has *the color of the sky.*

S_{shape} The figure + has *the shape of a cross.*

---

[1]Much of this chapter recapitulates ideas set forth in Adams, 1993, which is especially concerned with the abstractions that enter into physical theories, as contrasted, say, with types of persons, governments, and so on. These ideas are only sketchily developed both here and in the paper, and they call for much more detailed study. The author also acknowledges his limited acquaintance with the literature, both classical and current, on abstraction in general, or on the closely related subject of universals. Only occasional remarks will be made, especially in section 4.8, on contrasts between the present approach to abstraction and other recent approaches.

$S_{direction}$ He started in *the direction of the town.*

S_{time} Richard the lion-hearted ruled England at *the time of the crusades.*

Each sentence is of the form "X stands in relation R to the ϕ," where X is the name or other expression referring to a concrete particular, Marco Polo, Richard the lion-hearted, etc., and ϕ is a definite description of an abstract class, place, color, shape, direction, or time. According to Russell, prefixing ϕ by the definite article presupposes that one and only one such place, color, shape, direction, or time fits the description.[2] Our primary concern in this chapter will be with what guarantees that these presuppositions are valid, or, more generally, what guarantees the 'degrees of validity' that they have. Along the way, however, we will consider several subsidiary matters, including the following: (1) The *factual basis* of these presuppositions, which are associated with fact that they rest on *a priori* linguistic practices that in turn presuppose certain synthetic generalities. (2) The *modal* character of these generalities, which is associated both with possibilities of production, and with the idea that abstract entities of a given category generally form a kind of 'container space' (though that is not true in the case of classes). (3) The *instantiation relation*, R, between the *concretum*, X, and *abstractum*, ϕ, that it 'exemplifies',[3] and the associated *coincidence relation* that holds when two *concreta* instantiate the same abstract thing, and the possibility of reducing that to a binary empirical relation between the *concreta*. Other things will also be considered, but let us begin with instantiation.

4.2 Instantiation, Individuation of Abstracta and the Dual Interpretation of Coincidence

Instantiation relations vary with and are characteristic of the abstract category. Concrete things *belong* to classes, they are *at* places, they *have*

[2] We are not presupposing that the Russellian analysis applies to all expressions of definite description form. E.g., one questions its application to the sentence "He is repairing the hole in the roof." But it is hoped that the reader will find it worthwhile to explore the consequences of the hypothesis that the analysis applies in the contexts under consideration here.

[3] It is not easy to choose right term for the relation R, and other authors have used other terms for what we are calling instantiation. Thus, Donagan, 1963, uses 'exemplify' for the relation between a particular thing, say a turquoise, and the abstract color that it exemplifies. But while this seems appropriate for colors, it is less so for the relation between Marco Polo and the place to which he returned. 'Incident in' might be better for the latter, but that would be totally inappropriate for the relation between the turquoise and its color. However, *coincidence* seems reasonably apt for all cases in which two or more *concreta* instantiate the same abstract thing. Thus, X and Y are coincident in color if they have the same color, and they are coincident in place if they are at the same place.

colors and shapes, they *go in* directions, and happenings are *at* times.

The kinds of things that instantiate *abstracta* of a given category vary with the category. All 'things' including other classes can belong to classes, but only spatially localized things can be at places. Colored things have colors, although they need not be spatially localized since the sky is blue but it isn't at a place. Only shaped things have shapes, and the sky isn't one of them. Not things but *motions* have direction, and *happenings* are at times.

Abstracta are *individuated* by reference to *concreta* that instantiate them. The color of the turquoise is individuated as that of the sky: it exemplifies the blueness of the sky. The shape of the figure + is individuated as that of a cross: it exemplifies the shape of a cross. And so on. In general, an *abstractum*, ϕ, individuated in this way is described as "the ϕ to which Y stands in relation R," where R is the instantiation relation characteristic of the abstraction, and the whole statement "X has relation R to ϕ" has the form "X has relation R to the Φ to which Y stands in relation R." I.e., X and Y are coincident in respect to Φ.[4] "That turquoise has the color of the sky" says that the turquoise and the sky are coincident in respect to color.

The fact that statements like "That turquoise has the color of the sky" can be paraphrased as "the turquoise and the sky have the same color" suggests that they can be interpreted as simple empirical claims, in spite of the fact that they seem to refer to universals like the color of the sky. Specifically, "That turquoise has the color of the sky" appears to be equivalent to a coincidence statement, and such statements commonly have dual interpretations.[5] I.e., that two things, X and Y, have the same color can be interpreted either as an abstract existential claim "There exists a color that is instantiated by both X and Y," or as the binary relational claim "X matches Y in color," where matching in color is a seemingly straightforward empirical relation. More generally, if "X has relation R to ϕ" is equivalent to "X has relation R to the Φ to which Y stands in relation R," then it can be interpreted either as "there exists a Φ that both X and Y instantiate," or as "X is the same as Y in respect Φ," where sameness in respect Φ seems to be a binary empirical relation—sameness of color, shape, direction, and so on. But these relations may not be as simple as cursory consideration might make them seem.

[4]Note that ϕ stands for a *particular* color, say sky blue, while Φ stands for the category, *color*, to which ϕ belongs.

[5]Frege noted this in §64 of *The Foundations of Arithmetic* in the case of direction. One of Euclid's important contributions was to define *sameness* of direction, namely parallelism, as a binary 'empirical' relation. The reduction of co-directionality to a binary relation will be returned to in section 4.4.

Consider sameness of color.[6] The simplest color comparisons would seem to be between objects that aren't parti-colored, e.g., between two pure turquoises that can be compared 'side by side', under good illumination. Comparing them with chips or patches in Munsell color-charts might perhaps be better and spectroscopic examination might be still better, although that might have to be corrected to allow for the fact that it could show up differences between metameric indistinguishables. Each of these refinements could be said to refine the abstract category of color, perhaps in such a way that in place of one abstract color, say turquoise blue, a number of *shades* are ranged under it. But different complications arise in the case of parti-colored objects, which are important here because of their parallels with the category of places and the things that instantiate them.

We might say that two houses were painted the same color if their colors matched in most places, in spite of the fact that one had a green trim and the other a brown one. Of course the houses' color-sameness would only be 'approximate',[7] but then that is what is found in almost all practical color comparisons. In this case we are inclined to speak of *distributions of color* rather than of differences in shading. And, this suggests that speaking of *the color* of a house, as though it had only one color, is imprecise, and to speak precisely we should specify its *range* of colors. [8] And, if we do this we must be more careful in our use of the terms 'instantiate' and 'individuate'. The house may instantiate a given color if that color occurs in its color-distribution, but it only individuates the color if no other colors occur as well. Of course we may still speak loosely of "the color of the house" when some color predominates in the color-distribution.[9] Also, now rather than saying that two houses

[6]What follows in the remainder of this section and in much of the next concerning colors will focus on a rather narrow aspect of the enormously complicated subject of color concepts and measurement, and it will ignore 'dimensions of color' that are of the greatest importance in that science, especially variations of context and illumination. It should be said, too, that the approach to color-individuation and abstraction that will be developed here is the author's own, and very little will be said by way of comparing it to other approaches.

[7]Chapter 1 of Price, 1953, makes closely related points about exactness of 'resemblances'.

[8]On the other hand, we might be inclined to dismiss talk of precision as misguided, and speak instead of 'family resemblances' (§67 of Wittgenstein's *Philosophical Investigations*; Wittgenstein goes more deeply into color in *Remarks on Colour*, but those focus largely on aspects of color that are ignored here). However, we want to stress the likeness between color-coincidence and place- or time-coincidence, which seem less appropriately representable as family resemblances.

[9]Of course we may also choose to say that "the color of the house" designates a vague or fuzzy color, but that will be put aside here.

have the same color we might say that some abstract color-shade occurs or predominates in the color-distributions of both of them (this will be represented diagrammatically in the following section). Of course, putting matters this way makes the relation of approximate or partial color-coincidence more complicated than it seems to be when it is regarded as a matter of simple color-comparison, but it does invite us to look more carefully at this relation. This cannot be entered into in detail here, but we can now profitably consider how the complications just noted are reflected in *processes* of individuation. We will continue with color.

4.3 Processes of Individuation

One reason for not wanting to say that two houses individuate one abstract color if they are approximately coincident in color is that approximate color-coincidence is not a transitive relation. This non-transitivity implies that abstract colors cannot be identified with *equivalence classes* of things that are approximately the same in color, in the same way that numbers are commonly defined as equivalence classes of sets that are the same in number, because their members can be placed in one-one correspondence.[10] But the fact that parti-colored things are conceived to have color-distributions shows that abstract colors are still in the picture, because these are what are distributed over them. Moreover, these abstract colors are still related to color-coincidences, though not so simply that they can be represented as equivalence classes of color-coincident things. What follows briefly outlines one plausible way in which they may be related to color-coincidences, and even to a kind of coincidence-transitivity, which is significant because of its analogies to both place and time abstraction.

We can speak of color-matching as being *locally transitive at* a concretum, C, if any two things that are match C in color match each other in color,[11,12] and if *C* has this property then we can say that it is *uni-*

[10]Adams and Carlstrom, 1979, and Adams and Adams, 1986, discuss various practical means of meeting the difficulty of non-transitivity. The latter focuses on scientific classification, and the importance of taking into account the *purposes* for which classification systems or typologies are constructed. But the case of color fits in particularly well with that of place.

[11]Note the similarity to Euclid's *common notion* (Axiom) 1, "Things that are equal to the same thing are equal to one another," (Heath, Vol. I, p. 222). Proposition 30 of Book I of *The Elements* proves this kind of transitivity of the relation of parallelism (Heath, *op. cit.*, p. 314).

[12]This can be nuanced as a *degree of transitivity* at C, or as the degree of truth of the propositional function "if any *X* and *Y* are both coincident with C then they are coincident with each other." A 'degree' of this kind of transitivity at C is defined in

formly colored. Thus, we say that a house is *not* uniformly colored if it doesn't satisfy the transitivity condition. For instance, if one thing matches its trim while another matches its walls, but they two don't match each other, then the house is not uniformly colored. Conversely, to the extent that Munsell chips satisfy the transitivity condition we can speak unambiguously of 'their colors'. [13] Given this, the chips would *individuate* colors, and if all colors could be individuated in this way we would have a complete account of color-abstraction. But this is a big 'if', the detailed examination of which cannot be entered into here, and we will end the discussion of processes of color-abstraction with four brief comments.

First, it may help to picture the relations that are under consideration in a diagram that represents the 'color space', in which the colors of things like houses X and Y are represented by the regions in the space that their colors are distributed over, and a Munsell color-patch C is represented as an unextended 'color-point':

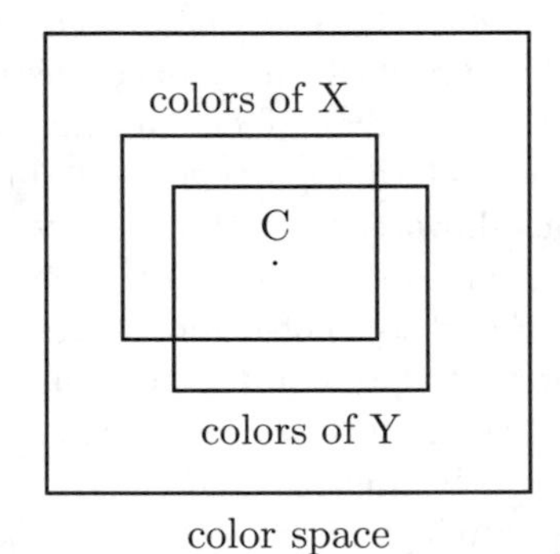

color space

X and Y are represented above as being approximately coincident in color because most of their colors are the same, though obviously neither

Adams and Carlstrom, 1979, as the proportion of pairs, $< X, Y >$, that satisfy the formula '$(XRC\&YRC) \rightarrow XRY$', where '$\rightarrow$' is the material conditional. A better measure would be to define it as the proportion of these pairs that satisfy 'XRY' out of all those that satisfy '$XRC\&YRC$'. Adams and Carlstrom use the material conditional measure in proving that any 'system' $< D, R >$, where R is a binary relation over a finite domain, D, in which the axioms of reflexivity, symmetry, and transitivity are 'highly true' must be 'close to an idealized system', $< D, R' >$, in which these axioms are exactly true, where the 'distance' between $< D, R >$ and $< D, R' >$ is defined as 1 minus the proportion of 'exceptional pairs', $< X, Y >$, that don't satisfy the biconditional '$XRY \leftrightarrow XR'Y$'.

[13] However, section 4.5 comments on more fine-grained individuation that can distinguish visually indistinguishable colors. It is also worth noting that Munsell chips are *color fast*, i.e., they are unchanging in color, hence they satisfy the conditions discussed in section 4.7 for being *color standards*.

of them individuates a unique color. But they instantiate a *common color*, C, and they would *individuate* C if C were the only color in the intersection of their color distributions.[14]

Second, Munsell color chips are, so to speak, 'prototype samples' of colors, which have a logical function akin to prototypes that are utilized in other systems of classification. Certain of these are discussed in Adams and Adams, 1991, (e.g., pp. 255–257), which makes the key point that these systems and their prototypes are selected or constructed for particular purposes. This is reflected in the fact that the Munsell system is also constructed for a purpose: namely that of determining *surface* colors. Thus, Munsell chip 'prototypes' can be placed in close juxtaposition to surfaces and portions thereof, to determine their colors, but they cannot be placed against the sky, a rainbow, reflected image, etc., to determine *their* colors.

Third, if all colors were located in the 'space' of perfectly uniform Munsell-chips, and all things were as uniformly colored as these chips, then the relation of color-matching would be perfectly transitive. This is the idealized scenario in which everything would individuate a unique surface color, and the 'naive conception' of color would be ideal.

But, fourth, even Munsell chips are not perfectly color-transitive, i.e., perfectly uniform in color, and therefore we are not entitled to suppose that they individuate unique colors. Given this, color-coincidences and processes of color-abstraction must be even more complicated than we have heretofore imagined. However, the smallness of Munsell chips, e.g., in comparison to houses, plus the fact that surface color-distributions are distributions over surface *spaces* in which ideally small things are points, suggests that going farther into surface color-abstraction should take into account the relation of surface colors to surface *places*.[15] We

[14]Note that the partial color coincidence pictured above shouldn't be perfectly transitive: the reader can easily imagine a series of regions, X_1, X_2, X_3, and so on, where each member of the series approximately coincided with the following one, but the first and last members were totally disparate—though the closer to exact coincidence the members were the longer the series would have to be for this to happen. In contrast, coincidence at Munsell patch C would be perfectly transitive because, given that C individuates a single color-point, any two things coincident with it would have to be at least partially coincident with each other. But, it must be recognized that Munsell patches aren't 'everywhere' in the space, and individuating colors that don't exactly match any patch is a further problem.

[15]Of course, the close connection between color and form or 'figure' has long been recognized (*viz.* Plato's *Meno*, 75, where Socrates says that figure is the only thing that always follows color). One problem that arises in a detailed examination of the connection is that while there are ideally small surface *points*, the nature of which is commented on below, these cannot easily be associated with colors (e.g., 'color points', by analogy with the 'mass points' that Maxwell theorized about and Quine

are not prepared to do this here, and we will turn next to brief comments on place- and point-abstraction.

The place to which Marco Polo returned is instantiated by Marco Polo, at least at the time he returned to it, but this place is analogous to the color of the house in not being an exact description, and coincidences in places of this kind are not perfectly transitive. Now, we do have an empirical test for a more exact kind of coincidence, namely that bodies are coincident in this sense if they touch each other. But this kind of coincidence isn't transitive either, and it doesn't necessarily individuate a single place since two bodies can touch a third body without touching each other.[16] This is analogous to the problem of individuating colors that arises from the fact that color-coincidence is not perfectly transitive. The problem was provisionally solved in that case by constructing uniformly colored Munsell chips, C, at which color-coincidence is *locally transitive* in the sense that any two things that are coincident with C in color must be coincident with each other in color. Now we must ask whether there exist spatial concreta, C, that are analogous to Munsell chips in having the property that any two things that are spatially coincident with C are spatially coincident with each other. I.e., are there objects that satisfy the 'local spatial transitivity condition', that any two things that touch them necessarily touch each other?

We saw in section 1.3 that an extended thing like the small dot "•" cannot satisfy the local transitivity condition, because it is possible to draw things that touch it without touching each other—as in the figure "‖•" Therefore, the only things that could satisfy the condition would have to be unextended, i.e., they would have to individuate spatial *points*. These would be the analogues of Munsell chips, which are in a sense 'unextended in color'. But do spatially unextended things exist? Section 1.3 has already argued that if they do exist they cannot be seen or otherwise discriminated by the senses. 'Non-empirical' grounds affirming the existence of such things will be discussed in section 4.5 and 'in depth'

comments on p. 249 of *Word and Object*). We would suggest that dealing with this and related problems requires bringing into the picture surface *topologies* and the *limit* and *continuity* concepts associated with them.

[16] Aristotle's characterization of the place a body, defined terms of the surface of the surrounding body (*Physics*, V, Chapter 4; commented on in footnote 21 of Chapter 1) characterizes the place that is coextensive with the body, and coincidence in that place is necessarily transitive. However this kind of coincidence can only be physical identity (at least when the coincidence is at one time), which we are not concerned with, and moreover not all places can be individuated in this way. In particular, we have already noted Aristotle's identification of a *point* with the place where it is (*Physics*, IV, Chapter 5), but unless we assume that these are individuated by 'mass points', which would be inconsistent with Euclid's view that the extremities of solids are *surfaces*, they cannot be individuated by bodies.

in Chapter 6. However, we now come up against a problem that greatly exercised the classical geometers, which modern axiomatic theorists side-step by treating *point* as an undefined or 'primitive' concept, but which cannot be side-stepped by us. For the present we will simply note the unsatisfactoriness of three attempts to individuate spatial points, as a prolegomenon to the theory to be developed in Chapter 6, according to which points are abstract objects that stand in clearly defined relations to concrete observables that individuate *places*, though not unextended points.

One theory is that points are the places of *material points*. Definition 1 of Book I of *The Elements*, "A point is that which has no part" could be interpreted in this way,[17] but this view has seldom been considered in modern times because of the doubtful provenance of such objects.[18]. Maxwell comments on them in Chapter 1 of *Matter and Motion*, and they are considered as possibilities in Continuum Mechanics (cf. Truesdell, 1977, p. 160). As said, whether or not material points do or could exist, they must be too small to be observed, and therefore an empirical account of point-individuation could not be based on them. This is one respect in which unextended spatial points are unlike unextended 'color points', since perfectly uniform Munsell chips might be small, but at least they could be seen.[19]

Another approach to individuating points is by way of the *extremities* of things that can be seen. This is also hinted at in one of the definitions in *The Elements*, namely "The extremities of a line are points" (Definition 3 of Book I).[20] For example, in the case of the cross in Diagram 1 of section 2.4, a single point might be individuated as follows:

[17]Heath's commentary, pp. 155–158 in Heath, 1956, is an excellent summary of ancient as well as fairly recent views on the concept of a point.

[18]But there are at least some theoretical grounds for it. Newton theorizes about material points in Book II of the *Principia*

[19]Another curious disanalogy between colors and places that is worth noting is that there is no chromatic analogue to the *exact place* that is coextensive with an extended body. For instance, even though a tricolor flag has an exact place it does not have an 'exact color'. This is related to the fact that while smaller places (and smaller *times*) can be *parts of* larger ones, colors like the blue in the tricolor are not parts of 'the color of the tricolor'—which doesn't exist.

[20]This is one of a trio of definitions that also include "The extremities of a surface are lines" (Book I, Definition 6), and "The extremity of a solid is a surface" (Book XI, Definition 2), which, combined with Definitions 5 of Book I and 1 of Book XI, "A surface is that which has length and breadth only" and "A solid is that which has length, breadth, and depth," approach the modern concept of dimension, to be returned to Chapter 8.

The arms and staff of the cross are two-dimensional but their edges are one-dimensional (cf. Chapter 8), and a place where the edges meet, like the one pointed at by the arrow above, looks as though it individuated a single point. But while this might be so, section 2.4 also noted that as yet we have no clear incidence test that applies to 'dependent' boundaries and their intersections, and therefore as yet we have no way of establishing that the intersection pointed at by the arrow satisfies the local transitivity requirement, that any two features that touch it must touch each other. Such a test will be described in Part II, but in the meantime we may note the third and recently most popular approach to characterizing spatial points.

This is in terms of *limits*, of which the best known example is Whitehead's theory of *extensive abstraction* (cf. Part IV of *Process and Reality*). The basic elements of that theory are *actual entities*, which are time-extended elements of subjective experience, but since the formalism is "... equally applicable to any scheme of relationship whatever...," (*Process and Reality*, p. 449), we will interpret them as regions of a physical space. The basic relation between these entities is that of *extensive connection*, but their diagrams (Diagrams I, p. 450 of *Process and Reality*) represent them as regions that have a common point. Of course, starting with spatial regions is unacceptable from our point of view, since such regions are themselves abstractions that may or may not be individuated by concrete observables. But even assuming that Whitehead's regions were coextensive with concrete observables, the theory that is based on of them does not seem to account for the kinds of facts that we are concerned with here. For instance, in section 1.3 we 'proved' that the segments forming the tepee, ⧖, do not meet in a common point because one its sides can be covered with ovals, no one of which touches all three of its sides (Figures 1A and 1B). On the other hand, the segments forming star, ⚹, cannot be 'separated' by such a test. Whiteheadian arguments to the same conclusions would have to show that an abstractive 'point class' (Definition 16, p. 456 of *Process and Reality*) is 'incident' in all of the segments forming the star, and no such class is incident in all of the ones forming the tepee.

In spite of the fact that Whitehead's theory purports to be a theory of experience, it is not clear how it can account for the facts just cited, both because it does not involve constructive *tests*, and because there is no obvious way of taking into account the difference between what one sees when looking at the tepee and what one sees when looking at the star. This is connected with the fact that very little effort is made to show that the theory's definition of a geometrical point satisfies intuitive preconditions for being a 'real' definition, e.g., that something is a point in the sense of the theory if and only if any two regions that are incident in it must be connected. But we ourselves have not yet succeeded in individuating places that satisfy this requirement, and, following lines sketched in section 1.3, we will postpone consideration of this subject to section 4.5, and turn for now to *times*, which raise similar problems.

Times are more closely analogous to places than to colors, at least in that a longer time can have a shorter one as a part (cf. footnote 19). The time of the crusades is a vague, longish time, and 13th century is a shorter and more exactly defined one that is part of it. These times are coincident, but processes of determining contemporaneity are difficult to spell out, and prior to Einstein they seem just to have been taken for granted. In any event, this kind of coincidence isn't transitive, and, as with colors and places, we may ask whether there are 'unextended' happenings that individuate times, i.e, that satisfy the temporal transitivity requirement that any two happenings that are contemporaneous with them are contemporaneous with each other. It seems intuitively that they would have to be instantaneous, and therefore no more perceptible or 'witnessable' than spatial points. We would argue that in a sense this is right: there are no publicly perceivable *independent* happenings that satisfy the temporal transitivity requirement and individuate instants. But we may briefly note two indirect ways of individuating them.

Certain *temporal extremities*, i.e., beginnings and endings, seem to be both witnessable and instantaneous. We may see the beginning of a race, and if two other happenings, e.g., the report of the starter's gun and the starting of a stop watch, are contemporaneous with this they are contemporaneous with each other. Of course Einstein's critique should cause us to examine this, but at any rate it would seem to be taken for granted by the man in the street. But we would only point out that the start of a race is an extremity in the temporal topology, which is dependent on the race itself, and the analysis of its relations of contemporaneity with other happenings presupposes an analysis of contemporaneity relations between primary, independent happenings, like the race.[21] The second approach deals with this problem.

This approach interprets a theory of *interval orderings* (section 16.2.2

[21] Of course Einstein's considerations are relevant to this, but they bear only on temporal relations between happenings that take place at a distance from one another, and not on local time-ordering or contemporaneity.

of Suppes, Krantz, Luce, and Tversky, 1989, hereafter referred to as 'SKLT') so as to apply it to the comparison of temporal intervals.[22] The interval ordering theory is formulated in terms of systems, $< E, \prec >$, where E is a set of 'elements', which we will interpret here as *events*, and where, given events X and Y, we will interpret $X \prec Y$ to mean that X precedes Y in time; i.e., X ends before Y begins. $< E, \prec >$ constitutes an *interval order* (Definition 4, p. 309 of SKLT), if $\prec$ is irreflexive and for all X, Y, Z, and W in E, if X$\prec$Y and Z$\prec$W then either X$\prec$ W or Z$\prec$ Y (this is a condition for topological linearity, to be discussed in Chapter 10). Then, among other things Theorem 4, p. 316 of SKLT implies the following. If $< E, \prec >$ is an interval order and an associated weak order has a countable order-dense subset, then there exist real-valued functions $b(X)$ and $e(X)$, which, with caution, can be interpreted as the times of the beginning and ending, respectively, of X, such that: (1) $b(X) \leq e(X)$; (2) if $e(X) < b(Y)$ then $X \prec Y$, and (3) if $X \prec Y$ then $e(X) \leq b(Y)$. In words: (1) X's beginning is not later than its end; (2) if X's end precedes Y's beginning then X precedes Y, and (3) if X precedes Y then X's end is no later than Y's beginning.[23] This demonstrates that $b(X)$ and $e(X)$ satisfy essential conditions for being *ordinal numerical representations* of the *times* of X's beginning and end. This does not give a complete analysis of ordinary time concepts,[24] but it does contribute towards one.

One of the main advantages of the interval ordering theory in this interpretation is that the only concrete 'things' it involves are *primary* events like a race, R, and not the beginning or end of R. It does not even imply the existence, much less the 'witnessability' of a beginning or end, although it allows the *possibility* of their existence. The beginning of R would be another event, B, having the properties: (1) that e(B) =

[22] This is not the application discussed in SKLT, which is to judgments involving 'thresholds of discriminability', but the axioms of the theory apply equally to the time-orderings of happenings. The notation in SKLT is altered to reflect this change in interpretation, but it will be left to the reader to verify that this change does not affect the mathematical content of the theory.

[23] These conditions might seem too weak, since the weak inequality in (3) allows the possibility of X preceding Y while X's end was simultaneous with Y's beginning. This could actually happen, if X or Y corresponded to *open intervals* of times, and did not include their beginnings and endings. Some possibilities related to this are considered in SKLT under the heading of strong and strong* representations, cf. Theorem 8, p. 317.

[24] E.g., $b(X)$ and $e(X)$ do not *measure* the times of X's beginning and ending because, if $f(\alpha)$ is an strictly increasing real function then $f(b(X))$ and $f(e(X))$ also satisfy conditions (1)–(3). For this reason and because b(X) is a 'pure number' like π and not a time in the ordinary sense like 12 o'clock midnight, we may prefer to interpret it as a number ordinally *correlated* with the time of X's beginning.

b(B) (B's end is simultaneous with its beginning), and (2) that b(B) = b(R) (B's beginning—and end—is simultaneous with R's beginning). But while the theory entails the existence of b(R), which we have interpreted as 'the time of R's beginning', it does not imply the existence of an event B having properties (1) and (2). This might be regarded as challenging the 'rightness' of interpreting b(R) as the time of R's beginning (cf. footnote 24), but at least a framework is provided in which beginnings and endings can be incorporated. This will be returned to in the following section, but now let us turn very briefly to shape and direction.

Shapes and directions are obviously abstract entities, but they have been much less 'theoretically regimented' than places, times, and colors. Frege made the not uncommon mistake of regarding both direction and shape as precise geometrical concepts, the first being what is common to parallel lines, and the second what is common to similar figures (*The Foundations of Arithmetic*, §65–§68).[25] But even cursory acquaintance with geometrical theory shows that neither of these views can be right. As to directions, any given straight line has *two* of them, as is implicit in Euclid's definition of parallel lines as "...straight lines which, being in the same plane and being produced indefinitely *in both directions*, do not meet one another..." (Heath, Vol. I, pp. 190–194, italics mine; cf. Heath's critique of the direction theory of parallels, which traces it to Leibniz). There are various reasons for not wanting to identify sameness of shape with geometrical similarity, the most obvious of which is the difficulty of defining the latter for solids (cf. the commentary on Definition 10 of Book XI of *The Elements*, pp. 265–267 in Heath, *op. cit.*, Vol. 1).[26] Moreover, there is no obvious reason for taking geometrical similarity to define sameness of shape, given that topology furnishes us with the equally precise concept of *homeomorphism*, which is much more general than similarity.

But these may be cases in which one is best advised to ignore theory, and look directly to the ordinary use of expressions like "the shape of a cross" and "the direction of the town." The only thing that is common to these usages is that they seem to have spatial reference. A thing has the same shape as the *place* of the thing. The direction in which it moves may be along a path, and the direction in which it turns can be clockwise or counterclockwise. But whether there exist morphological or directional analogues to spatial points, temporal instants, or chromatic standards

[25] "Aus der geometrischen Aenlichkeit geht der Begriff der Gestalt hervor," (*Grundlagen der Arithmetik*, p. 75).

[26] Note that the alleged similarity of so called 'incongruent counterparts' (e.g., right and left hands) is actual equality, i.e., congruence.

like Munsell chips is doubtful. If so, they are not widely dispersed in morphological or directional 'spaces'. This is evident by the prevalence of *shapelessness* in the case of shape, which reflects the fact that many things found in Nature can't easily be classified as to shape. Harking back to the point made earlier, that artificial classification systems serve purposes, what would be the purpose of such a classification? 'Paradigm shapes', circular, rectangular, pear-shaped, etc., are recognized by everyone, and using these terms allows us to communicate important information. But would it be useful to have a systematic shape-classification system into which all things could be pigeonholed?

Direction may be even more amorphous. North and clockwise are directions, even though things traveling continuously North may meet at the North Pole and continuous clockwise turns may end where they started. Again, it may be more profitable to consider the *use* of directional information than to try to formulate the laws of a 'logic of direction'. But we will now turn to another kind of logic, which justifies the transition from concrete things like races, pears, and turns to the abstract 'qualities' that we ascribe to them: of their being at such and such places and times, of their having such and such colors and shapes, and of their moving in such and such directions.

4.4 Principles of Physical Abstraction I: 'Principal Principles' and Their Grounds

Our thesis or hypothesis is that propositions like S_{class}–S_{time} can be analyzed as being of the form

S X has relation R to the Φ to which Y has relation R,

where X and Y designate concrete particulars, Φ designates an abstract category, class, place, color, etc., and R is the relation of instantiation that is characteristic of category Φ. Thus, in S_{color}, i.e., in "That turquoise has the color of the sky," X and Y are the turquoise and the sky, respectively, Φ is the category of color, and R is the relation between a concrete thing and an abstract color that Y 'instantiates'—the color of Y. Of course we have noted that there is an alternative to the realist analysis, which involves a relation between *concreta* X and Y and the abstract color that they instantiate. That is an 'empirico-nominalist' analysis according to which "That turquoise has the color of the sky" is equivalent to "The turquoise and the sky are the same color," where sameness of color is an empirical relational concept—albeit one that proves to be quite complicated on examination.

But we are concerned with the realist interpretation, which seems to be implicit in the Russellian analysis of the definite description "the color of the sky," the reason being that it presupposes that there is one and only one abstract color that the sky has, the concept of which may be fundamental to applied color theory. More generally, for any X and Y of the class to which abstractions of the kind Φ apply, statements of form S seem to presuppose

AB There is one and only one Φ to which Y has relation R,

and something like this seems to be fundamental to applications of theories of entities of kind Φ—classes, places, colors, times, etc.

In this section we will focus on a part of AB, namely the basic realist assumption:

EXIST = There exists a Φ to which Y has relation R.

Thus, in the cases of classes, places, etc., we have:

EXIST_{class} = There exists a class of persons born in Venice in the 13th century.

EXIST_{place} = There exists a place that Marco Polo came back to.

EXIST_{color} = There exists a color of the sky.

EXIST_{shape} = There exists a shape of a cross.

$\text{EXIST}_{direction}$ = There exists a direction in which he started.

EXIST_{time} = There exists a time at which Richard the lion-hearted ruled England.

EXIST_{class} follows by universal instantiation from Cantor's principle of abstraction:

UEXIST_{class} = For every property π there exists a class ϕ such that for all X, X has π if and only if $X \in \phi$.

With apologies to David Lewis,[27] this may be called the *principal principle* of class existence. And, it is clear that it is a direct generalization of the Russellian interpretation of ordinary language sentences like S_{class}.[28]

EXIST_{place} is not a direct consequence of explicit postulates of typical physical theories, but such theories presuppose that the 'things' that

[27]Lewis, 1978. Lewis's Principal Principle postulates a fundamental relation in which probabilities of probabilities (second-order probabilities) stand to the first-order probabilities to which they apply. Lewis's principle is not related to the principles that we are concerned with here.

[28]However, we pointed out in section 1.2 that Cantor regimented ordinary usage, among other things by ignoring the time-dependence that is implicit when persons are characterized as having being born at such and such a time.

they treat of, e.g., bodies in space, have places in space.[29] This might be called the principal principle of place-existence:

> UEXIST_{place} For every body X there exists a place π such that X is π.[30]

Again, this is a direct generalization of pre-theoretical assumptions, which are reflected in the Russellian interpretation of expressions like "the place that Marco Polo returned to." [31] And, to the extent that theories of physics and geometry assume this principle they build on the Russellian interpretation of statements like S_{place}.

EXIST_{time} is similar to EXIST_{place}, and it too follows by universal instantiation from a principle principal of time-existence:

> UEXIST_{time} For every happening X there exists a time τ such that X takes place at τ.

Something like this applies to color, though here we must be more careful:

> UEXIST_{color} For every uniformly colored body X there exists a color χ such that X has color χ.

This principle of color existence stands to ordinary usage in the same way that the principles of class, place, and time do, and to the extent that theories of color presuppose it they build on that usage.

Although $\text{EXIST}_{direction}$ and EXIST_{shape} have the same logical form as EXIST_{class}–EXIST_{color}, we have already pointed out that such theories as build on them are less clear. This is suggestive, since it raises the traditional metaphysical question: what justifies these propositions?

The fact is that while the man in the street is comfortable making assertions using sentences S_{class}–S_{shape} because, given their 'empirico-nominalist' interpretations, he can verify them empirically. But he is *un*comfortable with metaphysical questions concerning abstract exis-

[29]Many theories begin with *postulates of motion* that presuppose the existence of places in space, e.g., Newton's First Law "Every body continues in a state of rest, or of uniform motion..." Postulates of systems of Particle Mechanics (e.g., McKinsey, Sugar, and Suppes, 1953), and of the author's axiomatic formulation of Rigid Body Mechanics, 1959, treat locations in coordinatized spaces as properties of particles and rigid bodies.

[30]And, we should add that now we should make time-dependence explicit, and formulate the principle as

> UEXIST_{place} = for every body X and every time t, there exists a place where X is at time t.

[31]It is worth considering the extent to which Aristotle's *Physics* is a compendium of principles of ordinary use of terms involved in the physics of his day, which was very rudimentary.

tence propositions like EXIST_{class}–EXIST_{shape} and the universalizations that follow from 'Russellian-realist' interpretations of S_{class}–S_{shape}. Plausibly, EXIST_{class}, EXIST_{place}, EXIST_{time}, and EXIST_{color} are acceptable because generally accepted scientific theories accept them, but that is not the case with $\text{EXIST}_{direction}$ and EXIST_{shape}, which may be more doubtful because scientific theories of shape and direction are less widely accepted. But we must not forget that Cantor's principle, from which EXIST_{class} follows, initially derived *its* plausibility from ordinary usages that did *not* presuppose a theory of classes. What this suggests is that the mere ubiquity of usages of sentences like S_{class} that involve definite descriptions like "the class of persons born in Venice in the 13th century" predisposes if it doesn't logically compel persons to accept Russellian 'readings' of them. This is highly speculative of course, but we will end with one more speculation.

Whether something more fundamental than grammatical habit underlies the use of the definite description forms involved in S_{class}–S_{shape} is uncertain. And the extent to which special 'Kantian' impulses incline us to represent the world in terms of places and times, which don't operate in the cases of other abstractions like colors, is equally uncertain. But at least it is plausible that grammatical habits furnish the *immediate* impetus to accepting the existential consequences to which the Russellian readings of S_{class}–S_{shape} lead. And, to that extent these consequences rest on *a priori* grounds. Moreover, they are associated with other abstract grammatical habits that rest on synthetic 'facts', to which we now turn.[32]

4.5 Principles of Physical Abstraction II: Identity

Now we will consider the other half of

AB = There is one and only one Φ to which Y has relation R,

one half of which was discussed in the previous section. That is:

[32]This is not to deny the force of 'logical reconstructions' of mathematical abstractions, e.g., of the complex numbers as pairs of real numbers, which increases our confidence in theoretical claims such as that $i^2 = -1$, by showing that if the theory of the real numbers is consistent then so is the theory of complex numbers. Something like this might be said of Whitehead's theory of extensive abstraction, whose aim is to show how claims about abstract points can be replaced by ones about abstractive classes of regions. This will be returned to in section 4.6, but the point for now is that the confidence reposed in the most fundamental principles of the theories that concern us derives from the pre-theoretical, ordinary use of sentences like S_{class}, S_{place}, etc. Thus, the most fundamental principles of the theory of classes, that for every property there exists a class of objects that have the property, and the principle of extensionality, gained their initial acceptance from the everyday, pre-theoretical use of sentences like S_{class}.

UNIQUE = There is only one ϕ to which Y has relation R.

Here we will concentrate on

UNIQUE_{class} = There is only one class of persons born in Venice in the 13th century,

UNIQUE_{place} = There is only one place from which Marco Polo started,

and

UNIQUE_{color} = There is only one color that the sky has,

which appear to be presupposed by S_{class}, S_{place}, and S_{color}, respectively.

UNIQUE_{class} implies that the property of being a person born in Venice in the 13th century individuates a unique class, UNIQUE_{place} implies that at the time of setting out on his travels Marco Polo individuated a unique place, and UNIQUE_{color} implies that the sky individuates a unique color. Each of these assumes a principle of identity that is formulated in terms of the instantiation relation characteristic of the kind of abstraction it involves. UNIQUE_{class} is an immediate consequence of set theory's axiom of extensionality, that classes are the same if the same things belong to them. UNIQUE_{place} is more complicated in part because "the place from which Marco Polo started" is a very rough description—just as claiming that Marco Polo returned to that place is rough, but we can at least note that it is formulated in terms of Marco Polo's incidence in it. However, it also involves time-dependence because it claims identity over time, which will be returned to in section 4.7, and we shall put off comments on other aspects of it to later in this section. For the moment we will concentrate on UNIQUE_{color}.

That a thing has only one color would seem to hold only if it is uniformly colored. Let us assume this in the case of the sky; i.e., let us assume that at least to a first approximation it satisfies the color-matching transitivity condition—that any two things that match it in color match each other in color. Now, given two principles of abstract color incidence and identity, we can argue that the sky should satisfy a stricter color-individuating requirement: namely that it cannot have two colors that are distinct in the sense that something can have one of the colors but not have the other. The principles are: (1) if χ_1 and χ_2 are different abstract colors then there are concrete objects having those colors that don't match each other in color, and (2) two objects match each other in color if and only if there is an abstract color that they have in common. Now suppose that the sky had distinct colors χ_1 and χ_2. By (1) there would have to be non-matching objects X_1 having color χ_1 and X_2 having color χ_2. Then X_1 and the sky would have χ_1

as a common color, hence by (2) they would match in color, and X_2 and the sky would have χ_2 as a common color, and by (2) *they* would match in color. Therefore the sky would match both X_1 and X_2 in color, which, assuming the sky's color-transitivity, would entail that X_1 and X_2 matched each other in color, contrary to supposition.

Concluding our comments on abstract color individuation, two points may be made, both having to do with the principle that if two abstract colors are distinct then there are objects having those colors that don't match each other.

The first point has to do with the possibility of including in the 'color space' colors like infrared and ultraviolet that are outside the visible range, which are therefore not distinguishable by direct color matching methods. Clearly the inclusion of such 'colors' violates principle (1) above, although they might be distinguishable by nonvisual means or even by indirect visual means such as spectral decomposition.

The second point is that, assuming that there are at least two abstract colors, the principle that any two different colors are instantiated by objects that don't match in color implies that every abstract color must be instantiated by some object. But this is only plausible if it is interpreted in a modal sense: i.e., every abstract color *could* be instantiated by some concrete object. As well as reintroducing modality, which will be emphasized in what follows, this allows a place in color space for Hume's missing shade of blue.[33] An analogous possibility is conceivable in the case of spatial places, if there were hollow objects that were so impenetrable that nothing could be put inside them. But let us now consider spatial abstraction in general.

We have already argued that there are no bodies in space, or surface features in surface spaces, for which empirical spatial coincidence—touching—is transitive. Hence there is nothing that is to space as 'color points', such as we imagined the sky to be, are to color. Therefore we cannot argue that a concrete object can individuate an abstract spatial points in the way that we argued that the sky might individuate an abstract 'color point'. But we can still formulate abstract principles of spatial coincidence and identity that are the analogues of the principles of abstract color coincidence and identity. These are: (1) if π_1 and π_2 are distinct spatial points then there are concrete objects that instantiate them but which are separate from each other (i.e., they don't touch each other), and (2) two objects touch each other if and only if there is spatial point that they have in common. These imply that if an object did

[33] Page 19 of Open Court edition of *An Enquiry Concerning Human Understanding*, 1952.

instantiate exactly one abstract spatial point then coincidence would be transitive 'at that object'. Thus, if object X were at exactly one point, π, then by (2) any objects Y and Z that touched X would also have to be at π, and if that were the case then, by (1), they would have to touch each other; hence X would satisfy the transitivity condition. The following section will take up the thorny question of what justifies principle (2) if no concrete things individuate abstract spatial points, since the principal principle of spatial abstraction only guarantees the existence of places that are individuated by *concreta*. But we will conclude this section with some very unsystematic remarks.

One of them simply transfers to abstract spatial points a comment at the end of our discussion of abstract color individuation. Assuming that there are at least two abstract points, principle (2) implies that every abstract point must be instantiated by some object. But again this is only plausible if it is interpreted in a modal sense: i.e., every abstract point *could* be instantiated by some concrete object. This allows that there might not be 'actual objects' at some points, although objects *could* be at them. Our 'spatial spaces' are *container spaces*, and some points and regions in them may be empty, although things might move into them.

The other extremely cursory comments concern very difficult questions that relate to the use of the definite article in S_{place} to refer to *the* place from which Marco Polo started. However, in view of the fact that S_{place} asserts that Marco Polo was at the same place at different times, and we will postpone the discussion of 'temporal reidentification' of places to section 4.7, we will comment here on an atemporal variant:

S'_{place} = Marco and Carlo are going to the same place.

S'_{place} would not normally be taken to mean that Marco and Carlo are going to *exactly* the same place, in the Aristotelian sense that they are going (or intending) to occupy exactly the same region of space, and, being more careful, we might say something like "Marco and Carlo are going to the same neighborhood".[34] Given this, and granting that practical specifications of location have some measure of inexactness, we might say that the study of ordinary language spatial abstraction should concentrate on that. But we are concerned here with relations *between* pre-theoretical ordinary language and the theoretical, here we will only note the following.

[34]Perhaps 'vicinity' would be better than 'neighborhood', but the latter is more suggestive of the topological ideas to be noted below.

Neighborhoods are conceived to include points, and therefore the inexact and the exact are not unrelated.[35] One among various approaches to the very complex problem of analyzing the relation between them construes the exact as an *idealization* of the inexact, and it inquires into the question of how far the laws supposed to be valid for the former continue to hold for the latter.[36] For instance, generalizing the transitivity of coincidence condition that is assumed to be valid for ideal *points*, we may consider *the degree* of transitivity of coincidence in small neighborhoods C, as measured by the proportions of pairs of things X and Y in the neighborhood of C that are in the neighborhood of one another (cf. footnote 12 on this). This can be linked to topological ideas, e.g., that any two abstract points in a *Hausdorff* topological space have disjoint neighborhoods, which is valid for the surface spaces that are described in Part II of this study (cf. Theorem 7.3.2 of Chapter 7).

But these matters are too complicated to be discussed in detail here, and we can only hope to have made it plausible that even allowing for the universal inexactitude of practical applications, we still *conceive* these applications in abstract, theoretical terms.

But there are other abstraction principles to consider.

4.6 Principles of Physical Abstraction III: Other Abstraction Principles

We noted that because spatial points are not uniquely individuated by independent concrete objects, their existence is not guaranteed by the principal principle of place existence. Similar considerations apply to temporal instants, whose existence is not guaranteed by the principal principle of the existence of times, and we might even argue that it is only by idealizing and assuming that things like Munsell chips are perfectly uniform in color that the existence of abstract colors is guaranteed by the principal principle of color existence. So, if theory confidently postulates, e.g., that two objects touch each other if and only if there is spatial point that they have in common, it must presuppose other principles of existence for abstract entities beyond the 'principal' ones discussed in section 4.4. This section will comment very briefly on certain of these, as well as on limits that theory seems to impose upon the abstract.

To the extent that pure geometrical theory is abstract it does not formulate principles that *relate* the abstract to the concrete, but the

[35] This still holds in certain fuzzy-theoretical representations, in which points can have 'degrees of belonging' to neighborhoods.

[36] This approach, and its relation to Plato's view of the relation between the real world and the world of appearances, is discussed at length in Adams, "Idealization in applied first-order logic," 1999.

Euclidean constructive formulation is suggestive. Its own *constructibilia*, straight lines, circles, etc., are not *concreta*, but we may hazard the following: in the planar case they are idealizations of finite two-dimensional surface features like the ones discussed in chapter 2. Moreover, although a feature like the visible segment '——' is not perfectly thin or perfectly straight, and the surface it is on is not perfectly flat, if it is assumed that an ideal, abstract segment is *near* to it, geometrical theory can be applied to it. Thus, a principle of application of geometry is that the abstract entities that are objects of the theory are in the same space and near to the concrete objects to which the theory is applied.[37] We have already commented on this in the case of spatial points, e.g., that there exists such a point actually 'in' the small dot '•', but other examples include the boundary or 'outline' of a concrete observable like the bar '▬', and dependent things like the interior of the hollow square '◘'. In fact, geometrical theory even accords abstract existence to *loci* like infinite lines that are not approximated by anything concrete (e.g., Proposition 12 of Book I of *The Elements*, Heath, op. cit., p. 270).

Similar remarks apply to ideal times like instants, as well as to empty time intervals in which 'nothing happens'. And, to the extent that color spaces are objects of theoretical study, similar remarks may apply to them as well. An important thing that all of these theories allow for is the existence of abstract entities, points, instants, colors, etc., not only near to all actual concrete things to which they apply but also to anything the theory *could* apply to, and especially to things that may be constructible according to the tenets of the theory. Thus, the theories characterize container spaces of abstract 'places' in which concrete things might be found or fitted.[38]

The abstract existence assumptions that we are concerned with in this section seem to be more theory-dependent than are the principal principles discussed in section 4.4. For example, a large part of the acceptability of set theory's Axiom of Choice derives from the proof of its consistency with the more 'fundamental' axioms of abstraction and extensionality. Similarly, if certain geometrical theories countenance the existence of 'points at infinity', or of points in four or higher-dimensional spaces, it is because these theories can be shown to be consistent with

---

[37] According to Wedberg, Plato was ambivalent as to whether the ideal lines, triangles, circles, etc., that are the objects of geometrical theory according to him, are also in the 'world of appearances', and therefore they might be *near* to the imperfect appearances that are in this world.

[38] Such places are commonly referred to by *metrical* descriptions. E.g., "the point half way between the centers of gravity of the earth and the moon" refers to a place that may well be in empty space, though something *could* be there. Metrical characterizations of places in spaces will be commented on in Chapter 15.

more fundamental geometrical assumptions. So called 'representation theorems' have the same function (cf. Krantz, Luce, Suppes, and Tversky, 1971, sections 1.1 and 1.2). Thus, the theorems on interval orderings show that given extended events, X and Y, whose temporal precedence relations satisfy the axioms cited in section 4.3, the times of their 'beginnings and endings', b(X) and e(X), can be measured, and this comes close to proving that it is consistent to assume that X and Y *have* instantaneous beginnings and endings.

But theory also places constraints on the abstract. Geometrical theory transcends the constructible by countenancing the existence of infinite lines, but it assumes that they conform to its laws. E.g., any two lines that are parallel to a third line are parallel to each other even in the infinite case. Dimensional limitations are less rigid, but even though higher dimensional geometries are consistent, classical geometry only admitted three of them.[39]

4.7 Identity over Time: Standards of Constancy

This is an enormously complicated topic that generalizes to changes in color, shape, direction, and so on, and it is connected to the well known controversies involving absolute and relative motion in space. As in much of what has gone before, we will concentrate largely on analogies between places and colors.

The claim that Marco Polo was in the same place at an earlier time that someone else was in at a later one cannot be verified in the same way as a claim that the two persons are in the same place at the *same* time. When the times are the same the two persons can be seen 'side-by-side', so to speak, but Marco Polo cannot be seen standing side-by-side with another person standing in the same place at a different time. In fact, if the 'someone else' is Marco Polo himself, it is trivial that he should be at the same place as *himself* at any one time, but it is another problem entirely to verify that he is at the same place at a later time that he was at an earlier one.

Similar remarks apply to color comparisons. We can compare the colors of two pieces of turquoise by placing them side-by-side at one

[39]A reason for classical geometry not to have a fourth dimension is that nothing is in it; e.g., no five 'things' can be found, all of which are the same distance from each other. Note that Isaac Barrow seems to have avoided fourth powers because geometry has no fourth dimension (cf. comments by J.M. Child on p. 76 his edition of *The Geometrical Lectures of Isaac Barrow*, 1916). But it is also worth noting Poincaré's comment at the end of "Non-Euclidean Geometries" in *Science and Hypothesis*, that as long as contradiction is avoided the axioms of geometry are 'free conventions', which could be interpreted to mean that absence of contradiction was the only absolute requirement of geometrical theory.

time, but we cannot do this to verify that the color of one piece at a later time is the same as that of the other piece at an earlier time.[40]

Now, certain theories such as Newton's mechanics explicitly assumed the meaningfulness of absolute motion and rest, but that was contested by Leibniz, and we may fall back on a position more like Leibniz's and argue that motion or change of place is a relational concept. It is worth quoting part of Leibniz's most famous statement of a relativist position, in paragraph 47 of the 5th letter of his Correspondence with Clarke:

> I will here show *how* men come to form to themselves a notion of *space*. They consider that many things exist at once, and they observe in them a certain *order* of coexistence, according to which the relation of one thing to another is more or less simple. This order is their *situation* or distance. When it happens that one of these coexistent things changes its *relation* to a multitude of others, which do not change their relation among themselves; and that another thing, newly come, acquires the same relation to the others, as the former had; we then say it has come into the *place* of the former ...

Now, if we clarify the *relations* between the things involved, and what the 'multitude of others' is, a 'Leibnizian view' of change of place in color-space might be characterized somewhat as follows. In the case of places, the relations are the relative 'situations' or distances between the things related, which include physical coincidence relations, touching or separation, between them. In the case of color, the analogous relations are those of chromatic coincidence between the things compared—color matching or difference.

As to the 'multitude of others', although Leibniz does not give detailed examples, one is apt to think of the fixed stars in astronomical space, among which relatively few—the planets, and especially the moon and sun, which are very large—wander. And, of course, what makes the fixed stars 'fixed' is that at least to crude observation they maintain fixed relations of situation to one another. There is no close analogue to the fixed stars in color-space, but at least certain things are more constant in their color-matching relations than other things are. Certain crystals tend to be color-fast, and certain dyed textiles are more color-fast than others, and one thing that shows this is that they change relatively little

[40]Sense impressionists like Locke seem to have assumed that a person can compare a current sense impression with a memory image of an earlier one. No doubt certain crude color comparisons can be made in this way, but it is doubtful that this works for more precise comparisons, e.g., as to whether a turquoise seen today is the precise shade of blue as one seen yesterday.

over time in their relations of color-coincidence or matching to other color-fast things.[41] But note that in case of colors, the 'multitude' does not so far outnumber the rest, and it is not so well defined, as the fixed stars. This leads us to some concluding comments on *standards*.[42]

Standards are carefully selected things that are 'fixed' or 'fast' in that they have the property of maintaining constant coincidence relations to one another. But the fact that they are *selected* implies that the selection is 'reasonable'—it is *useful*. One use of standards in general, perhaps not always the most important one, is that 'locating' other things in relation to them is *informative* as to the relations of these things to each other. In the case of the stars, it is more useful to be told which constellations different planets are in than to be told which planets different constellations are near to.[43] In the case of colors it is more useful to be told the Munsell colors of tree leaves than to be told which tree leaves Munsell chips match. This is connected with the fact that Munsell chips are themselves color-fast and do not change over time in their color-matching relations.[44]

4.8 Summary

The key theses of the discussion of the abstractions that concern us are as follows. Geometry and certain other sciences, e.g., the science of color, are abstract, but the abstract entities that their variables range over are related to concrete objects by *principles of abstraction*. These have a logical status similar to that of Cantor's Axiom of Abstraction, in that they gain their acceptance from pre-theoretical, ordinary language usage, and to this extent they are linguistically *a priori*. Moreover, the principles of identity that apply to these abstractions are akin to Cantor's Axiom of Extensionality, and they rest on empirical, 'synthetic'

[41]Of course the spectroscopic methods that are now used to determine color also determine change in color. These have no obvious spatial analogues, and in view of this we might say that while change of place in physical space remains relative, change of color in color space is absolute.

[42]The very brief comments below on standards apply to places and colors the pragmatic, informational approach to classification and measurement developed in Adams, 1964, and in Adams and Adams, 1989. They apply equally to systems not considered here, including biological classification and typical systems of quantitative measurement like weight and length.

[43]But note that quite other 'standards of position' are used for the same purpose in other contexts. In particular, relation to the earth's surface is the standard for most terrestrial positions.

[44]But to the extent that sameness of shape is defined as geometrical similarity, that seems not to be relative to standards that are *chosen*. That is because *angle*, which is central to defining geometrical similarity, is measured on an 'absolute scale' (which is reflected in Euclid's Common Notion 4: all right angles are equal).

regularities—although they are not *a priori* synthetic in Kant's sense.

The entities whose existence is postulated by these principles of abstraction are more like *universals* than they are like the classes that are their extensions. In particular, they have a modal aspect that is akin the constructivity of many of the propositions of Euclid's *Elements*, which goes with the fact that the corresponding 'spaces' are *container spaces.* Thus, abstract points and other *loci* are 'wheres' that objects *might* be at, and the abstract colors of color theory are colors that objects *could* have. But delimiting these spaces is difficult, e.g., the *points* that Geometry theorizes about are not exactly coextensive with any physical things, and the same is true of infinite lines.

We now turn to details. Part II examines the topologies of the surfaces of bodies, starting with points on them, conceived as abstract entities, whose existence and relation to concrete observables is postulated by principles like those discussed in this chapter.

Part II

Surface Topologies

5

Overview

5.1 Introduction

In Parts II[1] and III we will narrow our concern with spaces in general, including classes, places, colors, and so on just to places, and specifically to places and points on the surfaces of bodies. For the sake of concreteness, we will concentrate here on flattish spaces with figures or 'features' in them like the following, taken from section 1.3:

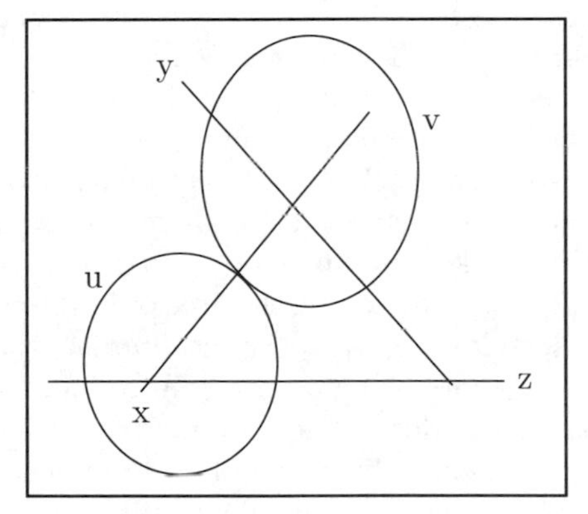

Diagram 1

The figure in the diagram is a composite system of concrete, visible marks, linear segments and ovals, which stand in relations of coincidence or separation, as discussed in Chapter 2. It is also evident that the

[1]The basic ideas discussed in Part II are set forth in Adams, 1973, 1987, and 1996. Beyond a few slight extensions, the present exposition gives detailed proofs of claims that are stated without proof in the earlier papers. The proofs in Chapter 6 are mathematically elementary, those in Chapters 7 and 9 make use of standard concepts and theorems of elementary topology, and Chapter 8 makes use of basic theorems in Dimension Theory.

diagram has empty spaces, in which other marks could be made, or *constructed*, as discussed in Chapter 3. And, that there are empty spaces makes it clear that there are abstract *places*, special cases of which are *points*, the totality of which form a kind of container space, since concrete things, marks, can be put in those places. The object of Part II is to describe in detail how these various things, the concrete observables, the constructibles, and the abstractions, are related, and the assumptions that are involved. Chapter 6 considers *points* on surfaces, and Chapters 7–10 extend the theory to *topologies* based on these points.

5.2 Theory of Points on Surfaces

Chapter 6 starts with an undefined class of *constructibilia*, closed under the primitive binary relation of coincidence, which is assumed to be reflexive and symmetric. Given this, the chapter next gives an account of so called *U-systems*, U, which are finite sets of constructible features such as U = {u,v,y},[2] which acts like the set-theoretical union of the regions covered by u, v, and y, and which can *cover* another feature or U-system V = {x,y}, if anything coincident with any member of V is coincident with some member of U.

U-systems permit us to define *separating covers*, an example of which is the set {u,v} that covers the left-hand segment, x, of the tepee formed of segments x, y, and z in Diagram 1, in such a way that *neither u nor v touches both of the other segments forming the tepee.* As in section 1.3, the reason for constructing such a cover is to demonstrate that the three segments of the tepee do not meet in a common point. The impossibility of constructing a cover like this can be taken to demonstrate the contrary, that there exists a point common to all three of x, y, and z. But of course the proof of this depends on *postulates of abstraction*, that relate concrete features to abstract points, which are not introduced until section 6.5. However, a rather lengthy formal development leads up to this, which is designed both to reveal empirical principles presupposed by the abstract representation, and to consider their plausibility.

Beyond the logical postulates of the formal development, namely that binary coincidence is a reflexive and symmetric relation between features, the pure, 'pre-topological' theory of points only requires one empirical postulate: the *Separation Postulate*, which among other things implies that if one segment of a system like the tepee can be covered by

[2]Throughout Parts II and III lower case letters, 'x', 'y', 'z', etc., will be used to denote individual surface features, or to act as variables ranging over them, and capital letters, 'U', 'V', etc. will denote U-systems, i.e., finite sets of features, e.g., we may write U = {x,y,z}. Underlined capitals, '$\underline{X}$', '$\underline{Y}$', etc., will be used to denote *I-systems*, as explained below.

a U-system that separates the system, then any other segment can be similarly covered. The postulate itself is rather complicated, but it is easily seen to be empirical in character, although it is hoped that it will also be seen to be empirically plausible, at least as an idealization. Moreover, it will be important that this is the *only* empirical postulate required by the theory of points.

Continuing the development, section 6.3 is concerned with *I-systems*, which are arbitrary classes of U-systems, e.g., $\underline{\mathrm{I}} = \{\{x,y\},\{y,z\}\}$. In a sense this acts like the set-theoretical *intersection* of the regions covered by the union of x and y together with the union of y and z. A particularly important kind of I-system is a *separable* one exemplified in Diagram 1 by the system $\underline{\mathrm{X}} = \{\{x\},\{y\},\{z\}\}$, which intuitively represents the intersection of the regions occupied by x, y, and z, which is therefore intuitively empty,[3] but which is 'shown' to be so by the fact that x is covered by the U-system, {u,v}, neither of whose members touches both y and z. Of course, so far we have only suggested that U-systems and I-systems 'act like' unions and intersections of regions covered by features, which does not prove that such abstractions—i.e., regions—exist and are individuated by these U- and I-systems.

Section 6.4 introduces a special kind of inseparable I-system, namely a *P-system*, like the three slices of a pie shown in Diagram 2:

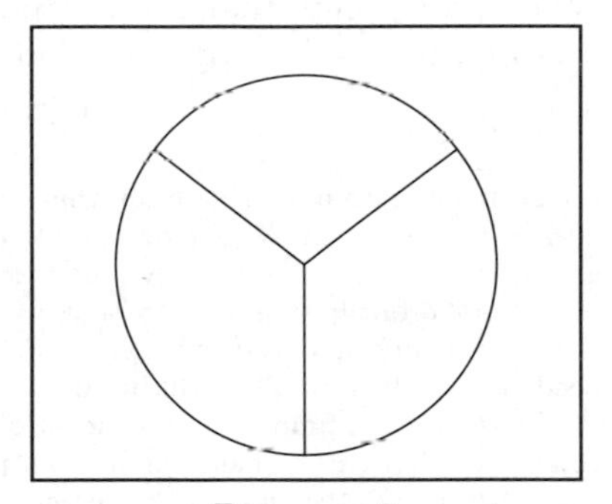

Diagram 2

The point at which the slices come together seems to satisfy the point-individuating requirement: namely that any two concrete things that 'include' this point would have to touch each other, and if we have not previously considered this way of individuating points, that is because it

[3]Note the difference between the I-system {{x},{y},{z}}, which corresponds intuitively to the intersection of the regions covered by x, y, and z, and the U-system {x,y,z}, which corresponds intuitively to the *union* of the regions occupied by x, y, and z, which is not empty.

is not obvious that *all* points can be individuated in this way.[4] But as already stressed, without bridge postulates that connect the ontologically abstract to the concrete, we cannot prove that the 'point' where the slices in the diagram meet does individuate an abstract point, and the last part of Chapter 6 deals with this connection, as well as making further comments on the relation of the present approach to point-individuation to other approaches.

Chapters 7–10 extend the inquiry to topology.

5.3 Basic Surface Topologies

Even cursory acquaintance with modern topology (cf. Kelley, 1955) informs us that the concept of a point is not a sufficient basis for topological theory, and further concepts are required that are usually treated as primitives in pure topological theory. The concept that is most commonly chosen as 'the' topological primitive is that of an *open set of points* (Kelley, p. 37), which have the property that finite intersections and arbitrary unions of them are also open sets. But neither *a priori* intuitions nor pure mathematical theories *define* these open sets in 'more primitive' terms, and the problem of applying topology to the concrete is to characterize these sets in an 'empirically appropriate' way. This is a serious problem, and section 7.2 proposes to solve it by stipulating that basic open sets are classes of points that are *not* points of U-systems shall be the open sets of the theory to be developed here. [5] This is illustrated in Diagram 3:

[4] A more fundamental problem is that it isn't clear that we are skillful enough to draw slice-shaped features that meet in exactly one point. Diagram 2 represents such features by drawing *outlines*, which are themselves lines of some thickness. But while we idealize and conceive the outlines as though they were infinitely thin, the gap between the actual finite outline and the infinitely thin 'conceptual' thing it supposedly represents needs to be filled in. This will be returned to in Chapter 8.

[5] The lack of an intuitive basis for the fundamental concept of an open set, and the wide variety of formal topologies that can be based on a given point set, is often not recognized. Given this 'indeterminacy', the choice of a particular topology cannot be argued for as 'right', but only as more or less appropriate.

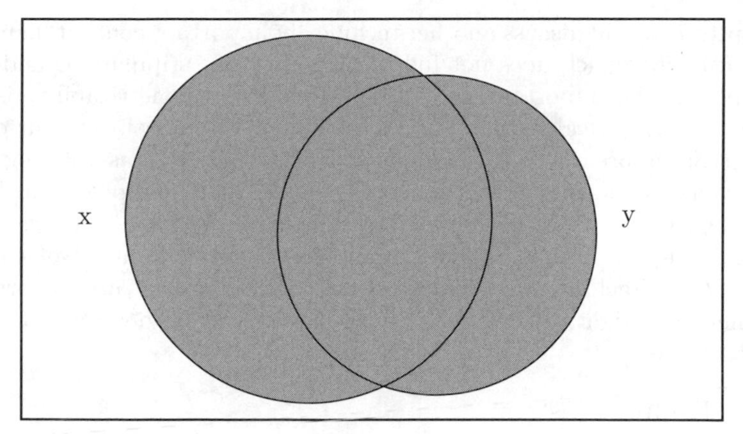

Diagram 3

Circles x and y in the diagram are concrete surface features, which define corresponding classes of points, as will be described in Chapter 6. The sets of points exterior to x and y are stipulated to be *basic* open sets, of which three are pictured in the diagram: (1) the points exterior to x, (2) the points exterior to y, and (3) the points exterior to both x and y and which is the intersection of sets (1) and (2). The significant thing about these sets is that *none of them contains its topological boundary*. This assumes that the complements of *these* sets, and surface features like circles x and y in particular, *contain* their boundary points, while their exteriors do not. But this is a *stipulation*, or 'convention' in Poincaré's sense, and it is not an empirical claim. *Ipso facto*, it can only be argued for in terms of its advantages and disadvantages as a foundation for topological theory.[6]

A major advantage of the stipulation emerges almost immediately. That is that the topology thus defined has two important properties. With the addition of one empirical *finiteness postulate*, it follows that it is *Hausdorff* (Kelley, p.67), and *compact* (Kelley, p.135). These properties are technically very important because they permit us to apply a wealth of mathematical results which themselves have important empirical applications, including all of those to be stated in Chapter 8 relating to connectedness and boundaries, in Chapter 9 on dimension, and in Chapter 10 on homeomorphism and lines. However, prior to that in

[6]However if no distinction were made between regions that contain their boundary points and those that don't, and, as intuition is apt to suggest, it were assumed that both a feature and its exterior had a *common boundary* (which they do in a 'refined', topological sense), this would lead to well known 'paradoxes' that will be commented on in section 8.2. It is the choice of *which* regions are to contain their boundaries that is the stipulation.

Chapter 7 we will discuss another technically important concept, namely *metrizability*, which does not follow from previous stipulations and assumptions. An important aim of the discussion of metrizability is to clarify its 'empirical meaning', which highlights its relation to our constructivist approach to the foundations of geometry. This will begin a program of 'operationalizing' concepts of mathematical topology, including in particular those of boundaries, dimension, and lines, in each case formulating operational methods of applying them that only involve *constructibilia*. Each of these topics involve sub-topics that are important in their own right, whose detailed discussion will be deferred to later chapters.

5.4 Boundaries

Now consider Diagram 4:

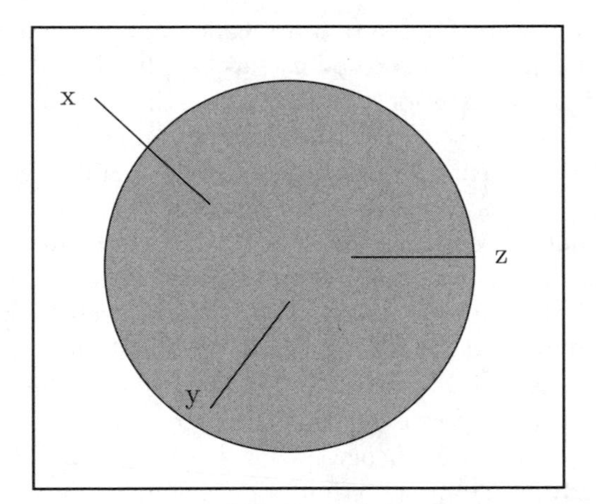

Diagram 4

We said that features like the circle in the diagram are supposed to include their boundaries, while their exteriors do not. But what empirical difference does it make whether the exterior or the circle, 'the exterior of the exterior', has the boundary? This distinction was discussed in section 2.4, where according to the 'genetic' conception of a feature, it was made to depend on how they are drawn. If the circle was drawn and the exterior was the 'background', then the circle has the boundary, whereas if the exterior was drawn and the circle was the background, then the exterior has the boundary. But which if either of these is right cannot be determined by inspection: one must know how the features were drawn in order to determine this.[7] However, there is an important conceptual difference. That is that the analysis of concepts like coincidence and related properties breaks down into 'cases', depending among other things on the way the things they apply to are formed.

Now focus on segments x, y, and z in Diagram 4. These exemplify things whose boundary relations to the circle will be analyzed separately, and the differences between the cases depend not only on how the segments are related to the circle but on how they were formed. It is clear that x touches the boundary of the circle and y does not, but z is a difficult 'borderline case' because determining whether it touches the circle's boundary would seem to require determining whether its boundary coincided with the circle's boundary, which could seem to be beyond powers of human discrimination. To deal with this problem we will operationalize boundary coincidence determination in a special way, in which coincidences between primary features are fundamental and all other determinations are reduced to them. Given this, it will not surprise the reader that the method of boundary coincidence determination is complicated and unintuitive in certain respects. For now we will confine ourselves to noting its salient features.

The segment in Diagram 4 whose relation to the boundary of the circle is perhaps easiest to determine is x. We know that it touches the circle's boundary because it touches the circle and passes outside of it, which can be observed because it only involves coincidence relations between primary features. Of course this assumes that x is itself *connected* in the topological sense (i.e., *continuous* in an intuitive sense),[8] since if it weren't then the part inside the circle might not be connected to the part outside, and it might not cross the circle's boundary. It follows that determining x's coincidence with the boundary of the circle involves more than just seeing the two of them: it requires (1) seeing x, (2) seeing that it is connected, (3) seeing that it and the circle are coincident, and (4), seeing that it isn't *covered* by the circle, i.e., seeing that part of it is outside the circle,[9] and (5) inferring from (1)–(4) that it crosses the boundary of the circle. All of these are taken in at a glance, but it is part of our enterprise to analyze the 'inner workings' of logical processes that

[7]Recall too that nothing in the printed diagram was *drawn*. The diagram only pictures or 'represents' things that might be drawn, but it does not picture them *being* drawn.

[8]Topological *connectedness* (Kelley, p. 52), which, as said, is related to intuitive continuity (e.g., that segment x is continuous), is also important because it is involved in the very idea of a concrete 'thing'. Cf. comments on this in 2.3.

[9]This isn't a simple observation either because 'inside' and 'outside' need to be analyzed. In fact, that x passes outside of the circle must be inferred from the fact that it touches something that is separate from the circle.

are taken for granted by the unanalytical observer. And, at least as they are approached here, the inner workings of the processes of determining the relations of segments y and z to the boundary of the circle are even more complicated. In fact, boundary crossings of the kind exemplified in the case of segment x are utilized as a stepping stone in the analysis of the relations of y and z to the boundary of the circle.

The idea that is fundamental to the analysis of all of these boundary concepts is that of a *boundary cover*, say of the circle in Diagram 4, which is defined as another primary feature or finite collection of features that touches all segments like x that cross the boundary of the circle in the sense that they satisfy conditions (1)–(5) above. For instance, the *boundary outline* of the circle in Diagram 4, which is shown in black, is a boundary cover of it.[10] Then we can say that y is *inside* the circle if it is covered by it, and furthermore some boundary cover of the circle is separate from y. And, we can say that z is coincident with the circle's boundary if every boundary cover of the circle is coincident with it. Both of these criteria are empirical because they are stated purely in terms of relations among primary surface features, but neither is obvious. Hence we will prove in Chapter 8 that they satisfy an intuitive condition for abstract boundary incidence: namely that features have common points with the abstract topological boundaries of other features if and only if they satisfy these criteria. Here we will only note that if these are the 'right' boundary coincidence criteria, then it is easy to explain why it is so much easier to verify that y *doesn't* touch the boundary of the circle than it is to verify that z does. Showing the former only requires constructing one boundary cover of the circle that is separate from y, while showing the latter requires establishing that all boundary covers touch it.[11]

The final point concerning boundaries, which will be discussed at length in section 8.2, relates to the cardinalities of our surface spaces. If they were finite then they would be *totally disconnected* (Steen and Zeebach, 1970, p. 31), and if segment x had more than one point it would also be disconnected (i.e., discontinuous), and it could not cross the

[10]It is part of the difficulty in interpreting a black-and-white diagram like this that without further stipulation it does not show exactly where in the boundary-cover the circle's boundary lies. This is related to the fact that it is more difficult to determine whether segment z touches the boundary than to determine whether x or y do.

[11]That it is much easier to determine that x touches the circle's boundary than that z does is because x passes outside the circle and z doesn't. Of course there is still much to explain, and we should examine the tacit assumption that we can *tell* that one primary feature covers the boundary of another, in the sense that any connected feature touching but not covered by the latter touches the former.

boundary of the circle—in fact, the circle wouldn't have a boundary.[12] Therefore if we are to apply the *topological* concept of a boundary to features on surfaces we must assume that our spaces are infinite,[13] which seems to transcend the empirical. This reflects two things.

One is that developing our theory within the framework of transfinite mathematics allows us to give a coherent, systematic account of boundaries in a way that avoids certain 'paradoxes' that will be discussed in section 8.2.[14] The other is that the transfiniteness of the present theory originates at the pre-topological level, with the *constructions* on which the theory of points is based. It is assumed that these constructions can be iterated *ad infinitum*, very much as is assumed by the Euclidean constructive postulates, and it follows from this that there must be infinitely many *constructibilia*. But, as the ancient disputes over infinite divisibility suggest, infinite iterability is a questionable idealization. This is still very real problem, and it has not yet been dealt with in a thoroughly satisfactory way, so we will only note it here, and comment later on certain aspects of it.

5.5 Dimensionality

Our surfaces and their spaces are intuitively two-dimensional, but still more than other topological concepts, dimensionality is much more complicated than intuition might make it seem. The aim of Chapter 9 is to relate topological concepts of dimension, which are normally applied to abstract mathematical spaces, to spaces of points on physical surfaces like those characterized in Chapter 8, and to describe operational tests for determining their dimensionality.

Beyond the fact that topological dimension theory makes dimension a non-metrical concept, and we do not want to presuppose metrical concepts at this stage,[15] the most important reason for applying it to our

[12] This assumes that surface spaces are Hausdorff.

[13] And if they are compact, as follows from our postulates, they must be uncountably infinite. However the uncountability is at the level of the abstract (i.e., there must be uncountably many abstract points), but that only requires a countable number of constructible *concreta*, which can be viewed as standing to abstract points much as the rational numbers stand to the real numbers.

[14] Cf. further remarks in section 17.8 on the advantages that our mathematical formulation provide in systematizing topological concepts, in contrast to the technically quite difficult but unsystematic 'empirical criteria' for boundary determination that Marr's theory of visual processing, 1992, sets forth.

[15] Thus, Euclid implicitly characterizes solids as three-dimensional because 'they have length, breadth, and depth' (Definition 1 of Book XI), surfaces as two-dimensional because they 'have length and breadth only' (Definition 5 of Book I), and lines as one-dimensional because they are 'breadthless lengths' (Definition 2 of Book I). These characterizations are implicitly metrical because length, breadth, and

'physical spaces' is that it demonstrates the equivalence of superficially quite different characterizations of dimension,[16] one of which is particularly suitable for elementary operationalization. Applied to surface spaces, which are intuitively two-dimensional, the 'test' illustrated in Diagram 5 below establishes that they must be *at least* two-dimensional. This involves two pairs of segments, horizontal segments x and y, and vertical segments z and w, and ovals u and v surrounding x and z, respectively:

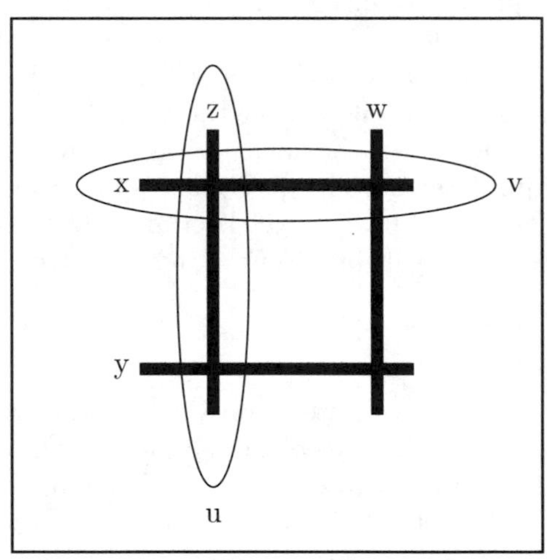

Diagram 5

The critical things about the segments and ovals above are: (1) x and y are separate, in fact oval v 'separates' them in the sense that x is inside it and y is outside of it, (2) z and w are separate and oval u separates them in the same sense, and (3) the ovals intersect each other. Now, given that any separator of x and y like v must intersect any separator of z and w like u, it follows from Condition C, p. 35 of Hurewicz and Wallman (1941), together with the abstraction postulates of our theory, that the surface space must be at least two-dimensional. A related test

---

depth are metrical concepts. Companion definitions, "An extremity of a solid is a surface," "The extremities of a surface are lines," and "The extremities of lines are points" (Definition 2 of Book XI and Definitions 6 and 3 of Book I) come closer to modern characterizations of dimension, and they will be commented on at greater length in the following section, as well as in section 9.4.

[16]The 'splitting' of the dimensionality tests to be described below, one for being *at least one-dimensional*, and one for being *at most two-dimensional*, is characteristic of modern dimension theory. The same thing holds for n-dimensionality for arbitrary n, including the interesting case n=0.

shows that it must be at most two-dimensional, although this test is somewhat more difficult to describe, as will be seen in section 9.3.

The practically important thing about the dimensionality tests described above is that, even if they presuppose the ideas of 'inside' and 'outside', which are explained in the theory of boundaries, they are comparatively easy to apply. Moreover, further reflection on them explains seeming paradoxes, including some curious fractal facts to be commented on in section 9.4, or the fact that a plane, the coordinates <x, z> of whose points are all rational numbers in the closed interval [0,1], is not at least two-dimensional—in fact it is zero-dimensional.[17] But if that were the case and ovals u and v had irrational coordinates, then they they might 'pass by' each other without intersecting.

The last point to make here concerning dimension is that the tests just described apply not only to entire surface spaces but also to features of them. For instance, the solid circle in Diagram 5 can be shown to be two-dimensional while its boundary is one-dimensional, and in a sense this 'vindicates' our intuitions. However while the boundary appears to be a circle, the *idea of circularity*, and of *linearity* in general, needs to be examined, since it by no means follows from the fact that the boundary is one-dimensional that it is like a *line*, as we intuitively conceive of lines.

5.6 Linearity

Modern dimension theory implies that the set of points in the unit square with coordinates <x, z> with real x and rational z is one-dimensional,[18] because it is a countable union of one-dimensional continua (cf. Theorem III 2, p. 30 of Hurewicz and Wallman, 1948). But this point set is not linear in any intuitive sense; in fact, it seems empirically indistinguishable from the set of all points with real coordinates, <x, z>, which is two-dimensional. One the other hand, Heath's discussion of Greek conceptions (Heath, 1956, Vol. I, pp. 158–165) makes it clear that the concept of a line is not easy to define. But the theme of this is interesting, and it is related to Aristotle's criticism of the allegedly Platonic definition of a line as 'breadthless length' (Definition 3 of Book I of *The Elements*).

Aristotle's criticism was that 'breadthless length' characterizes lines as a *species*, of which length is the *genus* (*Topics* VI. 6, cited on p. 158 of Heath's discussion), and the genus, length, already excludes breadth (elsewhere Aristotle characterizes length as "magnitude continuous in

[17]This space is not compact, and therefore it cannot satisfy the postulates of our surface spaces.

[18]Once again, this 'space' is also not compact, hence it couldn't be a surface space in our sense.

one dimension" (*Metaphysics* Δ , 1002a). Heath suggests that Aristotle advanced this criticism primarily as an argument against the Platonic theory of forms, which would imply that if length were a form it would be self-contradictory to 'divide' it in terms of having or lacking breadth. But now suppose, *contra* both Plato and Aristotle, that there were an abstract 'form of a line', and that concrete things were linear if they had (or 'partook') of that form? Modern topology suggests a way of making this precise, and Chapter 10 develops an operational theory both of the 'form of linearity' and of what it is to partake of it. The basic ideas are simple.

The interval of real numbers from 0 to 1 inclusive (the 'unit interval', symbolized as [0,1]), can be taken to be the form of ideal line segment, and it is not implausible to take it to epitomize Aristotle's "magnitude that is continuous in one dimension." Moreover topological theory tells us what it is to have the form of [0,1]; that is, it must be *homeomorphic* to it.[19] Of course, not all things that we think of intuitively as lines are homeomorphic to [0,1]. E.g., neither circles nor segments without their endpoints are homeomorphic to it, but at least we can say that these and almost all other curves that we are familiar with are 'composed of arcs', as circles are, and the *arcs* are homeomorphic to [0,1].[20] And, so are finite non-reentrant curves that include their endpoints, like the broken line:

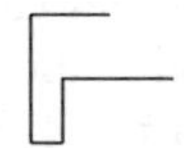

All that is left is to make operationally clear what it is for the boundary of a surface feature to be an arc in this sense, i.e., to be homeomorphic to [0,1], and carrying this out in detail is the main objective of Chapter 10. The problem of operationalizing this concept is that while homeomorphism is a precise topological concept, it is not 'elementary' in the way that the ideas of a boundary and of dimensionality are. To prove that a point set, Π, in a topological space is homeomorphic to [0,1], it must be shown that there exists a one-one, bicontinuous function—a homeomorphism—that maps Π onto [0,1]. But this is a *relation* between Π and [0,1], and not, apparently, an intrinsic *property* of Π. The question is: can the 'form' of Π be characterized intrinsically, in terms of properties, the possession of which is necessary and sufficient to guarantee the

[19] Kelley, 1955, p. 87. Note the etymology of the Greek derivative, 'homeomorphic': to have the same form.

[20] This is essentially the topological definition of an *arc*. Cf. Steen and Zeebach, 1970, p. 29.

existence of a homeomorphism to [0,1]?[21] Section 10.3 will show that this can be done, in terms of just two properties.

First, the numbers in the interval [0,1] are *ordered*, and therefore having points that are similarly ordered is one of the properties that anything homeomorphic to [0,1] must have. Second, there are uncountably many numbers in [0,1], and therefore anything homeomorphic to it must have uncountably many points. The problem is to describe 'operational tests' that can determine whether a physical continuum like the upper edge of the solid segment below has these properties:

Now, it may seem obvious that the points on the upper edge of the segment above are ordered, e.g., from left to right. If we picture them like beads on a string

we can even count them 'from left to right'. But the points on the edge of the segment are not really like this, like beads on a string, because the beads can be seen and the points cannot. Not only that, if the points are too small to be seen how can even a finite number of them be counted, and how can we establish operationally that an infinite set of them is uncountable? The answer this question is complicated, but the gist of it runs as follows.

We can infer that there are an arbitrary finite number of points on the edge by observing that it has arbitrarily many boundary-crossings, as represented in the outline below:

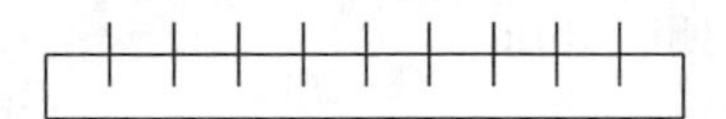

This suggests that we can 'confirm' that with sufficient skill, crossings can be interpolated between any two successive boundary-crossings—which is equivalent to infinite divisibility at the level of boundary-crossings. And, given this plus facts to be returned to concerning the ordering of these crossings, it follows that their ordering must be the same as that of the rational numbers in the interval [0,1]. And, again, this together the fact that the space of abstract points is *compact* entails that there are uncountably many abstract points between the ends of

[21]This is analogous to the question that Gauss answered by showing how to characterize a surface's curvature in terms of its own intrinsic properties, without assuming that it is embedded in a higher dimensional space.

the segment's edge.[22] Finally, that the boundary-crossings are ordered essentially as the rational numbers are follows from the fact that there is a crossing between any two other crossings, plus the fact that the 'betweenness relation' between these crossings satisfies the axioms of Group II of Hilbert's axioms for geometry (the 'betweenness axioms' on p. 6 of the First edition of the *Foundations of Geometry*, Hilbert, 1899). But, as said, carrying out these arguments in detail is tedious if not difficult, and that is the main program of Chapter 10.

There is one final comment to make before turning to details. Characterizing the 'form' of linearity is only the first of a many problems of operational characterization that confront the attempt to apply topology to the physical world. Unluckily, linearity is the only abstract concept that is defined 'relationally' that the author has made headway with, but two unsolved problems of this kind may be noted at this point, because they will be important later. One is to give an intrinsic characterization of the property of a whole surface space of being homeomorphic to a standard two-dimensional space, analogous to our intrinsic characterization of the homeomorphism of a boundary to the interval [0,1]. The ability to do this would be a criterion of adequacy of our theory of these physical spaces, but we have yet not been able to carry this out.

The other unsolved problem arises in attempting to extend the theory from elementary topological representations of the kind that we consider here to differential topological representations. This would allow us to account for all kinds of elementary facts, like the obvious difference between a solid rectangle and a solid circle, as well as much deeper facts such as the orientability of our surface spaces, which will be important when we come to superposition. These problems will be returned to, but it well to keep them in mind as another indication, if any more is needed, of the limitations of this work.

[22] This is essentially the same as the proof that the set of 'cuts' of the rational numbers, which themselves define real numbers, is uncountable.

6

Points on Surfaces

6.1 Introduction

What follows gives an informal axiomatic development of a theory of points on the surfaces or parts of surfaces of physical bodies such as the rectangle below:

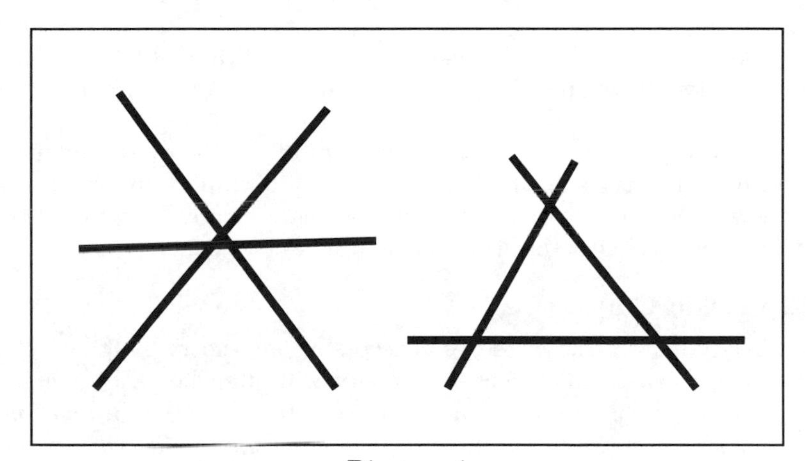

Diagram 1

As suggested in section 1.3, the *Leitmotif* of our approach is the separation test described in that section for determining whether three or more features have a point in common, as those composing the figure on the left in the diagram do but those on the right do not. But while the test is simple and its theory is elementary, its detailed development is somewhat lengthy. That is what this chapter is concerned with.

In doing this we will follow the steps briefly described in section 5.2, beginning in section 6.2 with basic properties and relations among concrete *constructibilia* that can be produced on the surface, followed by the theory of *U-systems*, which are finite sets of constructibilia that individuate unions of point sets. Section 6.3 introduces the concept of *separation*, in terms of which the test for having a common point is formalized, and it formulates an associated *separation postulate* and derives some of its consequences. Section 6.4 introduces the theory of *I-systems* which are arbitrary classes of U-systems that individuate *intersections* of point sets, and their separability properties, and section 6.5 introduces *PI-systems*, which are 'inseparable and indivisible' I-systems that intuitively individuate *points* on the surfaces we are concerned with. Section 6.6 develops the *theory of abstraction*, which is based on *abstraction principles* of the kind discussed in sections 4.4–4.6, that relate concrete features and *constructibilia* to abstract points, and which are essential to demonstrating incidence properties of U-systems, I-systems, and PI-systems. Section 6.7 will conclude this chapter with a brief commentary on relations between the conception of a spatial point developed here and other conceptions.

As with matters discussed in the later chapters of Parts II and III, most of the ideas and mathematical results stated in this chapter have been presented elsewhere, especially in Adams, 1973, and the main difference between this presentation and earlier ones is that proofs of the results are given.

The postulates, theorems and definitions of the theory will be numbered by the sections in which they occur. For example, our first postulate will be given as: '6.2.1. *Postulate*', which indicates that this is the first formal statement that is formulated in section 6.2.

6.2 Basic Concepts

As already stated, the primitive concepts of our theory are those of a class of *constructibilia* (which for brevity will often be called *marks*), that can be produced on a surface, and the binary relation of *touching*, or *coincidence* between them. To say that a *constructibilium* (mark) of such and such a kind *exists* is to say that it can be *produced*. We will assume that all of the marks that are actually on the surface are included in the class of *constructibilia*, and confirming generalizations about marks typically involves producing them.

Here is the first formal proposition of our theory, which is trivial:

6.2.1 Postulate. The relation of touching is symmetric and reflexive in the class of marks.

This postulate can be regarded as logical and not empirical in character, since it is part of the meaning of touching that if one thing touches another then the second thing touches the first. That the relation is held to be reflexive—that any mark touches itself—may be regarded as a convention that simplifies statements of the theory, but which we would argue has no substantive import.[1]

We next define three companion concepts that will play an important role.

6.2.2 Definitions

(1) A *U-system*, is a finite, possibly empty set of marks.

(2) A U-system X *touches* (or is *coincident with*) mark y if a member of X touches y; X touches U-system Y if it touches some member of Y. Two U-systems that do not touch are *separate*.

(3) A U-system X *covers* mark y if any mark that touches y touches some member of X; X covers a U-system Y if it covers every member of Y.

Note that lower case letters are used as variables ranging over marks and capitals are used to range over U-systems of them. Later we will introduce a third class of variable to range over another category.

Here are some elementary consequences of postulate 6.2.1. and the definitions in 6.2.2. They are given as theorems, but because they are trivial their proofs are omitted.

6.2.3 Theorems

(1) The touching relation is symmetric in the class of U-systems, and reflexive and in the class of non-empty U-systems.

(2) The covering relation is reflexive and transitive in the class of U-systems.

(3) U-system X covers U-system Y if and only if every mark or U-system that touches Y touches X.

(4) If X, Y and Z are U-systems and X covers Y and Y touches Z then X touches Z.

(5) If $X_1, \ldots, X_n$ cover $Y_1, \ldots, Y_n$, respectively, and $Y_1 \cup \ldots \cup Y_n$ covers Z then $X_1 \cup \ldots \cup X_n$ covers Z.

Interestingly, one of the things that does *not* follow from Postulate 6.2.1 and the definitions in 6.2.2 is that if a U-system X covers a mark y then the subset of marks in X that actually touch y also covers it. This is stated in Theorem 6.3.2 below, but proving it requires us to add an empirical postulate, which is central to the theory of separation to be discussed next.

[1]The 'empirical status' of axioms or postulates is not always easy to ascertain (cf. Adams, Fagot and Robinson, 1970), and justifying the claim that the assumption that the touching relation is reflexive has no substantive import is not entirely trivial. We must leave this aside here, however, merely observing that ordinary usage offers little guidance in the matter, which suggests that we feel intuitively that it makes no difference whether or not we assume it.

6.3 The Separation Test and Its Theory

To introduce this subject we will recapitulate in some detail remarks made in section 1.3 on the problem of distinguishing systems of marks that are coincident, or meet in common points, from systems that are not. Let us focus on the question of how we determine that the trio of segments on the left in Diagram 1 meet in a common point. As pointed out in section 1.3, we can exclude the possibility of directly *seeing* an abstract point that is common to the segments. What we see are the *segments*, and the problem is to make clear what we see *in* them that we don't see in the trio of segments on the right. This is like asking what we see in the pair of segments that form the equals sign '=' that we don't see in the system '-/', which leads us to conclude that the former are at least roughly parallel but the latter are not. Of course these properties might be left as primitive, unanalyzed 'gestalts', but if Euclidean geometry is a good guide we ought to try to reduce them to more fundamental concepts. That is especially true in the present case, because once three-mark coincidence is analyzed we still have to deal with that of four marks, five marks, and so on. Euclid's analysis of parallelism furnishes a model, since he defines parallelism by reference to a test for *non-parallelism*: namely, that it should be possible to construct extensions of the segments in question that intersect. The segments forming the equals sign are roughly parallel because they would have to be extended a fair distance before they intersected, while those forming the system '-/' are not because a very short extension of the horizontal segment would intersect the sloping one. Now we may consider an analogous test for non-coincidence.

What proves that the trio of segments on the right in Diagram 1 do not meet in a common point is the fact that it is possible to produce a collection of marks covering one of the figure's segments, no one of which touches all three of them. Diagram 2 below, which reproduces Figure 1b in section 1.3, illustrates this by adding two 'auxiliary ovals' that cover the left hand segment of the figure, though neither of the ovals touches all three of them:

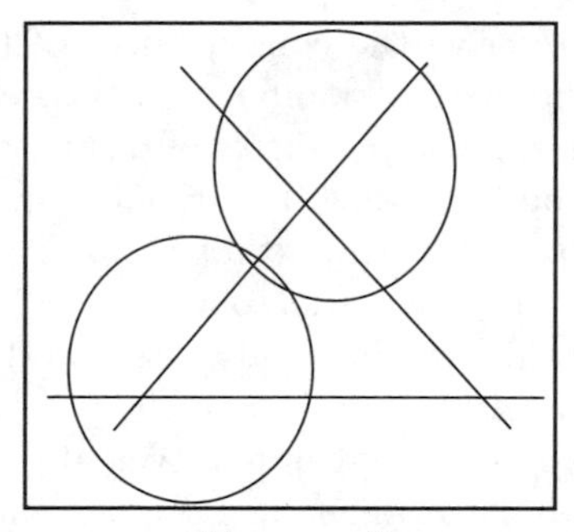

Diagram 2

Given that the pair of ovals in Diagram 2 cover one of the segments in the right hand figure Diagram 1, but neither touches all three of them, the three segments cannot have a point in common. If they had then one of the ovals would have to contain it, and therefore this oval would have to touch all of the segments, contrary to supposition. Therefore our constructive test demonstrates what it is supposed to: namely the non-coincidence of the segments.[2] Whether the failure of the test in application to other sets of marks would be enough to prove that they were *not* non-coincident—i.e., that they were coincident—is something that can only be considered after we introduce abstract points into our theory, in section 6.6. For now we will confine ourselves to analyzing the properties of the non-coincidence test. This is complicated, and the remainder of this section will consider a only few of its properties, an especially important one of which is that if *some* x in a U-system X can be covered with another U-system, no member of which touches all members of X, then this is true of all members of X. But to prove this we must introduce an empirical postulate that is fundamental to all that follows:

6.3.1 Separation Postulate. Let x and y be separate marks, and z be a mark that is separate from a U-system W. Then there exists a U-system that covers z and is separate from W, no member of which touches both x and y.[3]

---

[2]While this test gives theoretical characterizations of separability and inseparability of systems of marks like those in the figures in Diagram 1, it is practically very important that we seem to be able to 'see' *directly* that the marks in the left hand figure in the diagram are inseparable. We have not tried to analyze this sort of 'direct seeing', but because of the light it might shed on important epistemological matters, it would seem desirable to do so. Closely related matters will be taken up in sections 17.6 and 17.7.

[3]Saying that there *exists* a U-system, say $U = \{u_1, \ldots, u_n\}$, covering z, separate from W, and no member of which touches both x and y means that $u_1, \ldots, u_n$ can

Before starting to derive the consequences of this postulate, let us comment briefly on the justification for it. A theoretical justification derives from the fact that, given principles of abstraction to be formulated and discussed in section 6.6, it can be derived from standard topological assumptions. Another 'justification' that is *not* available to us assumes a correspondence between the marks to which the Postulate applies, and point sets. Assuming this, we might also assume the constructability of the 'intersections' of z with x and y, namely $z \cap x$ and $z \cap y$, as well as the 'remainder', $z \sim x \sim y$, that is left over after $z \cap x$ and $z \cap y$ are 'subtracted' from z, and in which case the postulate would be trivial. Thus, in that case $z \cap x$, $z \cap y$, and $z \sim x \sim y$ would exactly cover z, and their 'union', $(z \cap x) \cup (z \cap y) \cup (z \sim x \sim y)$, would contain nothing outside of z that would touch a U-system separate from z. But we cannot take for granted the possibility of constructing features, the point sets of which exactly coincide with the intersections $z \cap x$ and $z \cap y$, and the 'remainder', $z \sim x \sim y$. In fact, reasons for not postulating these things will be noted below. Instead we must make it plausible that the constructability of close *approximations* to $z \cap x$, $z \cap y$, and $z \sim x \sim y$ would guarantee the truth of the postulate.

For now we will try to make the Separation Postulate *empirically* plausible, by considering its application in another diagram, namely Diagram 3a, in which x and y are shown as separate vertical ovals, z and w are represented by separate, horizontal oval outlines, and w is the single member of the U-system $W = \{w\}$.[4]

---

be produced, so to speak 'all together', which is more than that each of them can be produced. This is plausible given that n is finite, but it would be implausible for infinite generalizations such as will be considered in section 6.4.

[4]Henceforth we will usually not bother to distinguish between singleton U-systems and the marks in them.

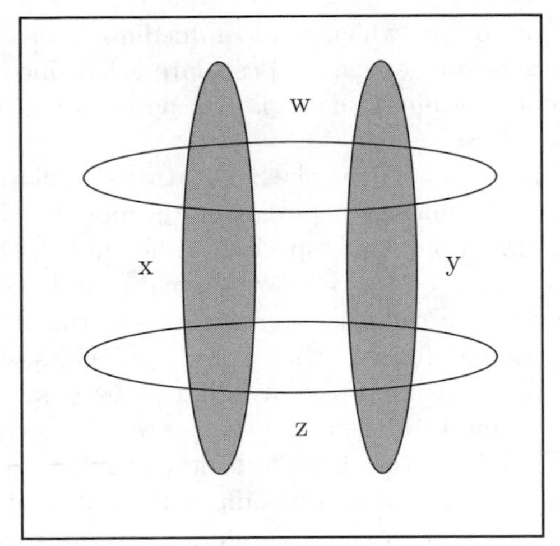

Diagram 3a

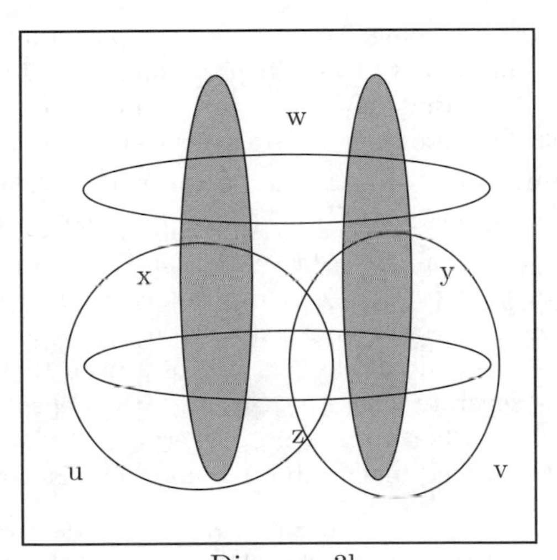

Diagram 3b

Diagram 3b confirms the Separation Postulate in application to the figure in Diagram 3a, by adding auxiliary ovals u and v, neither of which touch both x and y, both of which are separate from w, and whose 'union' covers z.

There are two reasons why this example is important in its own right. One is that because pure logic cannot guarantee the possibility of producing things that do not actually exist, the fact that the Separation Postulate requires producing new features shows that it is empirical in character.[5] The other, which is an immediate consequence, is that the new postulate is independent of Postulate 6.2.1, since that is 'purely logical', and it does not imply that it is ever necessary to make 'auxiliary constructions'.[6]

Concluding our comments on the Separation Postulate, we may note two reasons for not wanting to postulate the possibility of producing exact feature-intersections and especially differences.[7] One is that the 'union' of $z \cap x$, $z \cap y$, and $z \sim x \sim y$ would exactly coincide with z, but it is too much of an idealization to suppose that it is possible to draw a mark or union of marks that exactly coincides with an already existing one. Our present approach avoids this by postulating only the constructability of marks and unions that *cover* the preexisting one. A special reason for not postulating the constructability of the difference $z \sim x \sim y$ is that if that were non-empty it would not contain all of its boundary points, and this would conflict with an assumption to be stated in Chapter 8, to the effect that *all constructabilia* contain their boundary points.

In spite of the foregoing, however, and hoping that our examples give some initial plausibility to the Separation Postulate, we now want to show that it has intuitively the 'right' consequences. We will start with what is hardly more than a restatement of the postulate:

6.3.2 Theorem. If a U-system X covers a mark y then the U-system consisting of the members of X that touch y also covers y.

Proof. Suppose first that $X = \{x, w_1, \ldots, w_n\}$ and X covers y but x is separate from y. If $\{w_1, \ldots, w_n\}$ did not cover y there would exist a mark z touching y but separate from $\{w_1, \ldots, w_n\}$. Then z would touch x, since $\{x, w_1, \ldots, w_n\}$ covers y. Now, given that x is separate from y and z is separate from $\{w_1, \ldots, w_n\}$, the Separation Postulate implies that there exists a U-system U covering z that is separate from W, no member of which touches both x and y. This would imply: (1)

[5]This assumes that pure logic doesn't guarantee that there is more than one possible world, hence any postulate that entails the constructability of something not existing in this world must be non-logical—i.e., empirical.

[6]Note that if x, y, z and w were the *only* producible marks then x and y would cover z in the sense of Definition 3.2.4, and neither would touch both x and y, but they would not be separate from w.

[7]We put aside for the present more practical problems that can arise in producing marks satisfying the requirements of the Separation Postulate, such as the auxiliary marks in Diagram 2. If z and w were the horizontal segments in the figure '#', it would be practically very difficult to cover z with a pair of marks u and v, neither of which touched both of the slanting segments in the '#'. This 'problem of the microsurfacial' is not unrelated to problems of the microphysical, and it will be returned to in Chapter 16.

there must be a member u of U that touches y, since U covers z and z touches y, (2) u must be separate from x, since no member of U touches both x and y, and (3) u is separate from W. Hence, though u touches y, it must be separate from $\{x, w_1, \ldots, w_n\}$ since it is separate from x and from $\{w_1, \ldots, w_n\}$. But that contradicts the assumption that $X = \{x, w_1, \ldots, w_n\}$ covers y. QED

Now we will prove two generalizations of the Separation Postulate, starting by showing that it holds when the single mark y replaced by a U-system, Y.

6.3.3 Theorem. Let x and z be marks that are separate from U-systems Y and W, respectively. Then there exists a U-system that covers z and is separate from W, no member of which touches both x and Y.

Proof. Assume that x and Y are separate and so are z and W. It will be proved by induction on the number of members of Y that there exists a U-system, U, that covers z, is separate from W, and no member of it touches both x and y.

Suppose first that Y is empty. Then we can let $U = \{z\}$. Trivially then U covers z and it is separate from W, since z is separate from W. Furthermore no member of U can touch both x and Y since Y is empty.

Now suppose that the theorem is valid for all $Y = \{y_1, \ldots, y_n\}$ with n or fewer members, and that x and z are separate from Y and W, respectively. By the hypothesis of the induction there exists a U-system $U = \{z_1, \ldots, z_m\}$ covering z, separate from W, and no member of which touches both x and Y. Suppose that $z_1, \ldots, z_k$ touch x and are separate from Y, and $z_{k+1}, \ldots, z_m$ are separate from x though they may touch Y.

Now let y be separate from x, and let $Y' = Y \cup \{y\}$. We will construct a U-system U' covering z, separate from W, and no member of which touches both x and Y'; i.e., any member of U' that touches x is separate from y and all members of Y. Since y is separate from x and for $i = 1, \ldots, k$, z_i is separate from both Y and W, according to the Separation Postulate there exists a U-system U_i covering z_i and separate from $Y \cup W$, no member of which touches both y and x. Let $U' = U_1 \cup \ldots \cup U_k \cup \{z_{k+1}, \ldots, z_m\}$. It is easy to see that U' satisfies the required conditions.

First, U' covers z. That is because $U_1, \ldots, U_k$ cover $z_1, \ldots, z_k$, respectively, and therefore $U_1 \cup \ldots \cup U_k \cup \{z_{k+1}, \ldots, z_m\}$ covers $\{z_1, \ldots z_k, z_{k+1} \ldots, z_m\} = \{z_1, \ldots, z_k, m\}$, which itself covers z.

Second, U' is separate from W because each of the members of $U_1 \cup \ldots \cup U_k \cup \{z_{k+1}, \ldots, z_m\}$ is separate from W.

Finally, no member of U' touches both x and $Y' = Y \cup \{y\}$. All of $z_{k+1}, \ldots, z_m$ are separate from x, and for each $i = 1, \ldots, k, U_i$ doesn't touch both x and y, and moreover it is separate from Y.

This concludes the proof.

6.3.4 Corollary. For all U-systems X and Y, if X does not cover Y then there exists a U-system that covers Y, not all of whose members touch X.

Proof. Suppose that X and Y are U-systems and X does not cover Y. By definition, there is a mark z that touches Y but not X. Then z is separate from X and Y is separate from the empty U-system, $\emptyset$, hence by Theorem 6.3.3 there is a U-system, U, that covers Y and is separate from $\emptyset$, no member of which touches both z and X. Since U covers Y and z touches it, U touches z. Therefore one of the members of U touches z, and that member does not touch x. QED

The next theorem generalizes the Separation Postulate a step farther and begins consideration of arbitrary *classes* of U-systems, further properties of which are set forth in succeeding sections.

6.3.5 Theorem. Let X, Z and W be U-systems and let $\underline{Y}$ be a class of U-systems such that Z and W are separate and no member of X touches all members of $\underline{Y}$. Then there exists a U-system covering Z and separate from W, no member of which touches X and all members of $\underline{Y}$.

Proof. By induction on the number of members of X. If X is empty we can let $U = Z$. Then U covers Z and is separate from W, and no member of it touches both X and all members of $\underline{Y}$ since X is empty.

Now assume the theorem is valid for all X', $\underline{Y}'$, Z', and W' where X' has fewer than n members, and suppose that X, $\underline{Y}$, Z and W are such that no member of X touches all members of $\underline{Y}$, Z and W are separate, and X has n members. Let x be a member of X, which therefore does not touch some member Y_x of $\underline{Y}$, and let $X' = X \sim \{x\}$. No member of X' touches all members of $\underline{Y}$, and therefore X', $\underline{Y}$, Z and W satisfy the hypothesis of the induction. Hence there exists a U-system U' that covers Z, is separate from W, and no member of U' touches X' and all members of $\underline{Y}$. Let U_1' be the set of marks in U' that touch X', and let $U_2' = U' \sim U_1'$. Clearly every member u of U_1' must be separate from some member Y_u of $\underline{Y}$. Moreover, since all of the members of U_2' are separate from X', U_2' itself must be separate from X', and since it is also separate from W it is separate from $X' \cup W$.

Now, for each u in U' construct a covering of it, C_u, as follows. If u is in U_1', let $C_u = \{u\}$. If u is in U_2' then it is separate from $X' \cup W$, and since x is separate from Y_x it follows from Theorem 6.3.2 that there exists a U-system covering u that is separate from $X' \cup W$, no member of which touches x and all members of $\underline{Y}$. Let C_u be such a covering. Finally, let U be the union of the U-systems C_u, for u in U'. It will be shown that U covers Z, is separate from W, and no member of it touches X and all members of $\underline{Y}$.

That U covers Z follows from the fact that it is a union of coverings, C_u, of members u of U', and U' covers Z.

That U' is separate from W follows from the fact that all of the U-systems, C_u, for u in U' are separate from W. Thus, if u is in U then $C_u = \{u\}$, and since U' is separate from W so is $C_u = \{u\}$. If u is in U_2' then C_u is separate from $X' \cup W$, hence from W.

Finally, no member v of U can touch X and all members of $\underline{Y}$. If v is in U it must be an element of C_u for some u in U'. Suppose u is in U_1'. Then $C_u = \{u\}$, hence $v = u$. Since no member of U_1' touches all members of $\underline{Y}$ and $v = u$ is in U_1', v does not touch X and all members of $\underline{Y}$.

Suppose that u is in U_2', which is separate from X'. Then u and *ipso facto* v are separate from X', so they don't touch X' and all members of Y_x. To show that v can't touch x and all members of $\underline{Y}$, suppose that it touches x. Therefore v is separate from Y_x because v is in C_u, u is in U_2', and none of the members of C_u touch x and all members of $\underline{Y}$. Since v doesn't touch X' and all members of $\underline{Y}$ and it doesn't touch x and all members of $\underline{Y}$, it doesn't touch $X = X' \cup \{x\}$ and all members of $\underline{Y}$. Thus, whether v is in C_u for u in U_1' or U_2', it does not touch X and all members of $\underline{Y}$. This completes the proof.

We end this section by proving the following:

6.3.6 Corollary. Let X be a U-system that belongs to a class $\underline{X}$ of U-systems, and let U be a U-system that covers X, no member of which touches every member of $\underline{X}$. Then for every member of $\underline{X}$ there is a U-system, no member of which touches every member of $\underline{X}$.

Proof. Let X be a member of a class $\underline{X}$ of U-systems, and let U be a U-system that covers X, no member of which touches every member of $\underline{X}$. Let Y be member of $\underline{X}$. Y is separate from the empty U-system $\emptyset$, hence by 6.3.5 there is a U-system V that covers Y and is separate from $\emptyset$, no member of which touches U and all members of $\underline{X}$. No member of V could touch X and all members of $\underline{X}$, for if some member did then it would touch all members of $\underline{X}$ as well as some member of U, since U covers X, contrary to the assumption that no member of V touches U and all members of $\underline{X}$. QED.

Applied to the 'tepee' in Diagram 2, the foregoing implies that if marks like the two ovals can be drawn, covering one side of it but not touching all of the other sides, then the other sides can themselves be covered in a similar manner. As we will say, *separating* the sides in this way shows that they do not meet in a common point, and what the corollary shows is that this kind of separability is independent of the side chosen.

A final comment on the kind of separability just noted is that showing that a class $\underline{X}$ of U-systems is separable in the above sense shows that the intersection of the systems in the class is empty. Generalizing, a class of U-systems can be regarded as individuating the *intersection* of the classes of points in U-systems of the class, and separability establishes that this intersection is empty. In fact, plausible abstraction principles establish the converse: if a class of U-systems cannot be separated in this way then the spatial regions individuated by the systems belonging to it has a non-empty intersection. But prior to that we must define and establish some properties of these 'intersective systems'.

6.4 Intersective Systems

Henceforth we will follow the suggestion above that a class of U-systems, $\underline{X}$, individuates the intersection of the regions individuated by its members, and accordingly we will call these classes *intersective systems*, or I-systems. To be consistent with this way of interpreting classes like $\underline{X}$, we will define suitable criteria characterizing the conditions under which such a system touches, or is coincident with, a given mark, U-system, or other I-system. The key concept is that of member of $\underline{X}$ *separating* $\underline{X}$, already characterized as meaning that the member, say X, has a cover, Y, no mark of which touches all of the members of $\underline{X}$. Intuitively, this means that X is separate from the intersection of the members of $\underline{X}$, or equivalently, since X is itself a member of $\underline{X}$, the intersection of the members of $\underline{X}$ is empty.

Before starting to investigate the properties of I-systems, it should be noted that in moving from the consideration of finite U-systems to arbitrary and possibly infinite classes of them we also move from things that are in some sense 'observable' to things that may be 'merely theoretical'. A generalization like the Separation Postulate, which is about relations between finite sets of marks, can be inductively confirmed by observation, but it is not so obvious that a generalization about infinite classes of U-systems can be confirmed in this way. In fact, geometrical points will be characterized by reference to such classes, and claims about the latter could seem to be no more accessible to direct observational confirmation than are claims about points that are 'too small to be seen'. We may take some solace in the fact that the Separation Postulate is the only empirical assumption on which the theorems that will be stated in this chapter depend, and that is 'finite', but we shall have to reraise the question of empirical confirmability in subsequent chapters, whose *topological* theorems depend on further postulates. In the meantime, however, we will take a step towards simplifying the confirmation problem by proving a *compactness theorem* that shows how important claims about I-systems can be verified by reference U-systems that cover them. These theorems will come at the end of the present section, and the intervening theorems will lead up to them.

We begin with some definitions.

6.4.1 Definitions

(1) An *I-system* is an arbitrary, possibly empty class of U-systems. Let X be a U-system that is a member of the I-system $\underline{X}$.

(2) X *separates* $\underline{X}$ if there is a U-system Y that covers X, no member of which touches all of the members of $\underline{X}$. $\underline{X}$ is *separable* if it has a member that separates it; otherwise it is *inseparable*.

(3) An I-system $\underline{Y}$, U-system Y, or mark y is *separate* from $\underline{X}$, respectively, if $\underline{X} \cup \underline{Y}$, $\underline{X} \cup \{Y\}$, or $\underline{X} \cup \{\{y\}\}$ is separable; otherwise $\underline{Y}$, Y, or y *touch*, or are *coincident* with $\underline{X}$.

(4) An I-system $\underline{Y}$, U-system Y, or mark y *covers* $\underline{X}$ if it touches every I-system that touches $\underline{X}$.

(5) A U-system X *is a non-extraneous cover of* $\underline{X}$ if it covers $\underline{X}$ and every member of X is inseparable from $\underline{X}$.

Note that one mark or U-system is separable from another mark or U-system in the sense of part (2) of the above definition if and only if the first is separate from the second. Also, note that part (4) of the above definition, of an I-system being covered by a mark, U-system, or other I-system, guarantees that in the case of I-systems, covering stands to touching in the same way that it does in the case of marks and U-systems, which must be stipulated explicitly since I-systems are a new category.

The following theorems are trivial.

6.4.2 Theorems. Let X be a U-system and $\underline{X}$ be an I-system.

(1) If $\underline{X}$ is separable then every member of it has a cover that separates it; moreover, if X is a member of $\underline{X}$ it separates $\underline{X}$ if and only if it has a cover, no member of which touches all members of $\underline{X} \sim \{X\}$.

(2) If $\underline{X}$ is a subset of $\underline{Y}$ and $\underline{Y}$ is inseparable then $\underline{X}$ is inseparable.

(3) Every subset of $\underline{X}$ covers $\underline{X}$.

(4) Every member of an inseparable I-system is inseparable from the I-system, and touches every other member of it.

(5) X covers $\underline{X}$ if and only if $\{X\}$ covers $\underline{X}$.

(6) If X covers $\underline{X}$ and $\underline{X}$ is inseparable, then $\{X\} \cup \underline{X}$ is inseparable and X touches every member of $\underline{X}$.

(7) If X covers Y and Y is inseparable from $\underline{Z}$ then X is inseparable from $\underline{Z}$, or equivalently, if X covers Y and is separable from $\underline{Z}$ then Y is separable from $\underline{Z}$.

Proof. Part (1) is an immediate consequence of Theorem 6.3.6, parts (2)–(5) are trivial, part (6) follows immediately from part (5), and part (7) follows immediately from (6).

Combined with Definition 6.4.1(3), part (5) of the above theorem implies that singleton I-systems $\underline{X} = \{X\}$ can be treated as identical to their only members.

6.4.3 Theorems. Let X and Y be U-systems, and $\underline{Z}$ be an I-system.

(1) $X \cup Y$ is inseparable from $\underline{Z}$ if and only if either X or Y is inseparable from $\underline{Z}$.

(2) Let X be a member of an I-system $\underline{X}$, let U be a U-system covering X, and let V be the union of members of U that are inseparable from $\underline{X}$. Then V covers $\underline{X}$.

(3) Let X be a member of an inseparable I-system $\underline{Y} \cup \underline{Z}$, and let U be a U-system that covers X. Then there is a member of U that is inseparable from $\underline{Y}$ and which touches all members of $\underline{Z}$.

Proof. To prove part (1), suppose first that either X or Y were inseparable from $\underline{Z}$. Then by part (7), $X \cup Y$ would be inseparable from $\underline{Z}$ since it covers both X and Y.

Now suppose that both X and Y are separable from $\underline{Z}$, but $X \cup Y$ is not separable from $\underline{Z}$. Given that X and Y are separable from $\underline{Z}$ it follows from 6.4.1(3) that both $\{X\} \cup \underline{Z}$ and $\{Y\} \cup \underline{Z}$ are separable. Hence, by part (1) of the present theorem, there exist U-systems U and V covering X and Y, respectively, such that no member of U touches X and all members of $\underline{Z}$, and no member of V touches Y and all members of $\underline{Z}$. By Theorem 6.3.2, we can assume without loss of generality that all members of U touch X and all members of V touch Y, and therefore no members of either U or V touch all members of $\underline{Z}$. Hence no member of $U \cup V$ touches all members of $\underline{Z}$, and therefore $U \cup V$ is separable from $\underline{Z}$. Since the union, $U \cup V$, covers the union of the things covered, $X \cup Y$, and $U \cup V$ is separable from $\underline{Z}$, it follows from 6.4.2(7) that $X \cup Y$ is separable from $\underline{Z}$.

Part (2) is a trivial consequence of part (1).

Part (3) follows almost immediately from part (1) of the present theorem and parts (2), (4), and (6) of Theorem 6.4.2. If X is a member of an inseparable I-system $\underline{Y} \cup \underline{Z}$, and U is a U-system that covers X, then by (6) U is inseparable from $\underline{Y} \cup \underline{Z}$. If $U = \{u_1 \ldots, u_n\}$ and it is inseparable from $\underline{Y} \cup \underline{Z}$ then it follows easily from part (1) of the present theorem that at least one of $u_1, \ldots, u_n$ must be inseparable from $\underline{Y} \cup \underline{Z}$, say u_i is. Then $\{\{u_i\}\} \cup \underline{Y} \cup \underline{Z}$ is inseparable, hence by the converse of part (2) $\{\{u_i\}\} \cup \underline{Y}$ and $\{\{u_i\}\} \cup \underline{Z}$ are inseparable, since they are subsets of $\{\{u_i\}\} \cup \underline{Y} \cup \underline{Z}$. Therefore according to 6.4.2(4) $\{\{u_i\}\}$ and *ipso facto* u_i is inseparable from $\underline{Y}$, and it touches every member of $\underline{Z}$.

Part (1) of the foregoing shows that in a sense unions of U-systems like $X \cup Y$ 'interact' with intersective systems like $\underline{Z}$ in the way we would expect them to, assuming that the U-systems individuate the unions of the point sets of their members, and that the I-systems individuate the intersections of the point sets of *their* members. Of course, this is far from proving that U-systems and I-systems do individuate point sets in this way, but it is a step in this direction, and these will be carried to completion after we have introduced postulates of point-abstraction in section 6.6. But more groundwork needs to be laid before coming to that.

The next three theorems lead to demonstrating the compactness properties mentioned earlier.

6.4.4 Theorem. Let X be a U-system that is separable from an I-system $\underline{X}$. Then there is a U-system U that covers X, no member of which touches all members of $\underline{X}$.

Proof. If X is separable from $\underline{X}$ then $\{X\} \cup \underline{X}$ is separable, and therefore by 6.4.1 there is a U-system U that covers X, no member of which touches all members of $\{X\} \cup \underline{X}$. By 6.3.2 we can assume without loss of generality that all members of U touch X, hence no member of U touches all members of $\underline{X}$, as was to be shown.

6.4.5 Theorem. If $\underline{X}$ and $\underline{Y}$ are I-systems and $\underline{X}$ does not cover $\underline{Y}$, then there is a U-system Z that is inseparable from $\underline{Y}$ but not from $\underline{X}$.

Proof. Suppose that $\underline{X}$ does not cover $\underline{Y}$, hence there is an I-system $\underline{Z}$ that is inseparable from $\underline{Y}$ but not from $\underline{X}$. Then $\underline{X} \cup \underline{Z}$ is separable but $\underline{Y} \cup \underline{Z}$ is not. Let Z be a member of $\underline{Z}$, hence of $\underline{X} \cup \underline{Z}$ and of $\underline{Y} \cup \underline{Z}$. Since $\underline{X} \cup \underline{Z}$ is separable and Z belongs to it, by 6.4.2(1) there is a cover, U, of Z, no member of which touches all members of $\underline{X}$ and all members of $\underline{Z}$. But since U covers Z and Z belongs to the inseparable system $\underline{Y} \cup \underline{Z}$, by 6.4.3(3) there is a member, u, of U that is inseparable from $\underline{Y}$ and which touches all members of $\underline{Z}$. Since u does not touch all members of $\underline{X}$ and doesn't touch all members of $\underline{Z}$ it clearly doesn't touch all members of $\underline{X}$, and so by 6.4.2(4), $\{u\}$ cannot be inseparable from $\underline{X}$. Therefore {u} is inseparable from $\underline{Y}$ but not from $\underline{X}$, and the theorem is proved.

6.4.6 Theorem. Let X and Y be U-systems and $\underline{Z}$ be an I-system. If $X \cup Y$ covers $\underline{Z}$ but X is separable from it, then Y covers it.

Proof. Suppose that $X \cup Y$ covers $\underline{Z}$, X is separable from it, but Y does not cover it. Then there is a U-system, W, that is inseparable from $\underline{Z}$ but not from Y; hence Y is separate from W. Both $\underline{Z}$ and W cover $\underline{Z} \cup \{W\}$. Since X is separable from $\underline{Z}$ it is separable from $\underline{Z} \cup \{W\}$ which $\underline{Z}$ covers, and since Y is separate from W it is also separable from $\underline{Z} \cup \{W\}$, which W covers. Therefore by 6.4.3(1), $X \cup Y$ is separable from $\underline{Z} \cup \{W\}$. But $X \cup Y$ was assumed to cover $\underline{Z}$, hence it must cover $\underline{Z} \cup \{W\}$. $\underline{Z} \cup \{W\}$ must be inseparable since W is inseparable from $\underline{Z}$, hence $\underline{Z} \cup \{W\}$ is inseparable from itself. But it has been shown not to be inseparable from $X \cup Y$, and therefore $X \cup Y$ cannot cover $\underline{Z} \cup \{W\}$, which is a contradiction. This proves the theorem.

Parts (1) and (2) of the following theorem show among other things that no matter how much 'finer' intersectively interpreted systems are than the U-systems that belong to them, nevertheless any two separable I-systems are 'separable by U-systems' in the sense that there is a U-system that covers one of the I-systems but which is separate from the other one.

6.4.7 Theorems. Let $\underline{X}$ and $\underline{Y}$ be separable I-systems.

(1) There is a finite subset of $\underline{X}$ that is separable.

(2) If $\underline{X}_1, \ldots, \underline{X}_n$ are I-systems whose union is separable then there exist U-systems $X_1, \ldots, X_n$ which cover $\underline{X}_1, \ldots, \underline{X}_n$, respectively, such that $\{X_1, \ldots, X_n\}$ is separable.

(3) If $\underline{X}_1, \ldots, \underline{X}_n$ are I-systems, any two of which are separable, then there exist separate U-systems $X_1, \ldots, X_n$ that cover $\underline{X}_1, \ldots, \underline{X}_n$, respectively.

(4) Let $\underline{X}_1, \ldots, \underline{X}_n$ I-systems. Then there exist U-systems $X_1, \ldots, X_n$ such that for all $i, j = 1, \ldots, n$, X_i touches X_j if and only if $\underline{X}_i$ is inseparable from $\underline{X}_j$.

Proof. Suppose that $\underline{X}$ and $\underline{Y}$ are separable I-systems. Then $\underline{X} \cup \underline{Y}$ is separable, hence by 6.4.2(1) any member of it, Z, has a cover, U, no member of which touches all members of $\underline{X} \cup \underline{Y}$. If Z belongs to $\underline{X}$ then by 6.4.2(3) it covers it, and therefore so does $\{U\}$. Hence by 6.4.3(2) the union V of all members of U that are inseparable from $\underline{X}$ also covers it.

By 6.4.2(4), V touches all members of $\underline{X}$. Since no member of U hence no member of V touches all members of $\underline{X} \cup \underline{Y}$, and every member of V touches all members of $\underline{X}$, none of them touch all members of $\underline{Y}$. Therefore no member of V is inseparable from $\underline{Y}$, hence by 6.4.2.7, V itself is separable from $\underline{Y}$. V has already been shown to cover $\underline{X}$, and part (1) of the theorem is proved.

The proof of Part (2) is as follows. Suppose that $\underline{X}_1, \ldots, \underline{X}_n$ are I-systems whose own union is separable. Then $\underline{X}_1$ is separable from $\underline{X}_2 \cup \ldots \cup \underline{X}_n$, hence by Part (1) of this theorem there exists a U-system X_1 that covers $\underline{X}_1$ such that $\{X_1\}$ is separable from $\underline{X}_2 \cup \ldots \cup \underline{X}_n$. $\underline{X}_2$ is separable from $\{X_1\} \cup \underline{X}_3 \cup \ldots \cup \underline{X}_n$, and again there exists a union X_2 that covers $\underline{X}_2$, such that $\{X_2\}$ is separable from $\{X_1\} \cup \underline{X}_3 \cup \ldots \cup \underline{X}_n$. This procedure can obviously be iterated, ultimately yielding U-systems $X_1, \ldots, X_n$ covering $\underline{X}_1, \ldots, \underline{X}_n$, respectively, and equally clearly the set of these unions, $\{X_1, \ldots, X_n\}$, is separable.

Parts (3) and (4) follow easily from parts (1) and (2).

Theorem 6.4.7(1) is our key compactness theorem, and the next theorems follow easily from that.

6.4.8 Theorem. Let $\underline{X}$ and $\underline{Y}$ be I-systems. Then $\underline{X}$ covers $\underline{Y}$ if and only if $\underline{X} \cup \underline{Y}$ covers $\underline{Y}$.

Proof. If $\underline{X} \cup \underline{Y}$ covers $\underline{Y}$ then $\underline{X}$ covers it, since by 6.4.2(3) $\underline{X}$ covers $\underline{X} \cup \underline{Y}$ because it is a subset of it, and $\underline{X} \cup \underline{Y}$ covers $\underline{Y}$.

Now suppose that $\underline{X}$ covers $\underline{Y}$ but $\underline{X} \cup \underline{Y}$ does not. By 6.4.6 there is a U-system, Z, such that $\{Z\}$ is separable from $\underline{Y}$ but not from $\underline{X} \cup \underline{Y}$. Then it follows easily from 6.4.2(9) and 6.4.2(1) that there is a U-system, U, covering Z, all of whose members touch Z, and no member of which touches all members of $\underline{X} \cup \underline{Y}$. By 6.4.4, since U covers Z, $\{U\}$ covers $\{Z\}$, and since $\{Z\}$ is inseparable from $\underline{Y}$, {U} is also inseparable from it. By 6.4.7 it follows that for at least one member, u, of U, such that $\{u\}$ is inseparable from $\underline{Y}$, and therefore by 6.4.2(4), u touches all members of $\underline{Y}$. Since $\underline{X}$ covers $\underline{Y}$, u must also touch all members of $\underline{X}$. But this contradicts what was earlier shown: that no member of U can touch all members of $\underline{X}$ and all members of $\underline{Y}$.

6.4.9 Theorem. For any I-systems $\underline{X}$, $\underline{Y}$ and $\underline{Z}$, $\underline{X} \cap \underline{Y}$ covers $\underline{Z}$ if and only if $\underline{X}$ and $\underline{Y}$ both cover $\underline{Z}$.

Proof. If $\underline{X} \cap \underline{Y}$ covers $\underline{Z}$ clearly $\underline{X}$ and $\underline{Y}$ do since by 6.4.2(3) both of them cover $\underline{X} \cap \underline{Y}$.

Now suppose that $\underline{X}$ and $\underline{Y}$ both cover $\underline{Z}$. $\underline{X} \cap \underline{Z}$ and $\underline{Y} \cap \underline{Z}$ cover $\underline{X}$ and $\underline{Y}$, respectively, and both of latter cover $\underline{Z}$, therefore both $\underline{X} \cap \underline{Z}$ and $\underline{Y} \cap \underline{Z}$ cover $\underline{Z}$. Therefore by 6.4.2(3), $\underline{Z}$ covers $\underline{X} \cap \underline{Z}$. If $\underline{Y}$ covers $\underline{Z}$ and $\underline{Z}$ covers $\underline{X} \cap \underline{Z}$ then $\underline{Y}$ covers $\underline{X} \cap \underline{Z}$. Therefore, $\underline{Y} \cap \underline{X} \cap \underline{Z} = \underline{X} \cap \underline{Y} \cap \underline{Z}$ covers $\underline{X} \cap \underline{Z}$, and therefore they cover $\underline{Z}$ since $\underline{X} \cap \underline{Z}$ covers $\underline{Z}$. To show that $\underline{X} \cap \underline{Y}$ covers $\underline{Z}$, suppose that system $\underline{W}$ is inseparable from $\underline{Z}$, hence $\underline{W}$ is inseparable from $\underline{X} \cap \underline{Y} \cap \underline{Z}$. This means that $\underline{W} \cap \underline{X} \cap \underline{Y} \cap \underline{Z}$ is inseparable, and therefore its subset $\underline{W} \cap \underline{X} \cap \underline{Y}$ is also inseparable, and it follows that $\underline{W}$ is inseparable from $\underline{X} \cap \underline{Y}$. Therefore any system inseparable from $\underline{Z}$ is inseparable from $\underline{X} \cap \underline{Y}$, so $\underline{X} \cap \underline{Y}$ covers $\underline{Z}$, as was to be proved.

Now we turn to another concept.

6.5 Indivisibility

The property that this section is concerned with is one that we would expect that features (marks, U-systems, or I-systems) that individuate single points should possess. We have already argued that even the smallest marks, and *ipso facto* the U-systems that are unions of them, do not possess this property, among other things because they are perceivable, but, as we have supposed, anything small enough to individuate or 'single out' just one point would be too small to be perceived. Sections 1.3 and 4.3 even gave a theoretical argument to this end. A mark, say a small dot '•', which is big enough to be seen must also be big enough that two other features could be produced, say the vertical lines in '⫲', which both touched the dot, but which were separate from one another. This would prove that the dot 'contains' more than one point, since if it contained just one point then the two vertical lines would also contain that point, and if they did they would not be separate. In fact, this property, which the dot lacks, is the one that this section will be concerned with. Thus, we posit:

6.5.1 Definition. A mark, U-system, or I-system is *indivisible* if any two marks, U-systems, or I-systems that are inseparable from it are inseparable from each other.

Of course we have just argued that neither marks nor U-systems are indivisible, but we can ask whether I-systems might be. These intersective systems are, intuitively, smaller than the systems whose intersections form them, hence they might be small enough to be indivisible. In fact, trivially, certain I-systems *are* indivisible: namely *separable* ones like the system of segments forming the tepee, 'X'. Because this is separable no marks or other systems are *in*separable from it, and therefore no two marks that are inseparable from it at separable from each other. But let us put that aside, and inquire whether *inseparable* indivisible I-systems might exist.

It is not inconceivable that a pair of tangent circles x and y, as in the diagram below, might form an I-system that is both inseparable and indivisible:

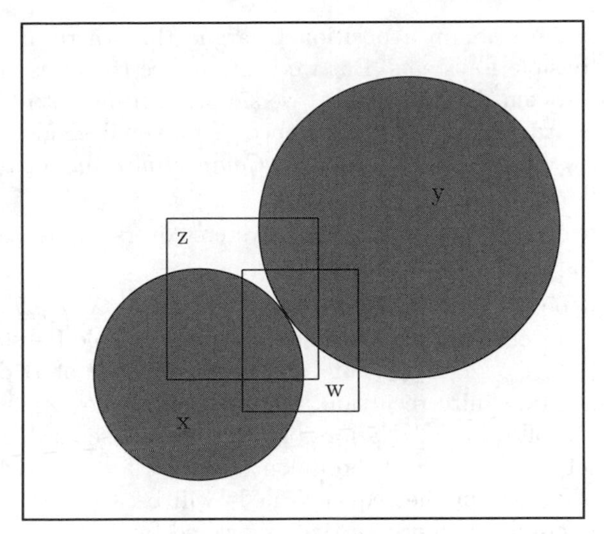

Diagram 4

The circles above touch each other, hence they are inseparable, and it looks as though it would be difficult to bring two things like rectangles z and w into coincidence with the pair of them, without bringing them into contact with each other: hence they are inseparable and seemingly indivisible. But as the diagram itself suggests, producing *perfectly* tangent circles may be a practical impossibility, and therefore we ought to avoid postulating it.

Rather than focus on intersections of circles, we might consider intersections of *boundaries*, such as the intersection of two edges pointed at in the diagram below, which is copied from section 4.3:

It seems intuitively that the edges and the intersection pointed at are not 'empty' in the way that the intersection of the segments forming the tepee is, hence they are inseparable. Moreover, it is hard to imagine how two separate things could be produced, both of which touched the intersection of the edges. Hence it seems to be indivisible, and, intuitively, it seems to individuate or 'locate' a unique point. But this *is* intuitive. We do not yet have an exact characterization of edges or boundaries, much less of intersections of them. And pending such a characterization we cannot even be sure that they correspond to I-systems. Thus, it might seem that we are not in a position to argue that there are insepara-ble and indivisible I-systems. But we will now see that this appearance would be mistaken. In proving this we simply transfer into the present setting standard proofs of the existence of maximal ideals in lattices, which are closely related to *Wallman Compactifications* of topological spaces (Kelley, 1955, p. 167).

6.5.2 Theorem. Every inseparable I-system covers an inseparable and indivisible I-system.

Proof. Let $\underline{X}$ be an inseparable I-system, and let $B = \underline{X}, \underline{Y}_1, \underline{Y}_2, \ldots, \underline{Y}_\omega, \underline{Y}_{\omega+1}, \ldots$ be a well-ordering of the set of all inseparable I-systems. Now let $\Gamma = \underline{Z}_1, \underline{Z}_2, \ldots, \underline{Z}_\alpha, \underline{Z}_{\alpha+1}, \ldots$ be the subsequence of B defined by the following transfinite recursion. For each ordinal α, $\underline{Z}_\alpha$ is the first member of B following all $\underline{Z}_\beta$ for $\beta < \alpha$, that is inseparable from the system $\underline{X} \cup \Gamma_\alpha$, where Γ_α is the union of all $\underline{Z}_\beta$ for $\beta < \alpha$. Let $\underline{Z}$ be the union of all $\underline{Z}_\alpha$ in the sequence Γ. It will be shown that $\underline{Z}$ is an inseparable, indivisible I-system that is covered by $\underline{X}$.

To show that $\underline{Z}$ is inseparable note first that any two members of Γ must be coincident. This follows trivially from the construction of Γ. If $\underline{Z}$ is separable then by 6.4.7(1) there must be a finite subsystem of $\underline{Z}$ that is separable, and since $\underline{Z}$ is the union of the members of Γ there must be a finite sub-sequence of Γ, the union of whose members is separable. But by the construction of Γ, any member of a sub-sequence of it must be inseparable from the union of all prior members of the sequence, hence the last member of any finite sub-sequence must be inseparable from all prior members of Γ, so the union of members of the subsequence cannot be separable. Hence $\underline{Z}$ must be inseparable.

That $\underline{X}$ covers $\underline{Z}$ is trivial, since by construction $\underline{Z}_1 = \underline{X}$, hence by 6.4.2(2), $\underline{X}$ covers $\underline{Z}$, since it is a subset of the set-union $\underline{Z}$.

To show that $\underline{Z}$ is indivisible we show first that any I-system $\underline{W}$ that is inseparable from $\underline{Z}$ covers it. If $\underline{W}$ is in the sequence Γ, the union of whose members forms $\underline{Z}$, then $\underline{W}$ is inseparable from and covers $\underline{Z}$, by 6.4.7(1) again. If $\underline{W}$ is not in the sequence Γ it occupies some position in the sequence B, say $\underline{W} = \underline{Y}_\alpha$ for some ordinal α, but $\underline{W}$ is separable from the union of all the $\underline{Z}_\beta$ in the sequence Γ that precede $\underline{W}$ in sequence B, hence $\underline{W}$ cannot be inseparable from $\underline{Z}$.

Now suppose that $\underline{U}$ and $\underline{V}$ are inseparable from $\underline{Z}$. By what was just shown $\underline{U}$ and $\underline{V}$ both cover $\underline{Z}$. Then by 6.4.9, $\underline{U} \cap \underline{V}$ covers $\underline{Z}$, and since $\underline{Z}$ is inseparable, by 6.4.2(6), $\underline{U} \cap \underline{V}$ must also be, hence $\underline{U}$ is inseparable from $\underline{V}$. This concludes the proof.

So far we have not systematically explored logical relations between indivisibles and abstract spatial points; we have only argued informally that they satisfy a necessary condition for individuating single points—namely that any two things that are inseparable from them must be inseparable from each other. In the next chapter we will investigate these connections more systematically, but we will conclude this one by deriving some further properties of indivisibles that confirm the hypothesis that they individuate points.

6.5.3 Definitions

(1) A PI-system is an inseparable and indivisible I-system.

(2) If $\underline{X}$ is an I-system then $[\underline{X}]$ is the set of PI-systems that are inseparable from it.

(3) If X is a mark or U-system then $[X]$ is the set of PI-systems that are inseparable from $\{X\}$.

The next theorem shows that PI-systems satisfy many of the conditions satisfied by abstract points in topological spaces. In particular, parts (3) and (4) are 'translations' of the two Postulates of Incidence relating concrete observables and abstract points that will be stated in the following section, and parts (5) and (6) show that the relations of coincidence and covering between marks, U-systems, and I-systems correspond in the expected way to set-theoretical relations between the PI-systems that correspond to the marks, U-systems, and I-systems. Other parts of the theorem show that PI-systems have other properties that one would expect abstract points to have, though, as will be noted at the end of this section, it is not to be supposed that they *are* abstract points. All parts of the theorem follow easily from previous definitions and theorems, and their proofs will be omitted.

6.5.4 Theorems

(1) For every mark x, $[x]$ is non-empty, and every PI-system belongs to $[x]$ for some mark x.

(2) For any marks x and y, x touches y if and only if $[x] \cap [y]$ is non-empty.

(3) For any mark x and U-system Y, Y covers x if and only if $[x] \subseteq [Y]$.

(4) An I-system $\underline{X}$ is inseparable if and only if $[\underline{X}]$ is non-empty.

(5) An inseparable I-system $\underline{X}$ is indivisible if and only if any two PI-systems in $[\underline{X}]$ are inseparable.

(6) If X is a U-system then $[X]$ is the union of the sets $[x]$ for all marks x in X.

(7) If $\underline{X}$ is an I-system $[\underline{X}]$ is the intersection of the sets $[x]$ for U-systems X in $\underline{X}$.

(8) Two I-systems $\underline{X}$ and $\underline{Y}$ are inseparable if and only if $[\underline{X}] \cap [\underline{Y}]$ is non-empty.

(9) One I-system $\underline{X}$ covers another I-system $\underline{Y}$ if and only if $[\underline{Y}] \subseteq [\underline{X}]$.

(10) Any two PI-systems are inseparable if and only if they are inseparable from all of the same marks.

While the foregoing establishes that PI-systems act *like* abstract points in certain important respects, it does not prove that they *are* abstract points. In fact, they lack one obvious property of abstract points: namely that of satisfying the criterion of identity that applies to them. This will be returned to in the following section, but in concluding this one we may note two important 'negative' qualities of PI-systems.

One is that PI-systems are not essentially *extremities*, of lines or of anything else, although once lines have been characterized topologically it will turn out to be a theorem that certain points are extremities of them—and plausibly all points are. However, so far as 'pure' point theory is concerned, PI-systems are abstract in the sense of being essentially 'formless', and they can be only be said to be 'small' in the technical sense of satisfying the indivisibility requirement.

The second negative quality of PI-systems is that while they involve constructible *concreta*, they are abstract to the extent of being infinite classes of finite classes of these entities. They are not concrete or constructible in any ordinary sense, and since they are arrived at by a transfinite recursion that depends on the Axiom of Choice, there is no more a 'recipe' for constructing, or better, *re*constructing them than there is for constructing the result of an infinite series of random coin-tosses.

But there is a sense in which abstract spatial points are even less 'structured' than PI-systems, and we want now to consider geometrical abstraction 'proper'.

6.6 Abstract Points and a Problem

Reviewing, we have seen that the separability test shows intuitively that marks like those forming the tepee X do not have a point in common, and the divisibility test shows intuitively that a mark like the block █ does not individuate a unique point. But because abstract spatial points have not yet entered our formalism, we cannot yet prove that the test for point-individuation does what it is designed to do. To do this we must first formulate 'bridge-principles' like those noted in Chapter 4, that link the abstract to the constructible concrete. We shall consider three such principles, or *postulates of abstraction* in this chapter, which are made plausible by intuitive conceptions, as well as a more questionable 'hypothesis' that will require special consideration, but which, together with the other postulates, validates the separability and indivisibility tests.

The first postulate or *principle of abstraction* is that two marks touch if and only if they have a point in common. In fact, this a way of formulating the 'dual interpretation' of the touching relation discussed in section 4.2, either as a binary empirical relation between the marks, or as a ternary and partially theoretical relation between the marks and a third thing, namely the abstract point they have in common:

6.6.1 First Abstraction Principle: The Dual Interpretation of Coincidence. Two marks touch if and only if there exists a point that is incident in both of them.

The second and third principles bear on the *identity* of abstract points:

6.6.2 Second Abstraction Principle. Two points are the same if and only if they are incident in the same marks.

6.6.3 Third Abstraction Principle. If a point is not incident in a mark, then it is incident in another mark that is separate from the first.[8]

Before turning to the consequences of these principles, and, following that, to the 'hypothesis of abstraction', let us comment briefly on the status and justification of the principles. As stated, none of them can be derived from the postulates of the theory outlined in the earlier sections of this chapter—though a very important partial justification is that it can be shown not only that they are *consistent* with the theory developed previously, but they are 'non-creative' in the sense that they have no *empirical* consequences that do not follow the 'purely empirical' theory (Theorem 6.6.8). Of course, important as it is, this kind of non-creativity is not enough: we must also worry that the addition of other plausible abstraction principles might *make* them creative. That is something that can only be ascertained *a posteriori*, by considering plausible additions, which is something which the present study aims to contribute to. For now it is enough to note that in their limited form, something like these principles seem to be assumed in everyday reasoning.

As to the first principle, we have already noted that it essentially formulates the 'dual interpretation' of the touching relation. However, the example of the hollow rectangle, ◘ , shows that we cannot not assume this principle in full generality. In some sense the hole touches the rectangle that surrounds it, but if the hole is only what is left over after the rectangle has been drawn, then it and the rectangle have no

[8] In the presence of other assumptions principles 6.6.2 and 6.6.3 are equivalent to the properties of a topological space being *normal* and of its being *regular* (Kelley, 1955, p.113). These are quite strong assumptions and we shall have to inquire how far they are justified by ordinary use.

points in common. But this should remind us of the distinction made in section 2.4, between 'primary' features and dependent features like 'backgrounds' or boundaries. And, if our first abstraction principle is to hold at all it must only be applied to primary features, and it will have to be reconsidered when we come to secondary or dependent ones.

The second abstraction principle implies that distinct points are individuated by distinct concreta,[9] which also rules out the possibility of distinct points being in exactly the same place—of one being directly 'on top of' another.[10] The third principle also implies this, but it appears to be independent and the theorems to follow show that it is required for important purposes. Both of these capture part of the idea that points are 'pure locations', i.e., they are what things that things coincide spatially have in common. Perhaps this requires further consideration, but let us now turn to the principles' consequences.

The following definitions are required in order to apply the abstract theory to U-systems and I-systems as well as to marks.

6.6.4 Definitions. Let x be a mark, let X be a U-system, and let $\underline{X}$ be an I-system.

(1) $\Pi(x)$ is the set of points that are incident in x.

(2) $\Pi(X)$ is the union of the sets $\Pi(x)$ for x in X.

(3) $\Pi(\underline{X})$ is the intersection of the sets $\Pi(X)$ for X in $\underline{X}$.

The theorems below show some of what follows from postulates 6.6.1–6.6.3 and the definitions. Of these theorems, part (1) is a trivial restatement of 6.6.1, and the proofs of parts (2)–(4) are nearly as trivial. These 'results', which will sometimes be used in combination without special proof, are significant partly because they follow purely from the definitions in 6.6.4 and the abstraction postulates so far stated, and partly because they help to bring out some limitations as to what can be derived from these principles alone, which will be commented on following the proofs of the parts (2)–(4) of the theorem.

6.6.5 Theorems

(1) x touches y if and only if $\Pi(x) \cap \Pi(y)$ is non-empty.

(2) X covers y if and only if $\Pi(y) \subseteq \Pi(X)$.

[9] Of course we cannot postulate that distinct points are points of distinct *actual* things, for that would imply that there could not be distinct points in 'empty space'—i.e., on featureless regions of surfaces.

[10] This rules out interpreting being at the same point as meaning that things were not only in the same place but are the same in other respects as well. If we said that things like the red spot and the bump mentioned in section 2.5 were at different points because the bump, *qua* bump, had no color, we would implicitly be treating one point as locating the bump in 'physical space' and the other as locating the spot in 'color space'.

(3) If $\underline{X}$ is separable then $\Pi(\underline{X})$ is empty.

(4) If $\underline{X}$ is indivisible then $\Pi(\underline{X})$ is empty or a singleton set.

Proof of (2). Suppose first that $\Pi(y) \subseteq \Pi(X)$ but X doesn't cover y. Then there exists z touching y but separate from X. Then by part (1) there exists $\pi \in \Pi(y) \cap \Pi(z)$, while $\Pi(X) \cap \Pi(z)$ is empty. But if $\Pi(y) \subseteq \Pi(X)$ and $\pi \in \Pi(y) \cap \Pi(z)$, then π must also be in $\Pi(X) \cap \Pi(z)$, and that must not be empty.

Suppose conversely that X covers y but $\Pi(y) \not\subseteq \Pi(X)$. Then there exists $\pi \in \Pi(y) \sim \Pi(X)$. If $\pi \notin \Pi(X)$ then by 6.6.3 there exists z separate from X such that $\pi \in \Pi(z)$. Since $\pi \in \Pi(z)$ and $\pi \in \Pi(y)$, z touches y by part (1). But since X covers y, z touches X.

Proof of (3). Suppose that $\underline{X}$ is separable. Then for some $X \in \underline{X}$ there exists a U-system, $U = \{u_1, \ldots, u_m\}$ covering X, no member of which touches all members of $\underline{X}$. On the other hand, suppose $\Pi(\underline{X})$ is non-empty, hence there exists $\pi \in \Pi(X')$ for all $X' \in \underline{X}$. Then $\pi \in \Pi(X)$ and since $U = \{u_1, \ldots, u_m\}$ covers X it follows from part (2) that $\pi \in \Pi(U)$, and therefore $\pi \in \Pi(u_j)$ for some $j = 1, \ldots, m$. But if $\pi \in \Pi(u_j)$ and $\pi \in \Pi(X)$ for all $X \in \underline{X}$, some member of U does touch all members of $\underline{X}$.

Proof of (4). Suppose that $\underline{X}$ is indivisible but $\Pi(\underline{X})$ has at least two distinct members, say π_1 and π_2. Then by 6.6.3 there exist two separate marks, x_1 and x_2, that contain π_1 and π_2, respectively. Then both and $\Pi(\underline{X} \cup \{\{x_1\}\})$ and $\Pi(\underline{X} \cup \{\{x_1\}\})$ would be non-empty because they contain points π_1 and π_2, respectively. Hence $\underline{X}$ is inseparable from the separate marks x_1 and x_2, and therefore it is not indivisible.

Parts (3) and (4) of the foregoing theorem partially validate the separation and divisibility tests. Part (3) shows that if an I-system is separable according to the separability test then the U-systems belonging to it do not have an abstract point in common, and part (4) shows that if the system is indivisible according to the indivisibility test then the U-systems belonging to it can have at most one abstract point in common. However, completely validating these tests requires proving the converses of the foregoing, and it is easy to see that these do not follow from the principles stated so far.

The following can be shown about *normal* topological spaces, $< X, \Omega >$, (Kelley, 1955, p. 112), which are ones in which any two disjoint closed sets are subsets of disjoint open sets. If closed sets in these spaces are interpreted as marks and non-empty intersection is interpreted as touching, then the spaces satisfy Postulates 6.2.1 and 6.3.1,[11]

[11] The proof that, given this interpretation and assuming normality, the Separation Postulate is satisfied goes as follows. Suppose that x, y and z are marks, which are

and if the members of Ω are interpreted as abstract points and membership is interpreted as a point being a point *of* a mark, then the three abstraction principles are also satisfied. In other words, appropriately interpreted normal topological spaces are *models* of our theories of both the concrete and the abstract superficial. But not all normal spaces are compact (Kelley, 1955, p. 135), and unless a space is not only normal but compact, the members of inseparable systems do not always meet in common points. *Ergo*, it does not follow from the postulates stated so far that inseparable systems meet in common points. This leads to our 'topological hypothesis':

6.6.6 Hypothesis. If $\Pi(\underline{X})$ is empty then $\underline{X}$ is separable.

Assuming this, the validity of the separation test follows immediately, and Theorem 6.6.7 will show that the hypothesis also entails the validity of the indivisibility test. But prior to that, something should be said about the legitimacy of formulating the hypothesis, seemingly on a par with the abstraction postulates. However, it is not *on a par* with the postulates, since the postulates rest on independent foundations, but so far as the author knows the hypothesis has no such foundation. If it were a *postulate* then Russell's comment would apply with a vengeance: "The method of 'postulating' what we want has many advantages; they are the advantages of theft over honest toil" (*Principles of Mathematical Philosophy*, p. 71). The next theorem brings out the reward that derives from 'stealing' the hypothesis, but before stating that there are two other points to make in partial justification.

One is that Theorem 6.6.8 below shows that, along with the abstraction postulates, the hypothesis does not entail any consequences concerning touching, covering, and separation that are not entailed by the 'empirical' assumptions of sections 6.2–6.5 alone. In effect, while the

---

interpreted as closed sets, and W is a finite union of marks, which must be another closed set, and x and y are disjoint—i.e. they don't touch—and z and W are disjoint. Since x and y are disjoint, by normality there must be disjoint open sets, Γ_x and Γ_y, that contain x and y as subsets. Let x' and y' be the closures of Γ_x and Γ_y, respectively. x' must be disjoint from y and y' must be disjoint from x. Now let U be the system $\{z \cap x', z \cap y', z'\}$, where z' is the closure of $z \sim x' \sim y'$. This system obviously covers z, and none of its members touches W. $z \cap x'$ and $z \cap y'$ are clearly disjoint from W, since z is disjoint from W. That z' is disjoint from W follows from the fact that $z \sim x' \sim y'$ is a subset of z, hence its closure is a subset of the closure of z, which is z itself since z is closed, and z is disjoint from W. It remains to show that no member of U touches both x and y.

Clearly neither $z \cap x'$ nor $z \cap y'$ touches both x and y, since x' does not touch y and y' does not touch x. Since x' is the closure of the open set Γ_x and x is a subset of Γ_x, it follows that x is interior to x' and it is disjoint from the closure of $\sim x'$. Therefore x is also disjoint from the closure of the smaller set $z \sim x' \sim y'$, and by the same reasoning y is also.

hypothesis lacks external support, it is like a non-creative definition in not entailing new empirical consequences.

The other point is that the hypothesis links point-theoretical and topological concepts of *compactness.* This is related to our earlier remark, that even the first abstraction postulate only applies to marks that *contain* their boundaries. We now see that, while the principles and hypothesis of abstraction have not been explicitly topological, nevertheless they have topological presuppositions. This said, however, we do not pretend that the hypothesis is above criticism, although we cannot enter farther into this here.

Now let us look at a major consequence, namely the converse of 6.6.5(4), whose proof is simple:

6.6.7 Theorem. If $\Pi(\underline{X})$ is empty or a singleton set then $\underline{X}$ is indivisible.

Proof. Suppose that $\underline{X}$ is not indivisible. Then by 6.5.4(5) there must exist separable PI-systems, say $\underline{Y}$ and $\underline{Z}$, both of which are inseparable from $\underline{X}$. Therefore by 6.5.4(8), $[\underline{X}] \cap [\underline{Y}]$ and $[\underline{X}] \cap [\underline{Z}]$ must both be non-empty. 6.5.4(7) then implies that $[\underline{X} \cup \underline{Y}]$ and $[\underline{X} \cup \underline{Z}]$ must also be non-empty, and therefore according to 6.6.4(3), $\underline{X} \cup \underline{Y}$ and $\underline{X} \cup \underline{Z}$ are not separable. Hence, by 6.6.6 $\Pi(\underline{X} \cup \underline{Y})$ and $\Pi(\underline{X} \cup \underline{Z})$ must be non-empty, and there would exist abstract points π_1 and π_2 in $\Pi(\underline{X} \cup \underline{Y})$ and $\pi(\underline{X} \cup \underline{Z})$, respectively. Then 6.6.4(3) would imply that π_1 and π_2 would both belong to $\Pi(\underline{X})$. Hence $\Pi(\underline{X})$ is not empty.

Nor could $\Pi(\underline{X})$ be a singleton set, since that would imply that π_1 and π_2 were the same point. But by 6.6.4(3) again, the former belongs to $\Pi(\underline{Y})$ and the latter to $\Pi(\underline{Z})$, and if π_1 and π_2 were the same point then $\Pi(\underline{Y}) \cap \Pi(\underline{Z}) = \Pi(\underline{Y} \cup \underline{Z})$ would be non-empty and therefore, $\underline{Y}$ and $\underline{Z}$ would be inseparable according to 6.6.5(3), contrary to hypothesis.

Theorems 6.6.5–6.6.7 formalize an 'extended dual interpretation' of relations of incidence, touching and separation, not only among marks but also among U-systems, I-systems, and PI-systems. The final theorem of this section provides the partial justification alluded to above, which shows that the postulates and the hypothesis on which the dual interpretation rests are non-creative in the sense that they add nothing to the 'empirical' theory of touching and separation that does not follow from laws like the separation postulate that are themselves purely empirical. This follows from the next theorem, whose proof is trivial and will be omitted:

6.6.8 Theorem. Let $\mathfrak{C} = < M, S >$ be a system of elements, M, with a binary relation, S, satisfying postulates 6.2.1 and 6.3.1 (the separation postulate), and let Π be the class of PI-systems of $\mathfrak{C}$. If Ω is the class of equivalence classes of members of II under the relation of inseparability then the system $\mathfrak{C}II =< M, S, \Omega >$ satisfies postulates 6.6.1, 6.6.2, and 6.6.3, and hypothesis 6.6.6.

As said, this implies that any empirical data consistent with the postulates of sections 6.2 and 6.3 must also be consistent with the abstract principles stated in this section. Therefore it would seem that such objections as might be brought against the abstract principles or the 'hypothesis' could not be raised on empirical grounds, although this question will be returned to in later chapters, when further concepts are introduced. We will close this one with brief remarks that compare the present account of the nature of points with other theories.

6.7 Other Views on the Nature of Points[12]

Detailed comparison of our present account of abstract points with other views lies beyond the scope of this study, but it will not be amiss to comment briefly on similarities and dissimilarities to some of the better known views on this subject. Of course, our present concern is with points on the surfaces of bodies, and, with the exception of Euclid's plane geometry, this stands in sharp contrast to almost all classical and recent writings, and therefore close correspondences between the views put forth here and ones developed elsewhere are not to be expected. Nevertheless there are some interesting points of comparison.

Consider Definitions 5 and 2 of Book I of Euclid's *Elements*, "The extremities of a line are points" and "A point is that which has no part" (Heath, 1956, Vol.I, p.153). It was already noted that PI-systems and *ipso facto* abstract points are not essentially extremities, much less extremities of lines as Definition 5 might suggest—though the lines that will be considered in Chapter 10 have points as extremities and it will even be plausible that all points should be extremities of lines. But lacking a topological basis in terms of which boundaries and extremities can be defined, extremities do not exist in our 'point spaces'.

As to "A point is that which has no part," the abstract points of our theory are not supposed to have or to be parts in a 'mereological space'. Moreover, the PI-systems that individuate points in surface spaces can only be considered to have parts in the sense of having *members*—but then, the U-systems that are their members are larger than they are. If anything, as we conceive abstract points, they come closer to fitting Aristotle's characterization of a point as a *location*, i.e., as was mentioned in section 4.3, they are identical to the places where they are.[13]

[12]An essay would itself be required to discuss in detail the matters to be commented on in this section, but that would be beyond the scope of the present study.

[13]In fact this fits in with a view of Leibniz's. Putting aside our restriction to

But let us turn to recent 'constructive' characterizations of points.[14]

It could be argued that if we had not distinguished abstract points from the PI-systems that individuate them, and we had stopped short at the PI-systems, their 'construction' might be compared to Whitehead's method of extensive abstraction. The details of Whitehead's constructions differ greatly from our own, in large part because the objects—the so called 'actual existents'—on which his constructions are based differ greatly from the surface features on which our theory is based. But there is an equally important methodological difference between Whitehead's theory and the present one. Russell's previously quoted *mot* "The method of 'postulating' what we want has many advantages; they are the advantages of theft over honest toil" might be turned on its head. Our four principles of abstraction explicitly describe the relation between abstract points and the *concreta* that instantiate them, but part of the justification for postulating them consisted in showing in Theorem 6.6.8 that they add nothing to the only empirical postulate that has been stated, *viz* the Separation Postulate, because the abstraction principles *plus* the postulate have no empirical consequences that do not follow from the Separation Postulate alone. In contrast, Whitehead's theory, while not postulating any transcendental principles, postulates at least 20 principles like *Assumption 2*: "No region is connected with all other regions; and any two regions are mediately connected" (*Process and Reality*, p. 451). However the empirical status of this and related assumptions is questionable because, although regions are said to be *relata* of the relation of 'extensive connection', assumptions about them have the appearance of question begging, because *region* is itself an abstract spatial concept.

---

surface spaces, Leibniz held that *mathematical* points (in contrast to both *material* and *metaphysical* points) are "merely positions, i.e., possible terms for the relations of distance" (Russell, 1900, p. 115). Substituting relations of *touching* for Leibniz's relations of *distance*, we have essentially the dual, ternary interpretation of touching.

[14]Two views put forth between Leibniz's and modern times may be noted in passing. Berkeley held that points are *minima sensibilia*, e.g., that "The moon is only a round, luminous plane, of about thirty *visible points* in diameter" (*New Theory of Vision*, par. 44, author's italics). Of course this radical empiricism, maintaining as it does that *esse* is *percipi*, and points in particular must be perceivable, conflicts with our view that while they are *abstracta* that stand in well defined relations to *concreta*, they are too small to be discriminated by the senses. Other comments on Berkeley's views will be scattered throughout this work.

Kant's scant comments on points in the *Metaphysical Foundations of Natural Science* (1786) are interesting as an early, if not the earliest place in which Geometry is conceived as a part of Physics. Thus, Observation 2 on Proposition 8 of the second chapter, "Metaphysical Foundations of Dynamics" says in a comment "...if a point cannot directly drive by repulsive force..." (Ellington, 1985, p. 73), which suggests that points might be dynamical objects subject to laws like Newton's Laws.

The author would raise similar objections to other constructive theories, including ones of Poincaré, Russell, Nicod, Carnap, and mereologists of various stripes. This is not so much because these theories are wrong as because their authors, in their fussiness about axiomatic form, do not pay sufficient attention to the primitive concepts and postulates of their theories.

7

Towards a Topology of Physical Surfaces: Basic Definitions and Three Fundamental Properties[1]

7.1 Introduction: The Problem of Physical Topology

Although mathematical topology has widely disparate applications, to metrizability, dimensionality, orientability, the four-color problem, and so on, the problem of its application to the physical world is most clearly brought out in connection with the concepts of spatial *boundaries*, *interior*, and *exteriors*. These ideas have been foci of controversy from ancient times to the present, and though it is not our present purpose to comment on them in detail (recall the Stroll and Zimmerman works cited in section 2.3), we may just note the following. Consider rectangle R below, and the region of points that are exterior to it, E:

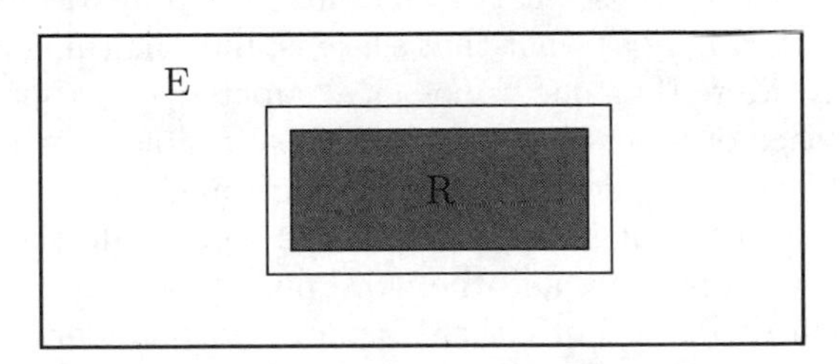

The question is: what are the relations between the inner boundary of E and the outer boundary of R, and between those and the regions whose boundaries they are? First, ordinary usage makes it clear that we intuitively conceive of these regions as *having* boundaries. We could

[1]This chapter develops in detail matters sketched briefly in sections vi and vii of Adams, 1973.

even point to them if asked. Second, we could argue that the boundaries of R and E must *coincide*, hence they shouldn't be pictured as separate, as they are here. If they were separate there would be a 'space' between them, and the points in the space would be exterior to R but not in region E, contrary our supposition that E contains all of the points exterior to R. And, finally, the boundary of each region ought to *belong* to the region it is the boundary of, since if it didn't it would be separate and some distance from it, which it clearly is not. But these three things are inconsistent. If region E includes its inner boundary and that coincides with the outer boundary of R, which in turn is included in R, then E includes points that are *in* R, and not exterior to it. To the extent that this argument depends only on our everyday ways of conceiving boundaries, interiors, and exteriors, this suggests that these conceptions are not wholly coherent.[2]

Now, modern mathematical topology rectifies the incoherence in our everyday concepts of boundaries, exteriors, and interiors, but the rectification necessarily involves a degree of arbitrariness. Whether it has to be as complicated as it will eventually be seen to be is another matter, but let us begin with the superficial way in which it modifies everyday conceptions. It gives up the idea that regions R and E both *have* boundaries in the strict sense. More exactly, if R contains its topological boundary then E only has a boundary in an extended sense, and that is *not* contained in E. Therefore, the contradiction in our intuitive conceptions is avoided by giving up one of the premises that leads to it. But this is done at a cost.

How are we to determine which of regions R and E has the boundary in the 'narrow' sense that it contains it? To answer this we might expect to look closely at what that sense is, but one thing is to be noted immediately. More than one 'topological superstructure' can be fitted to the same space of points, and mathematical topology, which applies to all of these structures, does not tell us which one is 'right'. Furthermore, *which* regions or point sets are stipulated to include their boundaries depends on the topology, and therefore pure topological theory cannot tell us which of R or E, if either, includes its boundary. If there were such a science as *Applied Topology*, standing to mathematical topology as Applied Physics stands to pure mathematical Physics, we might expact that the regions that contain their boundaries would be specified in the assumptions of that science. But there is no such science, and at best we may be guided by 'topological practice'.

[2]There are many other well known arguments that demonstrate the incoherence of everyday concepts of boundaries and related ideas, some of which are usefully summarized in Casati's and Varzi's recent book *Holes*, 1995.

We have already adopted one assumption that derives from 'proto-topological' practice, namely that objects that touch have a point in common. Now if the objects are solids, which cannot be in the same place at the same time, if they touch at all that can only be at their boundaries. Therefore, if solids touch and have a point in common, that must be a *boundary point* of both of them. *Ergo*, some points of the boundaries of touching solids are points of the solids themselves.

Euclid too, in theorizing about extremities, tacitly assumes that these extremities *belong* to the things, lines, surfaces, or solids, that they are extremities of. For instance, Proposition 10 of Book I (Heath, 1956, Vol. 1, p. 267) describes a method for bisecting a given line segment, say AB below, at point D:

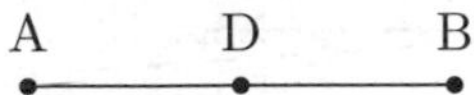

This tacitly assumes that segments AD and DB 'add up' to AB, which they wouldn't if they didn't include their end-points.

Furthermore classical physics, though like *The Elements* in having been developed without a foundation in mathematical topology, assumed that the surfaces of solids are not exterior to them. For instance, Routh's *Advanced Dynamics of a System of Rigid Bodies* describes the motion of a solid cone rolling on a plane surface as "... always touching the plane" (sixth ed., 1860, p. 111). And, that specialty that seeks to introduce modern topological and measure-theoretic rigor into classical physical theory, namely Continuum Mechanics, characterizes a body as "... a homeomorph of the closure of a regular region" (Truesdell, 1977, p.72). Putting aside what it is for a *body* to be homeomorphic to anything, this seems to imply that the region occupied by a body is a closed set, and this in turn implies that it includes its boundary points.[3]

With the foregoing examples to guide us, we too will assume, not that *bodies* include their extremities, but that *marks* do. Of course that requires us to be clear about what marks are, but we will rely here on the 'genetic' characterization of surface features that was discussed in section 2.5 According to this, if a hollow square, ◘ , is drawn by coloring in the part around the hollow, then that part is the *mark*, which will be assumed to contain its boundary, while the hollow is only 'background', and it will not be assumed to contain its boundary.

But this only 'scratches the surface', because along with abandoning the intuition that a thing and its exterior both have boundaries, much

[3]Perhaps it is unfortunate that physical bodies are assumed to occupy closed regions of space, which include their surfaces, since if they occupied *open* regions the troubling problem of impact would not arise.

deeper changes follow from the work of Cantor and others investigating Leibniz's 'labyrinth of the continuum' (Russell, 1900, Chapter IX). In the process of cutting the subject free of its metrical moorings, detailed exploration of the labyrinth requires the full resources of Cantorian set-theory. As already in the pure theory of points, we will require not only infinite sets, but the continuum, whose labyrinth Leibniz attempted to investigate.

That the finite cannot suffice is evident in the fact that finite point-sets do not have well-defined boundaries and interiors. For instance, while a rectangular array of dots like that below

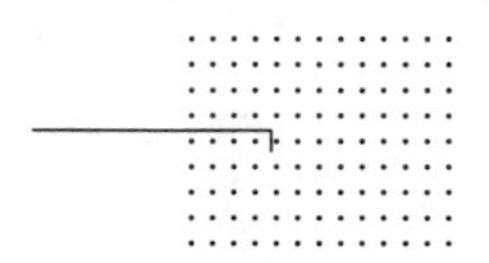

might seem to be bounded by its outer rows and columns, nonetheless any 'interior' dot can be reached by a line carefully drawn from outside the rectangle in such a way as to pass between the dots on the boundary, as shown.[4] Even if the array consisted of all points whose cartesian coordinates had rational values, it would still be possible to reach interior points from the outside by passing between the points with rational coordinates.

To bring the continuum into the picture we will use the fact that abstract points stand in one-to-one correspondence with equivalence classes of inseparable PI-systems, of which there are uncountably many.[5] Given this, we can define a *basic open set* in a surface space as a set of points that are *not* incident in some U-system, and an *open set* as an arbitrary union of basic open sets. The topology thus generated can be shown not only to be a standard *Hausdorff* topology,[6] but to be *compact* and to have various other important properties. These will be discussed in this and subsequent chapters, and this chapter will focus on the Hausdorff property, compactness, metrizability, and connectedness. This will begin the extension to the topological domain of our empiricist-operationalist

[4]This 'passing through' test will be very important in connection with the empirical account of boundaries and their relations to be given in chapter 9. Something close to this will also be commented on in Chapter 17, where Marr's (1982) theory of 'visual cues' for applying topological concepts like that of a boundary are discussed.

[5]However, the reader is warned that doing this in the simplest way requires us to add another empirical postulate, *viz*, the *Covering Postulate*, that the surfaces we are considering can be covered with finite numbers of marks, as well as two more 'hypotheses' akin to our Hypothesis 6.6.6.

[6]This appears to be the same as the topology *associated with the closure operator* constituted by the U-systems of our theory, Kelley, 1955, p. 43.

program.

7.2 The Basic Topology

We will begin by reminding the reader of some fundamental topological ideas. A *topological space* is defined by a point set, Ω, and a system, $\mathbf{N}$, of so called *open sets* that have the following properties: (1) the intersection of any two open sets is an open set, and (2) an arbitrary union of open sets is an open set (Kelley, 1955, p. 37).[7] A *neighborhood* of a point, π, is any point set that contains an open set containing π. A *basis* of a topological space (Kelley, 1955, p. 46) is a class of open sets that has the property that for every point π and every neighborhood of π, there is an open set that contains π and is a subset of the neighborhood. A *closed* set is the complement of an open set. A topology is *compact* (Franz, 1964, p. 67, condition [Kp']) if a system of closed sets has an empty intersection only if a finite subsystem of it has an empty intersection.

As points of our surface spaces we will take the abstract points characterized in the previous chapter, assuming that they satisfy not only the three abstract postulates of that chapter but also Hypothesis 6.6.6. This hypothesis is extremely important because many key properties of our topology depend on it.

Now we can define the open sets of our topology.

7.2.1 Definitions

(1) A *basic open set* is the complement of a point set $\Pi(X)$, for some U-system X.

(2) An *open set* is an arbitrary union of basic open sets.

Before considering the implications of this definition, a word should be said about our reason for defining the basic concept of our surface topologies in the above manner. In some ways it might seem more natural to define open sets as the sets of points that are *interior* to marks and U-systems, rather than as the sets of points *exterior* to them. In fact, Theorem 8.3.2 states that the sets of points interior to U-systems do constitute a basis for a surface topology. However, the reason for not *defining* our topologies in this way is that the idea of an *interior* presupposes other topological concepts; to say that a point is interior to a mark is to say that it is a point of the mark that does not lie on its boundary. But the idea of a *boundary* is itself a topological one, which will be taken up in the following chapter. On the other hand, the idea of

[7] Topological spaces can be defined in other ways, for instance in terms of *neighborhoods* of points, or, less directly, in terms of *uniform structures*, Bourbaki, 1966, Chapter 6. However, the approach in terms of open sets is most convenient for our purposes because of their close connection with U-systems.

an *exterior* is 'pre-topological'; to say that a point is exterior to a mark is only to say that it is not a point *of* the mark.

But now we will turn to the implications of our definition, starting with the fact that it does define a topology:

7.2.2 Theorems

(1) The set Ω and the class open sets form a topology.

(2) The class of basic open sets forms a basis for the topology.

(3) The closed sets of the topology are the classes of points, $\Pi(\underline{X})$, for I-systems $\underline{X}$.

(4) Let Γ_1 and Γ_2 be disjoint closed point sets. Then there exists a U-system X such that $\Gamma_1 \subseteq X$ and $\Gamma_2 \bigcap X$ is empty.

(5) The topology is compact.

Proof of (1). First we will show that intersections of open sets are open sets as we have defined them. Let Ω_1 and Ω_2 be open sets, i.e., for $i = 1, 2, \Omega_i$ is a union $\bigcup_{x\in\Sigma_i}(\sim \Pi(X))$ for U-systems X in an 'index set of U-systems', Σ_i. Then $\Omega_i = \sim \cap_{X\in\Sigma_i}\Pi(X)$ and

$$\begin{aligned}\Omega_1 \cap \Omega_2 &= [\sim \cap_{X\in\Sigma_1}\Pi(X)] \cap [\sim \cap_{X\in\Sigma_2}\Pi(X)]\\ &= \bigcup_{X\in\Sigma_1, Y\in\Sigma_2} \sim [\Pi(X) \cup \Pi(Y)]\\ &= \bigcup_{X\in\Sigma_1, Y\in\Sigma_2} \sim [\Pi(X \cup Y)].\end{aligned}$$

But since $X \cup Y$ is itself a U-system, this is a union of complements of point sets of U-systems and therefore it is an open set.

That the union of an arbitrary set of open sets is an open set follows easily from the fact that open sets are arbitrary unions of basic open sets, and therefore arbitrary unions of open sets are arbirary unions of arbitrary unions of basic open sets.

Proof of (2). If there is only one point in the space, π, then $\emptyset$ (the empty set) and $\{\pi\}$ are the only open sets, which are also basic open sets, and they are easily seen to be a basis for the topology.

If there is more than one point in the space, then for any point π_1 there exists a point π_2 that is different from π_1. Then by postulate 6.6.2 there exist separate marks x_1 and x_2 such that $\pi_1 \in \Pi(x_1)$ and $\pi_2 \in \Pi(x_2)$, and by 6.6.5(1) $\Pi(x_1)$ and $\Pi(x_2)$ must be disjoint since x_1 and x_2 are separate. $\sim \Pi(x_1)$ and $\sim \Pi(x_2)$ are basic open sets, and since $\pi_1 \in \Pi(x_1)$ and $\Pi(x_1)$ and $\Pi(x_2)$ are disjoint, $\pi_1 \in\sim \Pi(x_2)$. This open set is also easily seen to be a neighborhood.

Proof of (3). Let Π be a closed set of points. Then $\sim \Pi$ is an open set, and therefore it is a union, $\bigcup_{X\in\Sigma} \Pi(X)$, of basic open sets, $\sim \Pi(X)$, for U-systems X in a class Σ. Then

$$\begin{aligned}\Pi &= \sim \bigcup_{X\in\Sigma} \sim \Pi(X)\\ &= \cap_{X\in\Sigma}\Pi(X).\end{aligned}$$

If $\underline{X}$ is the class of U-systems X in Σ then $\underline{X}$ is an I-system, and by definition 6.6.4(3), $\Pi(\underline{X}) = \cap_{X\in\Sigma}\Pi(X)$. Therefore Π is the class of points of an I-system.

The argument goes in reverse, to show that if Π is the class of points of an I-system them it is a closed set.

Proof of (4). This follows trivially from part (3), that closed point sets are the point sets of I-systems, and from 6.4.6(4), that separable I-systems can be covered by separate U-systems.

Proof of (5). Suppose now that Σ is a set of closed sets, Π, whose intersection is empty. It must be shown that Σ has a finite subset, Σ', whose intersection is empty. By (3) above, each $\Pi \in \Sigma$ is the point set of an I-system, $\underline{X}\Pi$, i.e., $\Pi(\underline{x}\Pi) = \Pi$. Let $\underline{\Sigma}$ be the set of these I-systems for all $\underline{X}$ in Σ. Then it is easy to see that $\cap_{\Pi\in\Sigma\underline{X}\Pi} = \Pi(\underline{\Sigma})$, hence $\Pi(\underline{\Sigma})$ is empty. Hence by 6.5.4(4), $\underline{\Sigma}$ is separable and by 6.4.7(1) it has a finite subset, say $\{\underline{x}_1,\ldots,\underline{x}_n\}$, that is separable. Then the I-system $X' = \{\underline{X}_1 \cap \ldots \cap \underline{X}_n\}$ is separable, and by Theorem 6.6.5(3) $\Pi(\underline{X'})$ is empty. Now let Σ' be the set of members Π of Σ such that $\Pi = \Pi(\underline{X}_i)$ for some $i = 1,\ldots,n$. But $\Pi_{\Pi\in\Sigma'}(\Pi) = \Pi(\underline{X'})$, hence Σ', which is a finite subset of Σ, empty.

The foregoing, especially compactness, are significant technical properties of our topology, which are extremely important because they allow us to apply a great range of standard theorems of topology. But two other standard properties have yet to be considered, which are important for the same reason. These are the so called *Hausdorff* property, and the property of *metrizability*, which will be discussed in the following two sections. Neither follows from the principles postulated so far, and therefore assuming them requires us to extend our theory in important ways.

7.3 Finite Coverability and the Hausdorff Property

A topological space is *Hausdorff* if any two distinct points in it have disjoint neighborhoods (Kelley, 1955, p. 67).[8] This property is assumed in most elementary topological theory, but it is easy to see that it is not possessed by all spaces that conform to the principles heretofore assumed. It is consistent with everything assumed so far that our surface spaces are infinite in extent, although the marks that might be made on them are finite. In this case U-systems, which are finite sets of marks, must also be finite in extent, and therefore their *exteriors*, which define

[8] Such a space is also called a T_2, or *separated space*. Other topological separation properties will be encountered as we proceed, but these kinds of separation are to be distinguished from the purely 'point-theoretical' separation involved in the Separation Postulate, 6.3.1.

the basic open sets in our topology, must almost fill the space. Hence any two basic open sets, in fact any open sets and neighborhoods, must intersect in regions that almost fill the space. Obviously in these circumstances the topology defined above could not be Hausdorff.[9] To avoid this, we postulate the following:

7.3.1 Finiteness Postulate. There is a U-system that covers all marks.

This yields the Hausdorff property immediately, although first we will prove something slightly stronger:

7.3.2 Theorems. Let $< x, \Omega >$ be a basic surface topology.

(1) $< x, \Omega >$ is normal.

(2) $< x, \Omega >$ is Hausdorff.

Proof. Suppose that Γ and Θ are disjoint closed point sets. By 7.2.2(1) these are points-sets of I-systems, say $\underline{X}$ and $\underline{Y}$, which must be separable since they have no common points. By 6.4.7(4) with n=2, there exist separate U-systems x and Y, that cover $\underline{X}$ and $\underline{Y}$, respectively.

Now let V be a universal U-system. By an easy generalization of 6.3.3 there is a U-system U that covers V, no member of which touches both X and Y. Let U_X be the union of the marks in U that do not touch X, and let U_Y be the set of marks in U that do not touch Y. Since no mark in U touches both X and Y it follows that $U_X \cup U_Y = U$, and since U covers V and V is universal, so is $U_X \cup U_Y$. Therefore, $\sim [U_X \cup U_Y] = [U_X] \cap \sim [U_Y]$ is empty, and so the basic neighborhoods $\sim \Pi(U_X)$ and $\sim \Pi(U_Y)$ are disjoint.

Finally, the original closed sets Γ and Θ are subsets of the open sets $\sim \Pi(U_X)$ and $\sim \Pi(U_Y)$, respectively. Thus, we had $\Gamma = \Pi(\underline{X})$ and $\Theta = \Pi(U_Y)$, and disjoint U-systems X and Y cover $\underline{X}$ and $\underline{Y}$, respectively, hence $\Gamma \subseteq \Pi(X)$ and $\Theta \subseteq \Pi(Y)$. That $\Pi(X) \subseteq \Pi(U_X)$ and $\Pi(Y) \subseteq \Pi(U_Y)$ is trivial. If $\Pi(X)$ were not a subset of $\sim \Pi(U_X)$ then $\Pi(X)$ and $\Pi(U_X)$ would not be disjoint, and X would touch U_X (easy generalization of 6.6.4(6)). But U_X was defined to be the union of the marks in U that do not touch X. Similarly, $\Pi(Y)$ must be a subset of $\Pi(U_Y)$, which concludes the proof of part (1).

Part (2) follows from part (1) and the fact that points are closed sets in the basic surface topology. The latter follows from the fact that points correspond to maximal inseparable and indivisible I-systems, and by 7.2.2(3) all such systems are closed sets in the topology. This concludes

[9] Section 12.6 comments briefly on a modified *external component topology*, which is Hausdorff even though the underlying space may be infinite—although compactness is less easy to characterize in this case (cf. also p. 418 of Adams, 1973). Our present approach, which baldly posits finiteness, is therefore simpler, while still having considerable empirical plausibility in its application to the surfaces of actual physical bodies.

the proof.

The following theorem is a trivial consequence of the fact that the point sets of I-systems and *ipso facto* U-systems are closed (Theorem 7.2.2), and that in a normal space, if a finite union of closed sets is a neighborhood of a point then the union of the sets that contain the point is also a neighborhood of it.[10]

7.3.3 Theorem. If a U-system X is a neighborhood of a point π (i.e., $\Pi(X)$ is a neighborhood and $\pi \in \Pi(X)$),[11] then the class of members of X that contain π is also a neighborhood of π.

We end this section with brief comments on the empirical justification for the Finiteness Postulate. First, it is plausible that the surfaces that we are dealing with are of finite extent, because they are surfaces of *bodies* and ordinary bodies are of limited extent.[12] That a finite surface can be covered by a finite number of marks is also plausible. The paint that is applied by a single stroke of a paintbrush creates a single mark. Hence, if it is possible to paint over a finite surface with a finite number of brush strokes, it should follow that the surface could be covered with a finite number of marks, as the postulate holds.

The foregoing argument is not definitive, and among other things it can be objected to on the ground that it does not fit the microphysical picture. Thus, microscopic examination might be expected to reveal tiny unpainted regions of the surface, in which case we would have to say that the surface only *appeared* to be entirely covered. A related difficulty was already noted in connection with the Separation Postulate, and we can only reiterate what was said earlier: namely that our theory fits 'macrophysical appearance', but it must be reexamined when considering its applicability to the microphysical and micro-surfacial. Chapters 16 and 17 will return to the relation between macrophysical 'appearance' and microphysical 'reality', and the fact that the very idea of a *surface*

[10]However, I do not find this stated in a cursory survey of the topology texts that are available to me. In the case of n = 2 closed sets, Γ_1 and Γ_2, whose union is a neighborhood of a point π, if $\pi \in \Gamma_1 \sim \Gamma_2$ then π is separate from Γ_2, and by normality there are separate neighborhoods, Θ_1 and Θ_2, that contain π and Γ_2, respectively. Then since Γ_2 is a subset of Θ_2 and Θ_1 is separate from it, Θ_1 is separate from Γ_2. Hence $\Theta_1 \cap (\Gamma_1 \cup \Gamma_2) = \Theta_1 \cap \Gamma_1$, and since both Θ_1 and $\Gamma_1 \cup \Gamma_2$ are neighborhoods of π, so is $\Theta_1 \cap (\Gamma_1 \cup \Gamma_2) = \Theta_1 \cap \Gamma_1$, and this is a subset of Γ_1.

The case of any finite number of closed sets follows trivially from the case n = 2.

[11]In what follows we will often not distinguish between marks, U-systems, and I-systems and the point sets that they individuate.

[12]It is to be noted, though, that Chapter 12 will discuss *composite surfaces* that are formed when two or more or even infinitely many bodies are 'combined', e.g., as bricks in a wall, which will permit us to theorize about infinitely extended surfaces as 'sums' of finite ones.

might itself be held only to be a macrophysical appearance. But for the present we will confine attention to appearances, only remarking that the appearance-reality problem will be recurred to later in this chapter, where we discuss continuity, as well as in Chapters 16 and 17.

Now we turn to another property, in connection with which we introduce another 'hypothesis'.

7.4 Metrizability: A Hypothesis

A *metric* on a point set Ω (Kelley, 1955, p. 118) is a non-negative real-valued function, $d(\pi_1, \pi_2)$, on point-pairs π_1, π_2 satisfying the three conditions: (1) *symmetry*, i.e., $d(\pi_1, \pi_2) = d(\pi_2, \pi_1)$ for all π_1 and π_2, (2) the *triangle inequality*, $d(\pi_1, \pi_2) + d(\pi_2, \pi_3) \geq d(\pi_1, \pi_3)$ for all π_1, π_2, and π_3, and (3) $d(\pi_1, \pi_2) = 0$ if and only if $\pi_1 = \pi_2$.[13] The open neighborhoods of a point π_1 in the *topology associated* with $d(\pi_1, \pi_2)$ are sets of points π_2 such that for some fixed $\delta > 0, d(\pi_1, \pi_2) < \delta$, and a topology is *metrizable* if a metric can be associated with it. Of course the conditions defining a metric are very weak compared to ones satisfied, e.g., by a Euclidean distance measure, but nevertheless many important topological theorems, especially theorems on topological dimension, depend on this weak kind of metrizability. And, since not all topological spaces are metrizable we must inquire what properties our surface spaces must have if they are to be metrizable. This inquiry will be carried out from a point of view that brings in intuitive conceptions of measurement, but we will begin by noting a key result that states a necessary and sufficient condition for compact Hausdorff topological spaces to be metrizable. This is *Urysohn's Theorem*, which is stated in terms of the idea of a *basis* of a topological space (Kelley, 1955, p. 46), which, as previously stated, is a class of open sets in the space that has the property that for every point π and every neighborhood of π, there is a member of the class that contains π and which is a subset of the neighborhood.

Now we can state:

7.4.1 Urysohn's Theorem (Franz, 1964, p. 96). A compact topological space is metrizable if and only if it has a countable basis.

Our question now becomes: is there any reason for thinking that our surface spaces have countable bases? We know that the class of complements of point-sets of U-systems is a basis for a surface topology (Theorem 7.2.2(2)), and clearly the cardinality of this basis cannot exceed that of the class of U-systems that individuate its members. Moreover, given

[13] A *pseudometric* (Kelley, 1955, p. 119) is defined by omitting condition (3), but while metrics of this kind are important in many applications we shall not consider them here.

that U-systems are finite unions of marks, if the class of marks is infinite then the class of U-systems must have the same cardinality as that of the class of marks.[14] *Ergo*, the cardinality of the bases that we have used to generate the topologies of our surface spaces is no greater than that of the class of marks that can be drawn on the surface. But what do we know of that? So far all we know is what is said of constructible features in the informal remarks in Chapter 2, namely that they are things that can be drawn on bodies' surfaces. And there is no reason to exclude the possibility that uncountably many marks might be drawn on a surface. Hence we cannot exclude the possibility that the bases upon which our surface topologies have been constructed might be uncountable.

But the fact that a topology has an uncountable basis does not exclude the possibility of metrizing it, since it might also have countable basis. Let us now give a hopefully plausible 'metrological argument' that in fact our surface topologies do have countable bases.

Given that the open sets in the topology of a surface are defined as unions of exteriors of U-systems of marks on the surface, it is easy to see that any class $\mathbf{U}$ of U-systems is a basis for the topology if it satisfies the condition that for every point π and every U-system X that doesn't contain π, there is a U-system in $\mathbf{U}$ that covers X and doesn't contain π. In fact, this follows if it holds for all points π in a countable point-set, Π_0, that is *dense* in the space in the sense that every non-empty open set contains at least one point of Π_0 (this is the condition of *separability*, Kelley, 1955, p. 49). The diagram below pictures the relation between a U-system X, consisting of four ovals w, x, y, and z, a point π not contained in X, and an area Y shading in almost all of the 'space', in particular so as to cover all of the ovals in X, but *not* containing π:

[14]Because the cardinality of the class of finite subsets of an infinite set equals that of the set.

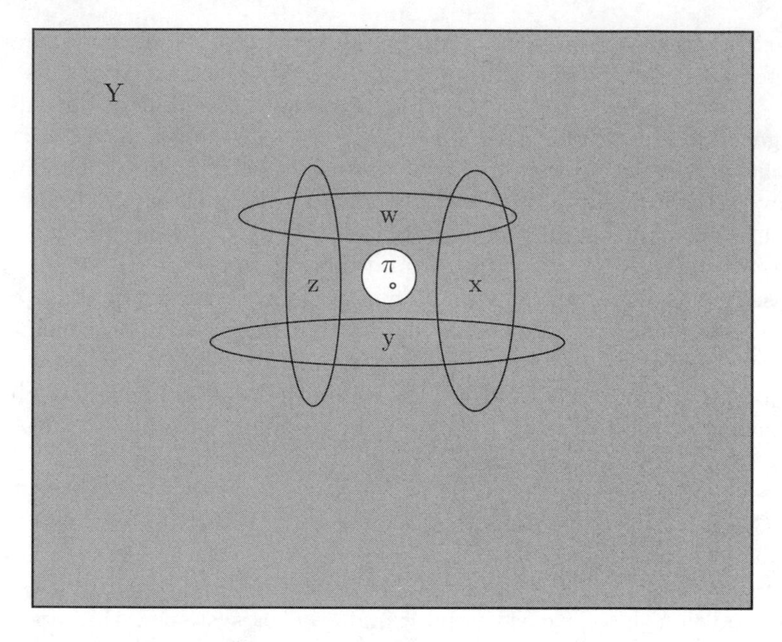

Now we suggest that countable classes of points, Π_0, and countable classes of U-systems, $\mathbf{U}$, that satisfy the conditions above ought to exist. If our surface is 'smooth' and two-dimensional, which, of course, we need to examine and have no right to assume at this point, then it should be possible to place a 'coordinate grid' on the surface, somewhat like lines of latitude and longitude on a sphere. The closer together the lines are drawn, the more dense their points of intersection will be, and if they are sufficiently dense then a point of intersection should be found in each open space of sufficient size. If the lines in the grid are as dense as the intersections of latitudes and longitudes whose degree measures have rational values, e.g., 36.7° north latitude and 126.75° west longitude, then we would expect to find one of these intersections in every open space of any size. Moreover, given that all of the degree measures have rational values, the grid will only have a countable number of latitude-longitude intersection-points, which would prove that the space was separable, though not yet that it had a countable basis.

To construct the basis we are seeking, each point in the intersection-set can be surrounded with a set of concentric rings of rational diameter and decreasing size.[15] If the sizes decrease to vanishing we can expect one of the rings will be small enough to fit inside any open space. And,

[15] Actually, to assure that the ring is an open set, we take it to be the exterior of a U-system that fills the entire space *except* for a circular 'hole', which itself forms the ring. Note that the rings need not be circles. All that is required is that they

assuming that their diameters have rational values, there will only be countably many rings surrounding each intersection point. Therefore, since there are only countably many intersection-points and each is surrounded by countably many rings, one of which is inside any open space, the set of all of these rings will constitute a countable basis for our surface topology. Thus, we have at least a plausible argument in support of the following hypothesis:

7.4.2 Countable Basis Hypothesis. There are countable classes, Π_0, of points and $\mathbf{U}$ of U-systems such that for every non-universal U-system X there exist $\pi \in \Pi_0$ and $Y \in \mathbf{U}$ such that Y covers X and $\pi \in\sim \Pi(Y)$.

Together with Urysohn's Theorem, this yields the following immediately:

7.4.3 Theorem. Surface topologies are metrizable and separable (Kelley, 1955, p. 49) in the sense that there is a countable set of points Π_0 that is dense in the space; i.e., such that every open set contains a member of Π_0.

Now we turn to our last fundamental property, which also has its associated hypothesis.

7.5 Topological Connectedness

A point set is topologically connected (Franz, 1964, p. 54) if it is not contained in the union of two non-empty open sets, which are *separated* in the sense that neither contains a point of the closure of the other. And, it follows easily that a *closed* set is connected if it is not the union of two disjoint non-empty closed sets. An example suggests the close connection between this concept and the intuitive idea of *continuity*.

This involves the two broad streaks, x and y, below:

Here each streak appears to be continuous in itself, but the two of them taken together seem not to form a continuous whole because they are separated by a small gap. This would show that the composite of the two streaks is not connected in the topological sense, because it is the union of the individual streaks x and y, and these are separate—disjoint. Conversely, the fact that the individual streaks appear to be continuous seems to show that they *are* connected in the topological sense because, it seems, two other closed regions, u and v, that 'covered', say x, as shown:

---

surround the intersection points, and that they have 'diameters' measured by the least upper bound of the distances between points in them.

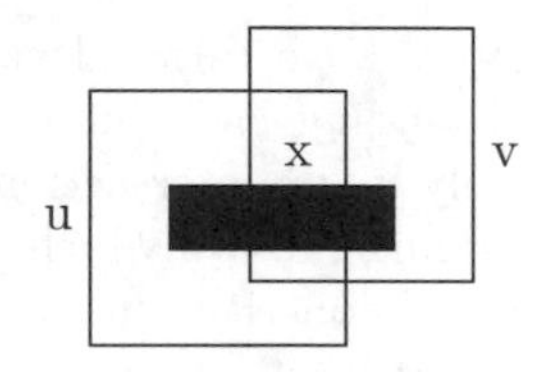

would inevitably intersect. But if true, these are empirical facts, and nothing so far stipulated in our formal theory tells us that when it is applied to concrete marks like x and y and the composite U-system $\{x, y\}$ formed from them, the former will turn out to be connected and the latter will not. If this is to be so, it can only be guaranteed by another 'hypothesis of application' which itself raises deep questions.

Let us begin by formulating perhaps the most obvious hypothesis, which, if true, would explain our intuitions about marks x and y and the U-system $\{x, y\}$ in the diagram above:

7.5.1 Connectivity Hypothesis. The point sets of all marks are connected.

We shall examine the grounds for this below, but trivial consequences can be noted immediately.

7.5.2 Theorems

(1) The union of two connected U-systems is connected if and only if a mark in one of them touches a mark in the other. (2) A U-system $\{x_1, \ldots, x_n\}$ is connected if and only for any $i, j = 1, \ldots, n$, there is a 'chain' of members of $X, x_i, x_p, \ldots, x_q, x_j$, each member of which touches the succeeding one.[16] (3) An I-system, $\underline{X}$, is connected if and only if there do not exist connected U-systems, x_1 and x_2, such that: (1) x_1 and x_2 are separate, (2) $x_1 \cup x_2$ covers $\underline{X}$, and (3) neither x_1 nor x_2 covers $\underline{X}$.

Proof of (1). Let $x = \{x_1, \ldots, x_n\}$ and $Y = \{y_1, \ldots, y_m\}$ be connected U-systems; i.e., suppose that $\Pi(X)$ and $\Pi(Y)$ are connected. By Definition 6.4.4(2) $\Pi(X) = \bigcup_{i=1}^{n} \Pi(x_i)$ and $\Pi(Y) = \bigcup_{j=1}^{m} \Pi(y_i)$. Suppose first that x_i touches y_j for some $i = 1, \ldots, n$ and $j = 1, \ldots, m$. Then by Theorem 6.6.5(1) $\Pi(x_i) \cap \Pi(y_j)$ is non-empty, and therefore $\Pi(X) \cap \Pi(Y)$ is non-empty. Then it follows from the fact that a union of connected sets whose intersection is non-empty is also connected (Franz, 1966, p. 41, Satz 7.5) that $\Pi(X) \cup \Pi(Y)$ is connected.

If no x_i touches any y_j for $i = 1, \ldots, n$ and $j = 1, \ldots, m$, then by 6.6.5(1) again, all of the intersections $\Pi(x_i) \cup \Pi(y_j)$ are empty, hence the sets $\Pi(X)$ and $\Pi(Y)$ are disjoint. Since X and Y are also I-systems it follows from 7.2.2(3) that $\Pi(X)$ and $\Pi(Y)$ are closed sets, hence

[16]This is clearly closely related to the *Finite Chain Theorem* (Kelley, 1955, pp. 60–61, Problem **R**).

$\Pi(X) \cup \Pi(Y)$ is a union of disjoint closed sets and therefore it is not connected.

Proof of (2). Let $X = \{x_1, \ldots, x_n\}$ be a U-system. Suppose first that some x_i and x_j in X are not connected by a chain $x_i, x_p, \ldots, x_q, x_j$, each member of which touches the succeeding one. Let x_i be the class of members of X that *are* connected to x_i by such a chain, and let $X \sim X_i$ be the class of all other members of X, which is clearly non-empty since it contains x_j. Then no member of x_i can touch any member of $X \sim X_i$, since if it did then that member would also be connected to x_i by a chain of the kind we are considering. Then $\Pi(X_i)$ and $\Pi(X \sim X_i)$ must be disjoint, and therefore $\Pi(X) = \Pi(X_i) \cup \Pi(X \sim x_i)$ must be disconnected.

The converse is proved by trivial induction. Suppose that $x_{n+1} = \{x_1, \ldots, x_n, x_{n+1}\}$, the theorem is valid for any $X = \{x_1, \ldots, x_n\}$, and every x_i and $x_j, i, j = 1, \ldots, n+1$ are connected by a chain. Then since the theorem is valid for all $X = \{x_1, \ldots, x_n\}, X$ must be connected. By hypothesis there is a member of X that touches x_{i+1} hence X touches x_{n+1}. By the connectivity hypothesis, x_{n+1} is connected, hence both X and x_{n+1} are connected. Therefore, since they are both connected and they touch, by part (1) of this theorem, their union is connected.

Proof of (3). If $\Pi(\underline{X})$ is disconnected there exist disjoint closed sets Π_1 and Π_2, both of which intersect $\Pi(\underline{X})$, such that $\Pi(\underline{X}) = \Pi_1 \cup \Pi_2$. By 7.2.2(3) again, there must be I-systems $\underline{X}_1$ and $\underline{X}_2$ such that $\Pi_1 = \Pi(\underline{X}_1)$ and $\Pi_2 = \Pi(\underline{X}_2)$. Since Π_1 and Π_2 are disjoint, $\Pi(\underline{X}_1) \cap \Pi(\underline{X}_2)$ is empty, and it follows easily from parts of 6.6.2 and 6.6.3 that $\underline{X}_1 and \underline{X}_2$ are separable. Then Theorem 6.5.4(10) implies that there exist separate U-systems X_1 and X_2 that cover $\underline{X}_1$ and $\underline{X}_2$, respectively. Hence by parts of 6.6.4 and 6.6.5 again, $\Pi(X_1) \cap \Pi(X_2)$ is empty, and $\Pi(\underline{X}_1) \subseteq \Pi(X_1)$ and $\Pi(\underline{X}_2) \subseteq \Pi(X_2)$. Therefore, while $X_1 \bigcup X_2$ cover $\underline{X}$, neither X_1 nor X_2 cover it, as was to be shown.

We will put these results directly to work in the next chapter, but we will conclude this one with inconclusive, unsystematic remarks on the justification of the Connectivity Hypothesis. The Hypothesis seems to be borne out in empirical examination of the two streaks ■■ ■■, which, in virtue of the gap between them, do not form a connected total, though, in virtue of the seeming absence of such gaps *within* them, the individual streaks appear to be connected—'continuous'. But closer examination could, probably *would*, reveal 'micro-gaps' within the streaks, which would prove that they were really disconnected. How does the seeming viewer-dependence of the marks' connectivity affect the present theory, built, as it is, on these marks as a foundation? This is closely related to issues discussed in sections 2.3 and 7.3, especially in relation to the individuation of marks. If we cannot use connectivity to determine what is and what is not 'part' of one streak in the pair , how are we to say whether there *is* a gap between it and the other one, and they do not touch? In other words, the very empirical *basis* of our theory seems to rest on the ability to pick out things without gaps—what we have been calling marks.

It might be argued that even if we are frequently or usually mistaken in our judgments of connectedness, if the ultimate 'surface atoms' should really be connected then there would be as much reason the think that they conformed to the postulates of our theory as classical physicists had for thinking that even if material atoms cannot be perceived, nevertheless they conform to the principles of their theories. And that being so, even if we should be mistaken about particular matters of fact, the *general theory* based on our principles might well be valid.

But this is a weak consideration, and would be so even if it were suggested that actual physical surfaces had fractal structures (cf. Mandelbrot 1983), wherein every portion of a real surface, no matter how small, exhibited gaps and fissures. What we now know of material microstructure suggests that the very idea of physical coincidence, contact or separation, which is presupposed in fractal theory just as much as it is in the present one, loses precise meaning in the microphysical. It is much more plausible that the principles of correspondence between microphysical reality and macrophysical appearance, of the sort that our theory is concerned with, must be more like correspondence principles of quantum-theory. In any case, the correspondence principles that might be appropriate in our case remain to be investigated, and we can only hope that the our macrophysical (or 'macrosurfacial') theory is an approximation that can be useful within the range of 'naive experience'. But this will be returned to in *Objection 6* in section 18.2, which concludes this essay.

8

Boundaries[1]

8.1 Introduction

Recapping the informal discussion in section 5.4, the key to our operational analysis of boundary concepts is that of a *boundary cover*. These are illustrated in Figures 3 and 4, in which boundary covers are 'drawn' over the boundaries of the two rectangles that appear in Figures 1 and 2.[2]

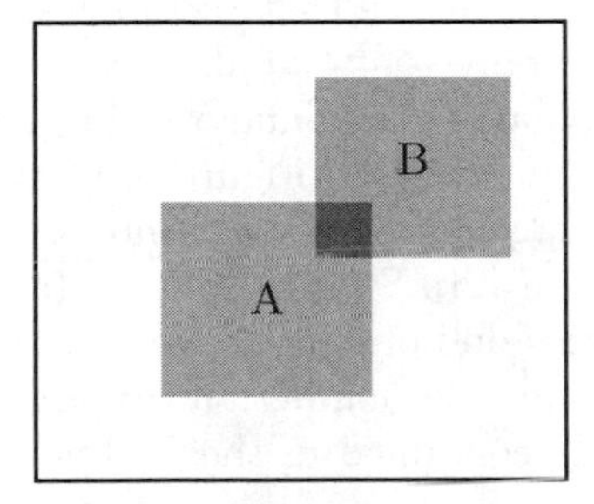

Figure 1

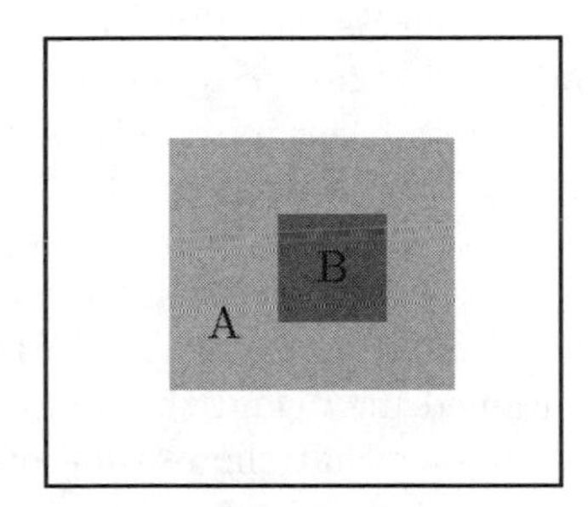

Figure 2

[1]Most of the ideas as well as a summary of relevant mathematical results in this chapter were originally set forth in sections VIII and IX of Adams, 1973.

[2]We do not mean to suggest that Figures 1 and 2 are to be regarded here as purely non-representational 'marks in themselves'. The so called 'rectangles' A and B are meant to represent marks made in some order, as discussed in section 2.3, and their region of overlap, which is more darkly shaded in, pictures where the second one to be drawn 'encountered' the other when it was drawn. Thus, the representation, the 'phenomenon' represented, and the relation between them are really quite complicated and deserve much more detailed analysis than we have entered into here.

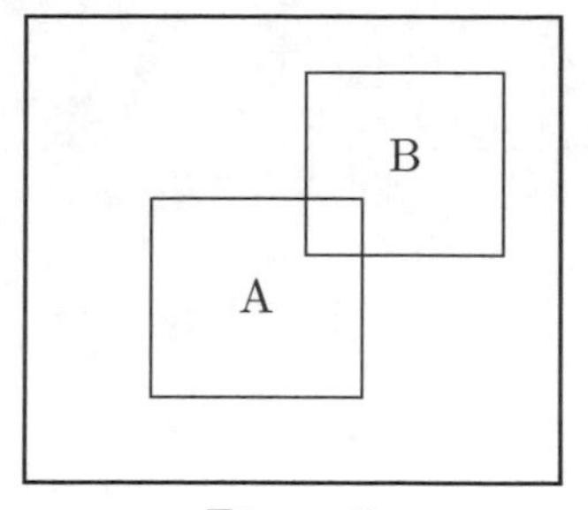

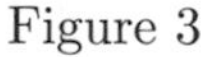
Figure 3

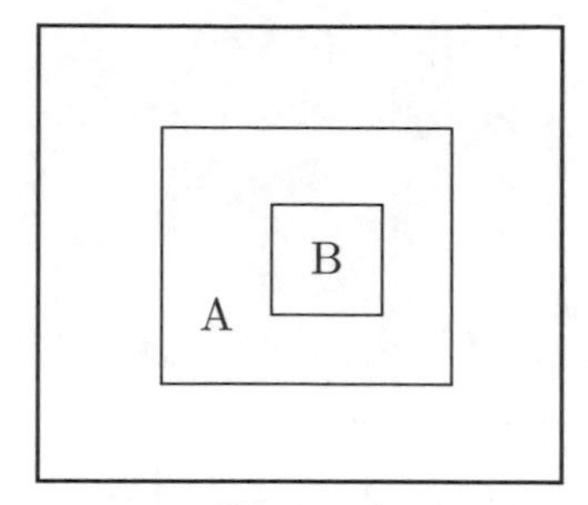

Figure 4

The hollow rectangles in Figures 3 and 4 can be conceived to outline the boundaries of the solid rectangles in Figures 1 and 2, much as visible lines in many maps outline the regions that they surround on the map. These lines are used to represent 'conceptual entities'—namely one-dimensional lines—by concrete, visible surface features, much in the way that small visible dots are used to represent 0-dimensional points.

What we want to do is use boundary covers in such a way as to represent not only the boundaries they cover but certain relations between them. That is also illustrated in Figures 1–4. We see that rectangles A and B in Figure 1 touch, or intersect. Not only that, we also see that their boundaries intersect. But the intersection of the boundaries in Figure 1 is 'conceptual', and it is only 'made manifest' in Figure 3, where the rectangles' outlines are made concrete in such a way as to intersect. Figure 2 stands in the same relation to Figure 4. In Figure 2, two rectangles, A and B intersect, but their boundaries do not, since B is entirely contained in A, and this non-intersection is made manifest in Figure 4 by the fact that the outline of B is entirely contained in the outline of A.

But the relation between the basic figures—the two rectangles in Figures 1 and 2—and their outlines is not as simple as it might seem. In contrast to the boundaries they outline, the outlines have some thickness, and this can distort or blur the representations. For instance, consider Figures 5 and 6 below, which also represent two rectangles, one without and one with its outlines. Figure 5 shows the rectangles intersecting:

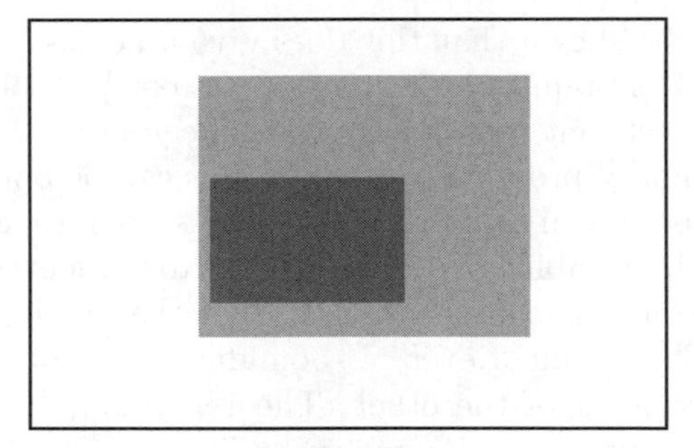

Figure 5

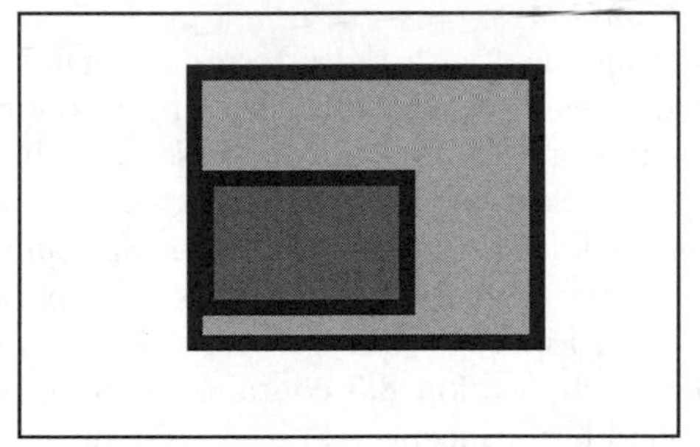

Figure 6

while their boundaries are separate, but when the boundaries are outlined, or 'covered' in Figure 6 they no longer appear to be separate. These outlines are too thick, but if they have to have some thickness, what is too thick?

We must now try to explain how we recognize without the aid of 'representational props' like the outlines in Figures 3, 4, and 6 that the boundaries of the bare rectangles in Figure 1 intersect, while those in Figures 2 and 5 do not. In effect, what we want to do is to extend the account of incidence relations for independently observable marks given in section 2.3 to dependent things like boundaries, which that section drew attention to as a separate problem. Still recapitulating section 5.4, we proceed as follows.

We can determine whether a U-system, X, covers the boundary of a point set Π, i.e., X is a *U-boundary cover of* Π, by determining whether X touches every connected U-system that intersects but is not covered by Π (cf., Franz, 1964, Satz 7.7, p. 42). Thus, the operational test for being a U-boundary cover utilizes the 'crossing the boundary' property illustrated in the relation between rectangles A and B in Figure 1. Of course, it must be shown that this does give a necessary and sufficient condition for a U-system to include all points on the topological boundary of the point set, but that is proven in Theorem 8.2.2. But given this test, other boundary properties and relations can be operationalized in terms of it. Thus, we can establish that one U-system touches the boundary of another by establishing that the first touches every U-boundary cover of the second (Theorem 8.2.4(1)), that the two U-systems have a common boundary point if every U-boundary cover of one touches every U-boundary cover of the other (Theorem 8.2.3(3)), and so on, up to *counting* the number of boundary points that the U-systems have in common if there are only finitely many of them (Theorem 8.2.3(6)), which is an important prologue to dimensionality. These ideas will be worked out in section 8.2.

Section 8.3 is concerned with the *interior*, as against the boundaries of point sets, and it ends by showing that the interiors of U-systems, as well as their exteriors, also constitute bases for our surface topologies (Theorem 8.3.2). Section 8.4 sketches extensions of these ideas to so called *relative topologies*, and especially to *topologies of boundaries*, where we consider 'parts' of boundaries, e.g., the part of the boundary of rectangle A in Figure 1 that is covered by rectangle B, and the boundary of that part. Section 8.5 contains an informal discussion of an 'alternative' boundary concept, according to which entire spaces may have boundaries, though they don't always have them. Section 8.6 adds further remarks on *boundary representations*.

But now we turn to formal details, starting with U-boundary covers.

8.2 Theory of U-boundary Covers

8.2.1 Definitions

(1) The boundary of a point set Γ is the set $\beta(\Gamma)$ that is the intersection of the closure of Γ and the closure of $\sim \Gamma$. (2) A point set Θ is *boundary cover* of Γ if $\beta(\Gamma) \subseteq \Theta$. (3) A *U-boundary cover* of Γ is a U-system X such that $\Pi(X)$ is a boundary cover of Γ.

In what follows we will constantly use the fact that whatever Γ is, $\beta(\Gamma)$ is *closed* set (cf. Kelley, 1955, p. 46, Theorem 10), hence by 7.2.2(3) it is the set of points of an I-system, and if Γ is itself a closed set then $\beta(\Gamma) \subseteq \Gamma$.

Now we prove a crucial *operationalization theorem*:

8.2.2 Basic Boundary Operationalization Theorem. Let Γ be a point set and let X be a U-system. Then X is a U-boundary cover of Γ if and only if it touches every connected U-system Y that intersects both Γ and $\sim\Gamma$.[3]

Proof. Suppose first that X is a U-boundary cover of Γ, and Y is a connected U-system that intersects both Γ and $\sim\Gamma$. Then by the elementary connectivity theorem of Franz cited in the previous section, Y must intersect $\beta(\Gamma)$, and since $\beta(\Gamma) \subseteq X$, X must intersect Y (Theorem 6.6.5).

Now suppose that X touches every connected U-system Y that intersects both Γ and $\sim\Gamma$, but it is not a boundary cover of Γ. Then there is a boundary point, π, of Γ that is not in X. π and X are disjoint closed sets (Theorem 7.2.2(3)) and since the space is normal (Theorem 7.3.2(3)), they have disjoint neighborhoods. Therefore π has a neighborhood, say Ξ, that is disjoint from X. Since the space is normal it is also regular (Kelley, 1955, p. 113), hence π has a *closed* neighborhood, say Ω, such that $\Omega \subseteq \Xi$, and therefore Ω is also disjoint from X. Since Ω and $\beta(\Gamma)$ are closed they are the point-sets of I-systems, and since they are disjoint, they separate according to 6.6.5. Therefore, according to Theorem 6.4.7(4), there exist separate U-systems, say U and V, that cover these I-systems, and it is easy to see that U is a closed neighborhood of π that is disjoint from X.

Now, remembering that U is a U-system that is a neighborhood of π, let U' be the class of marks in U that contain π. Then by Theorem 7.3.3, U' is also a neighborhood of π. Since all marks in U' contain π, all of these marks are connected, hence according to Theorem 7.5.2(3), U' itself is connected. Since U' is a neighborhood of π and π is a boundary point of Γ, U' must intersect both Γ and $\sim\Gamma$. But since X is separate from U and therefore from U', it does not intersect all connected U-systems that intersect both Γ and $\sim\Gamma$.

Before proceeding to other operationalization theorems, let us comment briefly on the sense in which the present theorem operationalizes the idea of a U-system X being a boundary cover of a point set Γ. According to the theorem what we have to *see* is that X touches any connected U-system that touches and then 'extends outside' of Γ. Of course this requires us to estimate something general, namely that X should touch *all* connected U-systems that touch and pass outside of Γ, but as long as Γ is itself sufficiently regular, persons seem to be able to judge this sort of thing. Moreover, and this is the crucial point, all that is required in making such a judgment is seeing that finite systems of marks are connected, and they touch and pass outside a region Γ, which

[3]As stated in footnote 10 of the previous chapter, it will usually do no harm to conflate U- and I-systems with the point sets that they individuate, e.g., simplifying "$\Pi(Y)$ intersects both Γ and $\sim\Gamma$" as "Y intersects both Γ and $\sim\Gamma$."

does not require a person to see an independent 'thing'—the *boundary* of Γ. A singular judgment seemingly about the infinitely thin boundary of Γ is replaced by a general judgment about the relation between Γ and visible U-systems that intersect Γ and its exterior.[4]

Note, incidentally, the fact that the operational test for a U-system to be a boundary cover of a point set Γ involves touching U-systems that extend *outside* of it, which reflects the fact that the test doesn't apply when Γ is an entire surface space, since nothing in the space is exterior to it. This shows that whatever operational criteria might apply to boundaries of entire spaces, they must differ in important respects from the one described here. But this will be returned to in section 8.5, and for now we will continue to discuss 'internal boundaries'.

The following theorems provide operational criteria, among other things for determining whether one thing intersects the boundary of another, and for whether two things have intersecting boundaries. Only part (1) of the theorem will be proved, since the proofs of all of the other parts follow the same pattern as that of part (1), and they are routine.

8.2.3 Boundary Intersection Operationalization Theorems. Let Γ and Θ be closed point sets, and let $\underline{X}$ be an I-system. (1) Then Γ intersects $\beta(\Theta)$ if and only if it intersects every U-boundary cover of Θ. (2) $\beta(\Gamma)$ intersects $\Gamma(\underline{X})$ if and only if every U-boundary cover of Γ touches every U-system that covers $\underline{X}$. (3) $\beta(\Gamma)$ intersects $\beta(\Theta)$ if and only if every U-boundary cover of Γ intersects every U-boundary cover of Θ. (4) $\beta(\Gamma)$ has at least n points if and only if there are at least n separate U-systems that touch it. (5) $\beta(\Gamma)$, $\beta(\Theta)$, and $\Gamma(\underline{X})$ have a common point if and only if for all U-boundary covers, U and V, of Γ and Θ, respectively, and every U-system X that covers $\underline{X}$, $\{U, V, X\}$ is inseparable (6) $\beta(\Gamma)$ and $\beta(\Theta)$ have at least n common points if and only if there are at least n separate U-systems that touch all U-boundary covers of both Γ and Θ.

Proof of (1). Suppose first that Γ contains a boundary point, π, of Θ, and X is a U-boundary cover of Θ. Then π is a point of X, and since it is a point of Γ, X and Γ intersect.

Now suppose that Γ intersects every U-boundary cover of Θ, but it does not intersect $\beta(\Theta)$. Since Γ and $\beta(\Theta)$ are both closed sets, they are the point sets of I-systems, say $\underline{X}$ and $\underline{Y}$, and since Γ and $\beta(\Theta)$ are disjoint $\underline{X}$ and $\underline{Y}$ are separable. Therefore, $\underline{X}$ and $\underline{Y}$ are covered by separate U-systems, say X and Y, hence $\Gamma \subseteq X$ and $\beta(\Theta) \subseteq Y$.

[4]This is like the judgment that line segments are parallel, which is seemingly about a singular relation between the segments, but which Euclid's theory 'unpacks' as a general judgment about arbitrary extensions of the segments.

Therefore Y is a U-boundary cover of Θ that is separate from Γ.

We conclude this section with informal remarks on the significance of the foregoing results. First, we now see why rectangle A in Figures 1 and 5 intersects not only rectangle B but rectangle B's boundary. That is because we can see that A would have to intersect any 'complete outline' or U-boundary cover of B, like the heavy line in Figure 5. Similarly, every complete outline of A would have to intersect every complete outline of B, as in Figure 3, and therefore according to Part (2) of the theorem, the *boundaries* of A and B must intersect. But that is not the case with the rectangles in Figures 2 and 4, which intersect each other, but whose boundaries do not intersect. Thus, once again we have replaced singular judgments about incidence relations involving the boundaries of A and B with general judgments about relations among visible U-boundary covers of the rectangles.

Parts (4) and (6) are especially important because of their bearing on *counting* boundary points and boundary-intersection points. Figure 6 on the left below illustrates part (4), by showing three separate U-systems, X, Y, and Z, that clearly would have to touch every U-boundary cover of rectangle A, thus proving that A's boundary has at least three points:

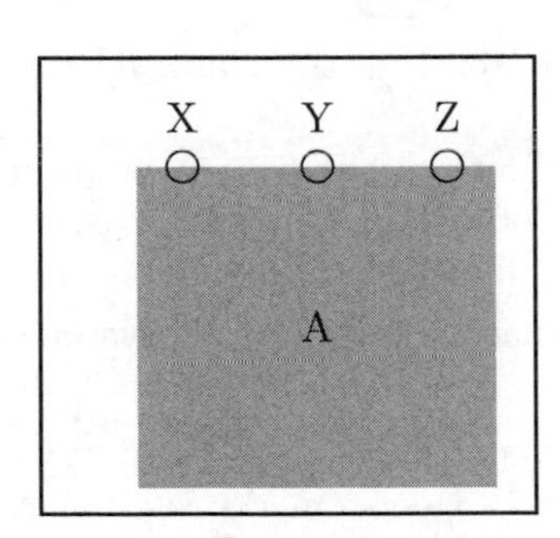

Figure 6

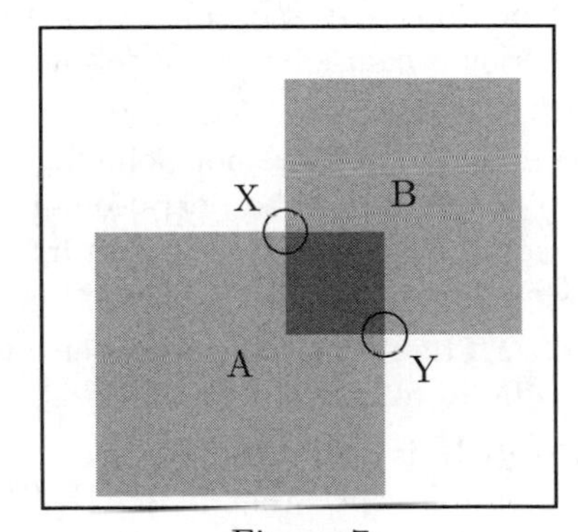

Figure 7

Moreover, the obvious fact that more such systems could be drawn in, separate from X, Y, and Z, is empirical proof that A's boundary has many more points—in fact, arbitrarily many if arbitrarily many separate marks could be drawn in in this way. This is suggestive of the intuition that boundaries are like *lines* that contain infinitely many points, which, moreover, are *ordered*, as will be returned to in section 8.6 and Chapter 10.

Figure 7 illustrates part (6) in showing that the boundaries of rectangles A and B intersect in at least two points, because the two separate circles X and Y would have to touch any U-boundary covers of A and B. Moreover, further consideration of the figure should make it clear that any *other* marks that touched all U-boundary covers of both of A and B would have to touch one of X or Y, which would constitute empirical proof that A and B had *only* two common boundary points. Modifying the figure slightly gives us empirical verification of the intuition commented on in section 4.3, that boundary-intersections individuate unique points, which is in turn related both to linear one-dimensionality, to be returned to in Chapter 10, as well as to Euclid's Definition "The extremities of a line are points," (Heath, 1925, Vol. I, p. 165). However, here these intuitions are not assumed as 'givens', but instead we have approached this 'plateau' by a long ascent, which aims to discover what lies behind the intuitions. The ascent will now be continued, to approach *interiors*, and then *boundary topologies*, which are essential preliminaries to surface dimensionality.

8.3 Interiors

Many theorems of classical geometry are formulated in terms of the interiors and exteriors of plane figures like triangles (e.g.,. Proposition 21 of Book I of *The Elements*, Heath, Vol. I, p.289). The idea of an interior is characterized as follows:[5]

8.3.1 Definition. The *interior* of a point set Γ is the class, $\iota(\Gamma)$ of points of Γ that are not points of $\beta(\Gamma)$.[6]

We shall need to operationalize this definition, but first we state a theorem that states an important property of interiors that could seem self-evident, but which, like so many similar facts, really requires proof:

8.3.2 Theorem. The set of sets of points that are interior to U-systems are bases for surface topologies.

Proof. It is sufficient to show that for every point π and every U-system, X, such that $\pi \in\sim \Pi(X)$, there is a U-system, Y, such that $\Pi(Y) \subseteq\sim \Pi(X)$, and π is an interior point of $\pi(Y)$. Then it will follow immediately that every basis set contains a closed set that contains π as an interior point.

Suppose that π is a point that is separate from a U-system X, i.e., $\sim \Pi(X)$ is a neighborhood of π. By Theorem 7.3.2(3), surface spaces are

[5] Unfortunately we must leave aside here what are probably even more important interior-exterior and 'side' concepts, such as the sides into which an extended straight line divides the plane (cf. the proof of Proposition 12, Book I of *The Elements*, Heath, Vol. I, p. 271), and the interior and exterior angles of a triangle (cf. Proposition 16 of Book I of *The Elements*, Heath, p.279).

[6] Thus $\iota(\pi)$ is an open set that is disjoint from the closed set $\beta(\pi)$, though the closure of $\iota(\pi)$ contains $\beta(\pi)$.

normal, hence *regular* (Kelley, 1955, p. 113), and therefore by condition [Rg'] of Franz, 1965, p. 62, $\sim \Pi(X)$ contains a closed neighborhood of π, say Γ. Hence Γ and X correspond to disjoint I-systems, which by Theorem 6.4.7(4) are covered by separate U-systems Y and Z. Therefore π is an interior point of $\pi(Y)$, which is itself disjoint from $\Pi(X)$, as was to be proved.

One interesting thing to note about this theorem is that it is not obvious that U-systems *have* interiors. For instance, *a priori*, the boundaries of rectangles A and B in Figures 1 and 2 might be supposed to be U-systems, and only the fact that U-systems are finite sets of marks seems to exclude this, because marks are assumed to be visible, and therefore they couldn't be 'infinitely thin'. But the assumption that marks are visible plays no part in our formal theory, and the only thing in the theory that guarantees that *enough* U-systems have interiors is the finite covering postulate. If all marks were as thin as boundaries, then indeed no finite number of them could cover any of our spaces. But now we see that the finite covering postulate guarantees a lot more: namely that there are 'arbitrarily small marks' with non-null interiors *everywhere* in the spaces.

Concluding this section, we will 'operationalize' four significant properties of interiors.

8.3.3 Operationalization Theorems for Interiors. (1) Let Γ be a point set and let Θ be a closed point set. Then $\Theta \subseteq \iota(\Gamma)$ if and only if $\Theta \subseteq \Gamma$ and there exists a U-boundary cover of Γ which is disjoint from Θ. (2) Let Γ_1 and Γ_2 be disjoint closed point sets. Then there exists a U-system X that *separates* Γ_1 and Γ_2 in the sense that $\sim \Pi(X)$ is the union of disjoint open sets Θ_1' and Θ_2' such that $\Gamma_1 \subseteq \Theta_1'$ and $\Gamma_2 \subseteq \Theta_2'$. (3) Let $\Gamma = \Pi(\underline{X})$ for some I system $\underline{X}$, and let X be a U-system. Then X contains an interior point of Γ if and only if it covers a U-system that is inseparable from $\underline{X}$, and which is separate from some U-boundary cover of X. (4) X covers $\iota(\Gamma)$ if and only if it touches every U-system that contains an interior point of Γ.

Proof of (1). Obviously if Θ is a subset of Γ and it is separate from a boundary cover of Γ then it is a subset of $\iota(\Gamma)$.

Conversely, suppose that $\Theta \subseteq \iota(\Gamma)$. Then it is separate from $\beta(\Gamma)$, and since both Θ and $\beta(\Gamma)$ are closed sets they are point-sets of I-systems, which are therefore separable. Therefore by 6.4.7(4) there are separate U-systems, say X and Y, that cover them. And clearly X covers the boundary of Γ; i.e., X is a U-boundary cover of Γ, and equally clearly Θ is disjoint from it.

Proof of (2). Suppose again that Γ_1 and Γ_2 are disjoint, closed point

sets. By part (1) there exists a U-system X such that $\Gamma_1 \subseteq \iota(X)$, and $\Gamma_2 \cap \Pi(X)$ is empty. Let $\Omega_1 = \iota(X)$ and let $\Omega_2 =\sim \Pi(X)$. Then both Ω_1 and Ω_2 are open sets. Moreover, it follows easily that (a) Ω_1 and Ω_2 are disjoint, (b) $\Pi(X)$ is their union, and (c) $\Gamma_1 \subseteq \Omega_1$ and $\Gamma_2 \subseteq \Omega_2$.
Proofs of (3) and (4). Part (3) follows trivially from part (1) and part (4) follows trivially from part (3).

Part (1) applies directly to *marks*. For example, assuming that labels A and B in Figures 3 and 4 are marks, they must be interior to the rectangles they label because they are covered by them, but they are separate from the U-boundary covers depicted. Parts (2), (3) and (4) will be used in succeeding chapters. Obviously many similar properties could be examined, but now we will turn to another topic.

8.4 Remarks on Boundary Topologies

Not only are boundaries defined topologically, but they themselves have 'intrinsic topologies'—what we will call *boundary topologies*. Thus, looking at the boundary of, say, the inner rectangle in Figure 5, one feels intuitively that it constitutes a 'space', regions in which have their own boundaries and interiors that stand to that space in the same way as regions and their boundaries in 'entire spaces' stand to those spaces.

Topological theory deals with the topologies of arbitrary subspaces of 'basic spaces' in terms of so called *relative topologies* (Kelley 1955, p. 51). A point set, Γ, in the basic topological space defines a topology *relativized to* Γ, call it the Γ-*space*, the open sets of which, the Γ-*open sets*, are simply the sets $\Gamma \cap \Omega$, where Ω is an open set of the basic space. This implies that neighborhoods in the Γ-space, the Γ-*neighborhoods*, are all point sets in it that contain Γ-open sets, and the closed sets in it, the Γ-*closed sets*, are the sets $\Gamma \cap \Omega$ in which Ω is a closed set of the basic space. The next theorem lists properties of Γ-spaces that follow trivially if Γ is a closed subset of a surface space:

8.4.1 Theorem. If Γ is a closed point set in a surface space then the Γ-space is Hausdorff, compact, separable, and metrizable, and a closed subset of Γ is connected in the Γ-space if and only if it is connected in the surface topology.

Now let Θ be a point set in the Γ-space, and consider the boundary of Θ in that space—call this the Γ-*boundary* of Θ, denoted $\beta_\Gamma(\Theta)$. This is the class of points in Γ-space, all Γ-neighborhoods of which contain points of both Θ and $\sim \Theta$. We must be careful to distinguish $\beta_\Gamma(\Theta)$, which is the boundary of Θ in the Γ-boundary space, from $\beta(\beta(\Theta))$. The latter is trivially identical to $\beta(\Theta)$, and we do not want to say that the boundary of a point set *in* a boundary space is necessarily identical to the point set itself.

The definition given above of a relative topology does not directly tell us how to operationalize concepts like that of connectedness, of boundary-incidence, etc., in a relative space, and in particular in a boundary topology. Concepts of boundary-incidence were operationalized in basic surface topologies, as described in Theorem 8.2.2, by the use of U-boundary covers, which are finite sets of visible surface features. But the analogue of a U-boundary cover in a boundary space would lie entirely in that space, and that is 'too thin to be seen'. If we are to operationalize boundary covers *in* boundary spaces we need to 'project them back' into the original, visible space, as an image seen in a microscope projects back into the visible realm what cannot be seen with the naked eye. This is not an entirely routine matter, but since it will prove important in the analyses of surface dimensionality and linearity, what follows will briefly sketch a way of doing it.[7]

Focus on the boundary of rectangle A in Figure 1, and the part of it that is covered by rectangle B, which corresponds to the point set $\beta(\Pi(A)) \cap \Pi(B)$. Theorem 8.4.2 below provides operational tests in terms of U-systems and their interiors, for determining whether such systems *touch* a point set of this kind, for determining whether they *cover* it, and related matters. Proofs are omitted since they are obvious.

8.4.2 Theorems. Let $\Gamma = \Pi(\underline{X})$ for an I-system $\underline{X}$, let $\Theta = \Pi(\underline{Y})$ for an I-system $\underline{Y}$, and let Z be a U-system.

(1) Z intersects $\beta_\Gamma \cap \Theta$ if and only if for every U-boundary cover, W, of $\underline{X}$, $\underline{Y} \cup \{Z, W\}$ is inseparable.

(2) Z covers $\beta_\Gamma \cap \Theta$ if and only if it touches every U-system that intersects $\beta_\Gamma \cap \Theta$.

(3) If Θ is closed then $\beta_\Gamma \cap \Theta$ is connected if and only if there do not exist separate U-systems X and Y, neither of which covers $\beta_\Gamma \cap \Theta$, but whose union covers it (i.e., $\beta_\Gamma \cap \Theta \not\subseteq \Pi(X)$ and $\beta_\Gamma \cap \Theta \not\subseteq \Pi(Y)$, but $\beta_\Gamma \cap \Theta \subseteq \Pi(X \bigcup Y)$).

(4) If Φ is closed then $\iota(X)$ contains a Γ-boundary point of $\beta_\Gamma \cap \Phi$ if and only if X covers some U-system Y, such that $\beta(\Gamma) \cap \Pi(Y)$ is connected, and it intersects but does not cover $\beta_\Gamma \cap \Theta$.

(5) If Θ is closed then $\beta(\Gamma) \cap \Theta$ has at least n boundary points if and only if there are at least n separate U-systems whose interiors contain Γ-boundary points of it.

[7]Relativizing a topology in a way that projects it into an 'invisible subspace' may rob it of its intuitive content. This is related to a quasi-Kantian view of the author's: his 'spatial intuitions', or 'anschauungen', are always two-dimensional (at least).

Let us comment briefly on these results and the operational procedures that are related to them. First, there is a sense in which the procedures described in parts (3)–(5) transpose the boundary-crossing methods for entire spaces outlined previously (section 8.2) into 'boundary-microspaces'. Doing this involves 'magnifying and projecting' them into the visible realm, which makes them more complex than they were in the macroscopic realm from which they are derived. Luckily, however, the complexity of our *descriptions* of these procedures does not necessarily correspond to the difficulty of applying them.

Consider first Figure 8 below, and circles x, y, and z, whose boundary covers are shown:

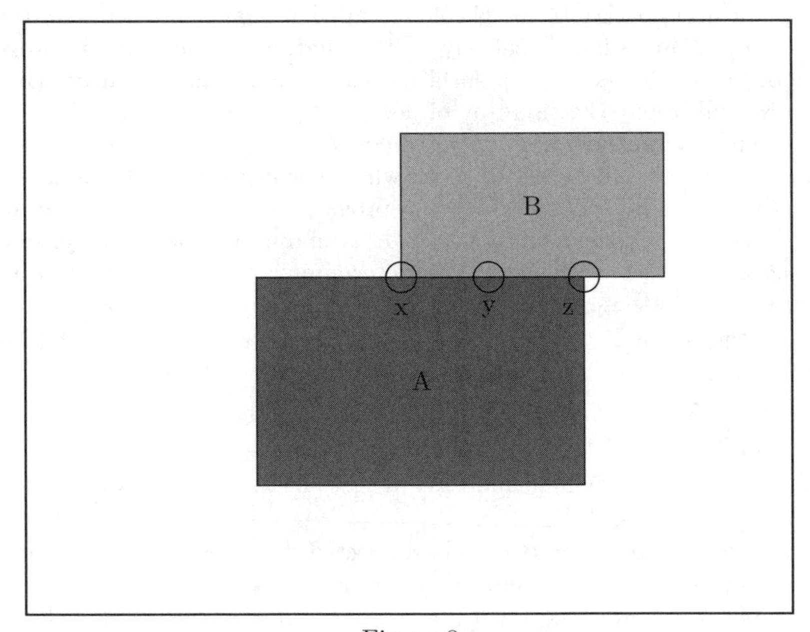

Figure 8

Obviously the circles contain common boundary points of rectangles A and B, since any U-boundary covers of these rectangles would have to touch them (Theorem 8.2.3(5)). However, only the interiors of circles x and z contain boundary points of $\beta_A \cap B$, i.e. these interiors contain boundary points of the part of the boundary of A that lies in B, but the part of the boundary of A that is in y is itself interior to the part that is in B.

Finally, two observations relate to applications of parts (3) and (5) of Theorem 8.4.2, to 'entire boundaries'. One is that with $\Theta = \emptyset$ and applied to a torus like Figure 9, part (3) of the theorem implies that its entire boundary is disconnected, because, as the figure shows, the union of the separate rings that compose the U-boundary cover of it do cover its boundary, but neither ring by itself does.[8]

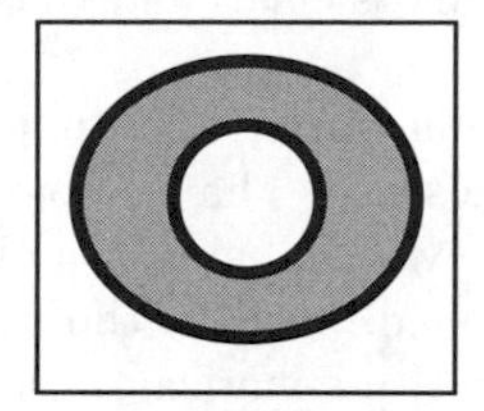

Figure 9

The other observation is that the fact that it is easy to see that circles like x and z in Figure 8 contain points on the boundary of the part of the boundary of rectangle A that is in B makes it easy to *count* the points that have this property. There must be at least two of them, since x and z both contain such points and they are separate, but there cannot be more than two since any other U-system that contained such points in its interior would have to touch either x or z. And, the possibility of counting 'extremities of boundaries' will turn out to be very important in connection with determining surface dimensionality, to be discussed in the following chapter. That is because for a surface to be no more than two dimensional it is sufficient that the boundaries of open sets in it should be no more than one-dimensional. And, it is sufficient for these boundaries to be one-dimensional that open sets in *their* topologies should be 0-dimensional—and, it is sufficient for a set to be 0-dimensional that it be finite.

Before turning to dimensionality, however, we will comment informally on the possibility of defining another boundary concept, according to which entire spaces might be said to have boundaries, as well as on the representation of boundaries.

8.5 Boundaries of Spaces

We noted in section 8.2. that entire spaces do not have boundaries in the topological sense. Nevertheless, we feel intuitively that many physical spaces and even possibly 'Space' itself has a boundary, or 'limit'. Of

[8]But it should be noted that having a disconnected boundary cover is neither a necessary nor a sufficient condition for the boundary itself to be disconnected. Thus, the entire torus its own boundary cover, and it is connected while its boundary is disconnected. And, a disconnected U-boundary-cover can cover a connected boundary if it includes 'extraneous' members that are separate from the boundary.

course, the operationalization of boundary concepts described in the previous sections followed the lead of mathematical topology, by treating boundaries as essentially 'two-sided', so that there is always something in the space 'beyond' them. But now we may ask whether the intuitive sense of a one-sided boundary is clear, and whether *that* can be captured topologically.

It should be noted immediately that not all physical spaces have boundaries in the intuitive sense. That is obvious in the case of Newtonian Space, which is conceived as being of unlimited extent. And more limited spaces are unbounded, such as the surfaces of spheres. Unless we want to say that an entire spherical surface is a boundary, escape from which would lead off the surface, it has no edges beyond which it is impossible to go. [9] So, the question arises: what is the difference between limited and unlimited spaces?

Part of the answer to the above question may be sought in considering the third of Euclid's triad of 'extremity definitions', namely "The extremities of a line are points." Recall rectangle A in Figure 6, the upper edge of which has the three small circles on it, x, y, and z. These are *arranged in order*, so that moving continuously from x to z or from z to x along this edge would necessitate going past or 'through' y. This order is suggestive of *direction of motion*, of *going through* y to get from x to z, which in turn suggests a kind of boundary. When no further motion in a given direction is possible, we are apt to say that a limit has been reached, a *ne plus ultra.* This kind of limit is one-sided, 'intrinsic', and it does not presuppose the existence of something beyond it.

Of course not all lines have extremities or limits in this sense. As Proclus pointed out long ago in criticizing "The extremities of a line are points" as a *definition*, circles are unbounded in this sense, since it is possible to move for ever on them in a given direction.[10] This is the kind of limit that the surface of a sphere lacks, while, because it is difficult to 'cross over' the edge of a sheet of paper, that may be conceived to have a limit of this kind.

Can this intrinsic, one-sided 'limit' sense of a boundary be described

[9] Of course it might be argued that the all physical surfaces, such as sheets of paper, are topologically equivalent to spheres, hence we shouldn't say that their edges are boundaries either. In fact, the surfaces of whole bodies are always homeomorphic to the surfaces of spheres, tori, and other more complicated solids. But that dodges the problem. If the surfaces of solids have no boundaries, what are we to make of Euclid's definitions "The extremities of solids are surfaces," and "The extremities of surfaces are lines"?

[10] Cf. Heath's comments on this on p. 165 of Vol. I. A recent translation of Proclus' *Commentary* uses the term 'limit' explicitly in translating Definition 3 as "The *limits* of a line are points" (Morrow, 1970, p. 93, my italics).

in topological terms? We will return to this in the case of lines in Chapter 10, where the ordering of points on them is characteristic of *linearity*—of *being* a line.[11] But it is not obvious how this might generalize to higher dimensional continua, in which points are not 'naturally' ordered. And, speculating, this might explain why, without importing extraneous concepts, the topology of spaces of arbitrary dimensionality confines itself to two-sided boundaries and their properties.

8.6 Remarks on Representing Boundaries

The operational account given in this chapter of boundaries and their relations has centered on boundary covers, and especially on U-boundary covers that in a sense 'manifest' the boundaries they cover by making them stand out. They might also be said to *represent* these boundaries, somewhat as boundary lines on maps represent boundaries between the regions represented. But 'represent' is vague enough to cover a multitude of sins, and our boundary covers do *not* represent the boundaries they cover in same the way that boundary lines on maps represent the boundaries they correspond to. Our boundary covers cover surface features, which are 'things in themselves' and not *maps* of anything else. Furthermore, that boundary covers *cover* the things they are boundaries of is a matter of fact, and not a convention. Of course, convention may enter in choosing from among the possible boundary covers of a feature one that manifests the feature's boundary properties and relations in a special way. For instance, not all possible boundary covers of the torus in Figure 9 make the disconnectedness of its boundary manifest. To do this the boundary cover must lie close to the boundary it covers—among other things it must be 'thin'. But let us put that aside, and consider another kind of representation that boundary covers might be thought to involve.

Our figures and the their boundary covers might be regarded as diagrammatic representations of abstract ideas, somewhat as diagrams are used to convey geometrical ideas. Plato's Cave analogy suggests something like this in picturing the relation between sensory experience and the 'eternal existents' of which geometrical theory treats, as being like the knowledge of the 'real world' that could be gained from seeing shadows thrown on the walls of a cave by objects in the outside world (*The Republic*, Cornford translation, p. 244). In a not unrelated comment, Berkeley speaks of "an idea, which considered in itself is particular, becomes general, by being made to represent or stand for other

[11] This will be described in 'Hilbertean' terms, in terms of an essentially topological *betweenness relation*.

particulars of the same sort" (*Principles of Human Knowledge*, Introduction, par. 12), which leads to a conclusion like Plato's, at least to the extent that what geometrical diagrams represent are 'in the mind'. Less philosophically, Firby and Gardiner speak of 'pretending' that an "actual space model" exactly realizes a corresponding "*ideal* model in our mind," which topological theory applies to but which we can only 'imagine' *Surface Topology*, 1992, p. 14). These 'Platonic' views will be discussed at greater length in Chapter 10, but the point to make here is that, while the boundary cover in Figure 9 is intended to illustrate something general, and this intention is in the author's head, nevertheless the boundary cover in the figure does cover the boundary of the torus in the figure, and it is not a mere 'resemblance' of something else in a world of ideas. This leads to a concluding remark about idealization and exactness in general.

Looking at the present work's figures is supposed to convey ideas to the reader's mind, but the theory developed here is not *about* the ideas. It is about surface features, including figures, but they are 'out there and not in the mind', and it involves 'idealization' only to the extent that the general propositions that it formulates, like the Separation Postulate, are 'approximately true' of those things. It is a collection of assertions like "Brazilians speak Portuguese," which are not exact truths pertaining to a realm of ideal beings, but rather ones that convey inexact but useful information about things or persons 'out there in the sensible world'. Our aim is to clarify *what* propositions of abstract topological theory, and ultimately geometrical theory, tell us about the world of the senses.

And now we will turn to dimensionality, and to how the abstract theory of topological dimension might apply to physical surfaces and things on them. Since the concept of a boundary is central to the modern theory, our 'operationalization' of boundary concepts will figure largely in our attempt to relate the theory of dimension to physical surface.

9

Surface Dimensionality[1]

9.1 Introduction

Of all of the geometrical concepts that have been in wide general use at least since classical Greek times, perhaps those associated with spatial dimensionality have been the most resistant to precise mathematical formulation. To illustrate the antiquity of dimensional concepts, we may note that three of the five definitions with which Book I of Euclid's *Elements* begins can be regarded as characterizing the dimensionality of points, lines, and surfaces. These are "A point is that which has no part" (Definition 1), "A line is breadthless length" (Definition 2), and "A surface is that which has length and breadth only" (Definition 5), to which should be added Definition 1 from Book XI: "A solid is that which has length, breadth, and depth." Though these do not say that points, lines, surfaces, and solids have the dimension *numbers* 0, 1, 2, and 3, respectively, dimension *concepts* are plainly involved. But it can be argued that what Euclid attempted to express, e.g., by saying that a surface has "length and breath only," was only properly formulated in the early years of this century. Moreover, this formulation was a work of genius in which the names of such giants as Poincaré, Cantor, Jordan, Peano, Lebesgue, Brouwer, Urysohn, Menger, and others figured largely, and which itself would not have been possible if it had not been for the foundation in analysis and the theory of sets that had been built up slowly throughout the 19th century. This alone should make it evident that the idea of dimension is a difficult one, as witness the fact that as late as 1930 we find a well known mathematician writing "In saying that a line is one-dimensional, a plane two-dimensional, and space three-dimensional, we mean simply that a line contains ∞^1 points, a

[1]This chapter recapitulates and slightly extends the discussion in sections 2.6–2.10 of Adams, 1986.

plane ∞^2 points, and space ∞^3 points,"[2] seemingly ignoring or being unaware of Peano's space-filling curve, that fills a two-dimensional region (Hausdorff, *Grundzüge der Mengenlehre*, p. 369).

In asking why dimension should be such a difficult concept to analyze, two points may be noted immediately. One is that Euclid did not intend his definitions to apply just to *straight* lines and to *flat* surfaces, and it is not clear, e.g., what is it for a spherical surface to have "length and breadth only." How is one to measure the length and breadth of such a surface, and why does it have *only* length and breadth? One perceives dimly what Euclid was driving at, but it is another matter to make this precise in a way that is applicable to all of the lines and surfaces that modern mathematics is concerned with, which, as the Peano curve suggests, is a far larger class than Euclid and his contemporaries were aware of.

The other point is that we should speak not so much of 'the' concept of dimension, as of a *system* of dimensionality concepts that are interrelated in often quite obscure ways. For instance, Euclid paired the characterizations of lines, surfaces and solids cited above with seemingly alternative characterizations: "The extremities of a line are points" (Definition 3), "The extremities of a surface are lines" (Definition 6), and "An extremity of a solid is a surface" (Definition 2 of Book XI). Thus, lines are what have points as extremities, surfaces are what have lines as extremities, and solids are what have surfaces as extremities. But what have those things to do with breadthless length and having length and breadth only? One feels intuitively that there are relations between them, but it is far from clear what they are. This is even more true of alternative characterizations of dimension considered in modern dimension theory, and one of that theory's great triumphs has been to demonstrate, often with great difficulty, equivalences between them—or more generally the conditions under which they are equivalent. This will be very important for us here, because the characterization of dimension that we have found easiest to operationalize is less well known than ones that are usually cited in the literature. A major contribution of dimension theory to the present enterprise is to demonstrate that what has a given dimension, say dimension 2, according to our operational criterion, is also two-dimensional according to more familiar criteria.

The complexity of modern dimension concepts will require us to give a longer exposition of dimension theory's basic ideas than has been given of the basic ideas of the more elementary topological theory with which the last two chapters have been concerned. This will be set forth in the

[2] Graustein, 1930, *Higher Geometry*, p. 178.

next section, though we will keep the exposition as short as possible.

A final point before starting concerns *fractals.* The inspiration of the fractal idea may be found in the 'Cantor discontinuum' (Franz, 1965, p. 47), and Hausdorff's fractional dimension measure, according to which the Cantor discontinuum has dimension $\log 2/\log 3 = 0.63093$ (Hurewicz and Wallman, 1948, p. 107). But Mandelbrot (1977) has developed these ideas spectacularly, and we will ultimately have to say something about them. But in the interval we will summarize key concepts and results of more 'standard' topological dimension theory (section 9.2), and then sketch a way of operationalizing them in order to apply them to physical surfaces.

9.2 Summary of Concepts and Results of Modern Dimension Theory

Here we will rely almost exclusively on formulations and results set forth in Hurewicz's and Wallman's classic *Dimension theory* (1948). The basic results in this work apply to *separable metric spaces*, which are properties of our surface spaces, as we have seen. Although our concern will be primarily with dimensions 0–3, the general theory is concerned with interconnections between alternative characterizations of *dimension* n, for arbitrary n. We will begin by quoting Hurewicz's and Wallman's basic definition of dimension *verbatim*, then we will comment on some of its features, and finally we will note connections with alternative definitions.

9.2.1 Definition. (Hurewicz and Wallman, p. 24) The empty set and only the empty set has dimension -1.

A space X has dimension $\leq n (n \geq 0)$ *at a point* p if p has arbitrarily small neighborhoods whose boundaries have dimension $n-1$.

X has dimension $\leq n$, $dim X \leq n$, if X has dimension $\leq n$ at each of its points.

X has dimension n at a point p if it is true that X has dimension $\leq n$ at p and it is false that X has dimension $\leq n-1$ at p.

X has dimension n if $dim X \leq n$ is true and $dim X \leq n-1$ is false.

X has dimension ∞ if $dim X \leq n$ is false for each n.

Among the features of this complex definition, the following are noteworthy. First, it is recursive, and it can be regarded as following Euclid in defining a 'space' or point set to have dimension n+1 if its boundaries have dimension n.[3] However, now the recursion starts with the empty set, which is the only set with dimension -1.

[3]Recall Euclid's trio of definitions "The extremities of a line are points", "The extremities of a surface are lines", and "The extremities of a solid are surfaces," which can be thought of as characterizing the dimensions of higher-dimensional entities in terms of the dimensions of their 'extremities' or boundaries. Of course, in the modern

Dimension 0 is particularly interesting: singleton sets of have this dimension because their points necessarily have empty boundaries. But it follows from a theorem to be cited below that all finite and even all countably infinite point sets are also 0-dimensional, and one only arrives at non-trivial dimensionality with sets of cardinality greater than $\aleph_0$, and especially with sets with the cardinality of the continuum.[4] That high cardinality is a prerequisite to positive dimensionality is a reflection of the facts that boundaries of neighborhoods are fundamental to the definition, and that sets of cardinality lower than that of the continuum have empty boundaries.

Another significant feature of the modern dimension concept is that it is *local*. Thus, it starts by defining the dimension of a space or point set *at a point*, and then it defines the dimension of the entire space to be its maximum dimension at points in it—or to be infinite dimensional if it has points at which it has arbitrarily high dimension. Figure 1 may help to explain why dimension should be defined locally, since it represents a circular disk with a 'tail', where, intuitively, the figure is two-dimensional at points in the disk, but one-dimensional at points in its tail. Therefore the entire 'space', disk+tail, is two-dimensional, because it is two-dimensional at some points and one-dimensional at all of the others.

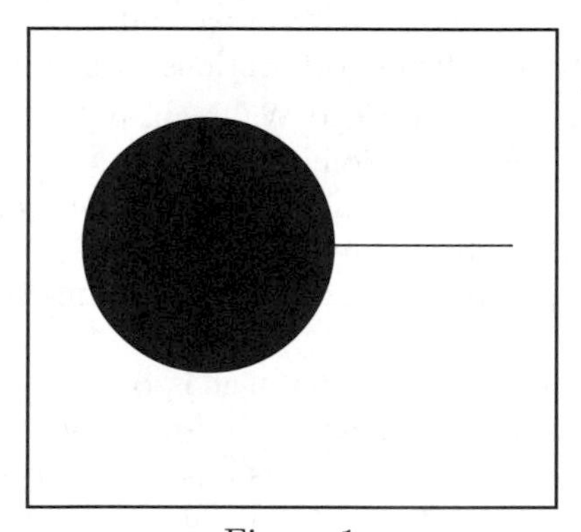

Figure 1

The locality of dimension is related to the fact that it is defined in terms of a *dimension ordering*. A space or point set is first defined to be

theory the 'entities' whose dimensions are defined are topological spaces and point sets or regions within them, and not, e.g., so called 'solids'.

[4]Though not even all of these have positive dimensionality, since the class of irrational reals, while obviously of the cardinality of the continuum, is 0-dimensional (Hurewicz and Wallman, p. 11, Example II 2).

at least n-dimensional, and then to be *exactly n-dimensional* if is at least n-dimensional but not at least n+1-dimensional. This ordering property will be made use of in describing operational procedures for determining surface dimensionality. Thus, it is convenient to operationalize one abstract dimensionality characterization for the purpose of establishing that a surface is at least 2-dimensional and to operationalize another characterization for the purpose of establishing that it is not at least 3-dimensional. Modern dimension theory makes this possible by offering a multiplicity of equivalent characterizations of dimension.

Certain consequences of the modern definition that will be used in the sequel may be noted immediately. Some things that one would expect intuitively, which nevertheless require proof, include the fact that standard Euclidean spaces, e.g. a straight line, a plane, the surface of a sphere, have the abstract dimensionalities that we would expect (Hurewicz and Wallman, Chapter IV). We will also use the fact that subsets of a given point set cannot have a higher dimension than the set (Hurewicz and Wallman, Theorem III.1, p. 26), and the less obvious fact that countable unions of closed sets cannot have a higher dimensionality than that of the set of highest dimension in the union (Hurewicz and Wallman, Theorem III.2, p. 30).[5] This tells us immediately that a countably infinite point set must have dimension 0, since singleton point sets are 0-dimensional. But seemingly much larger sets are also 0-dimensional, such as the set of points with rational coordinates in a finite dimensional Euclidean space. In Part III we will use this fact to argue that 'piecing together' countably many two-dimensional surfaces can never yield a space of dimension higher than 2.

Finally, two alternative abstract characterizations of dimension may be noted. One is the famous 'tiling characterization' due to Lebesgue, which is illustrated in the two-dimensional case by the figure below:

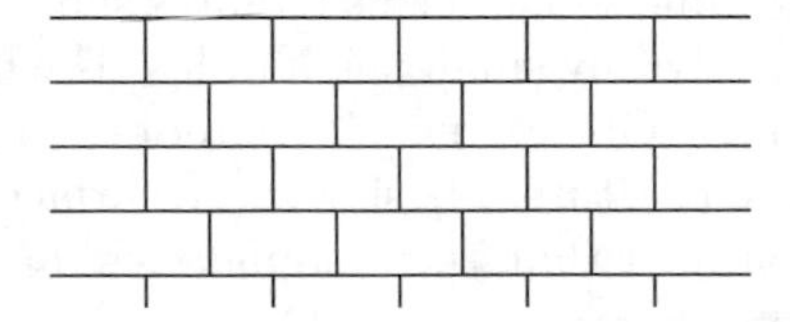

It is characteristic of two-dimensional surfaces that if they are 'tiled over' in the manner illustrated then there will always be points at which at least three tiles meet, and if the tiles are sufficiently small then it will

[5]This clearly doesn't apply to unions of point sets that are not closed, since the sets of rational and of irrational reals are both 0-dimensional, but their union is 1-dimensional.

always be possible to place them in such a way that no more than three of them meet at any point. Lebesgue's *covering theorem* (the 'Pflastersatz', cf. Theorem IV.2 in Hurewicz and Wallman, p. 42) makes the first part ('at least three') of this precise and generalizes it to n-dimensions (Hurewicz and Wallman, section IV.3), and Lebesgue and Brouwer generalized and proved the second ('no more than three') part (cf. Courant and Robbins, 1941, p. 251).

A less well known but for us extremely useful result follows from Proposition C on p. 35 plus a Remark on p. 78 of Hurewicz and Wallman 1948:

9.2.2 Theorem. (Hurewicz and Wallman, pages 35 and 78) Let X be a space of dimension $\leq n-1$ and let C_i and $C'_i, i = 1, \ldots, n$, be n pairs of closed, disjoint subsets of X. Then there exist n closed sets $B_i, i = 1, \ldots, n$, such that $B_i \cap \ldots \cap B_n$ is empty and for $i = 1, \ldots, n, B_i$ *separates* C_i and C'_i in the sense that $\sim B_i$ is the union of disjoint open sets A_i and A'_i, and $C_i \subseteq A_i$ and $C'_i \subseteq A'_i$. If there exist be n pairs of closed, disjoint subsets of X, C_i and $C'_i, i = 1, \ldots, n$, such any closed sets $B_i, i = 1, \ldots, n$, that separate C_i and C'_i in the foregoing sense have a non-empty intersection, then X is at least n-dimensional.

We will use this result in the case $n = 2$, to argue that the topologies of bodies' surfaces are at least 2-dimensional because they contain pairs of disjoint closed sets C_i and $C'_i, i = 1, 2$, that cannot be 'separated' by 'separate separators', $B_i, i = 1, 2$ in the manner described.[6]

Now let us consider how these results may be operationalized in such a way as to apply to our surface spaces.

9.3 Operationalizations

We will begin by applying the fundamental dimension definition, 9.2.1, to argue that the surface of a body like the page on which Figure 2 below appears is at most 2-dimensional. This requires showing that every point on the surface has arbitrarily small neighborhoods whose boundaries are at most 1-dimensional. Given that the exteriors of U-systems, i.e., of finite sets of marks, constitute a basis for the surface topology, this will follow if it can be argued that their boundaries, and in particular the

[6]What makes this especially useful for us is its linking the 'microsurfacial criterion of dimensionality' given in the fundamental recursive definition of dimension stated earlier, which is in terms of the dimensions of boundaries of arbitrarily small regions, to a 'macrosurfacial criterion' that is formulated in terms of closed regions C_i and C'_i of arbitrary size, which, being closed, are useful for us because they correspond to I-systems. On pp. 34–5 of *Dimension Theory*, Hurewicz and Wallman make the significant comment "It is the conjunction of properties of the small and the large which is largely responsible for the power of the dimension concept."

boundaries of the marks that compose them, are at most 1-dimensional.

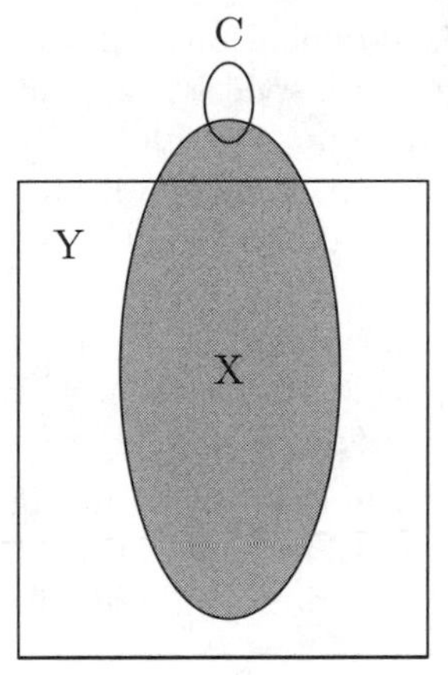

Figure 2

Consider first the oval X in the above figure. Its boundary *looks* 1-dimensional, but we must argue that according to Definition 9.2.1 it is at most 1-dimensional. Clearly this is an empirical matter, but it would be true if it were the case that, given any 'gap' in the boundary that is *not* covered by a second mark, say the rectangle Y, it is possible to place a mark like oval C, as shown, that has only a finite number of common boundary points with X. Applying the test for boundaries to have exactly n common points described in Theorem 8.2.3(6) to C and X shows that in fact they have *two* boundary points in common, and we hypothesize that something like this is the case for any U-system Y that does not cover the boundary of X:

9.3.1 Hypothesis. For any U-system Y that is not a boundary cover of oval X in Figure 2, it is possible to draw a mark, C, separate from Y, such that $\iota(C) \cap (\beta(X) \sim \Pi(Y))$ is non-empty and which has a finite number of common boundary points with X.

The empirical plausibility of related hypotheses will be returned to below, but let us note immediately that Hypothesis 9.3.1 implies that $\beta(X)$ is at most 1-dimensional:

9.3.2 Theorem. $\beta(X)$ is at most 1-dimensional.

Proof. It must be shown that every point of $\beta(X)$ has an arbitrarily small neighborhood whose boundary in the space of $\beta(X)$ is at most 0-dimensional. Basic neighborhoods in this space are its intersections with neighborhoods in the entire space, and the latter are exteriors of U-systems, Y; i.e., basic neighborhoods in the space of $\beta(X)$ are point-sets of the form $\beta(X) \sim \Pi(Y)$. It must be shown that if $\beta(X) \sim \Pi(Y)$ is non-empty then it contains a non-empty neighborhood whose boundary in the space of $\beta(X)$ is 0-dimensional. Since U-system interiors also constitute a basis of the entire space (Theorem 8.3.2), it is sufficient to show that there is a U-system, C, such that $\iota(C) \cap (\beta(X) \sim \Pi(Y))$ is a non-empty open set in the space of $\beta(X)$, and whose boundary in the space of $\beta(X)$ is 0-dimensional. By Hypothesis 9.3.1 there exists C such that $\iota(C) \cap (\beta(X) \sim \Pi(Y))$ is non-empty and which has only a finite number of common boundary points with X. But these are exactly the boundary points of $\iota(C) \cap (\beta(X) \sim \Pi(Y))$. This proves that $\beta(X)$ is at most 1-dimensional, since finite point-sets are 0-dimensional.

It has not yet been shown that oval X in Figure 2 is itself at most 2-dimensional, much less that all figures that can be drawn on the page are at most two-dimensional, but both of these follow from a generalization of Hypothesis 9.3.1:

9.3.3 First General Flatness Hypothesis. If a U-system is not universal then there is a U-system X separate from it, such that for any U-system Y that is not a boundary cover of X, it is possible to draw a mark, C, separate from Y, such that $\iota(C) \cap (\beta(X) \sim \Pi(Y))$ is non-empty and which has a finite number of common boundary points with X.

9.3.4 Theorem. Every surface space is at most 2-dimensional, and so is every U-system in it.

Proof. Combined with the foregoing hypothesis, Theorem 9.3.1 implies immediately that every non-empty open space on a surface is occupied by a non-empty U-system, X, whose boundary is at most 1-dimensional. Since the interiors of these U-systems constitute a basis for the surface space, it follows that every point of the space has arbitrarily small neighborhoods with at most 1-dimensional boundaries. Hence according to Definition 9.2.1, the space is at most 2-dimensional at every point and so is the entire space. Since subspaces of a given space cannot have higher dimensionality than the space, any U-system in a surface space must be at most 2-dimensional.

Now the dimensionality criterion implicit in Theorem 9.2.2 will be used to argue that surface spaces are at least 2-dimensional, which, combined with Theorem 9.3.4 will imply that they are exactly 2-dimensional. This depends on a second flatness hypothesis, which will be stated first in its application to Figure 3. This contains two pairs of parallel segments, horizontal segments C_1 and C_1', and vertical segments C_2 and C_2', where each horizontal segment intersects each vertical one. It also contains ovals B_1 and B_2, where B_1 'separates' C_1 from C_1' in the sense that C_1 is inside B_1 and C_1' is outside of it, and B_2 separates C_2 from C_2' in the same sense.

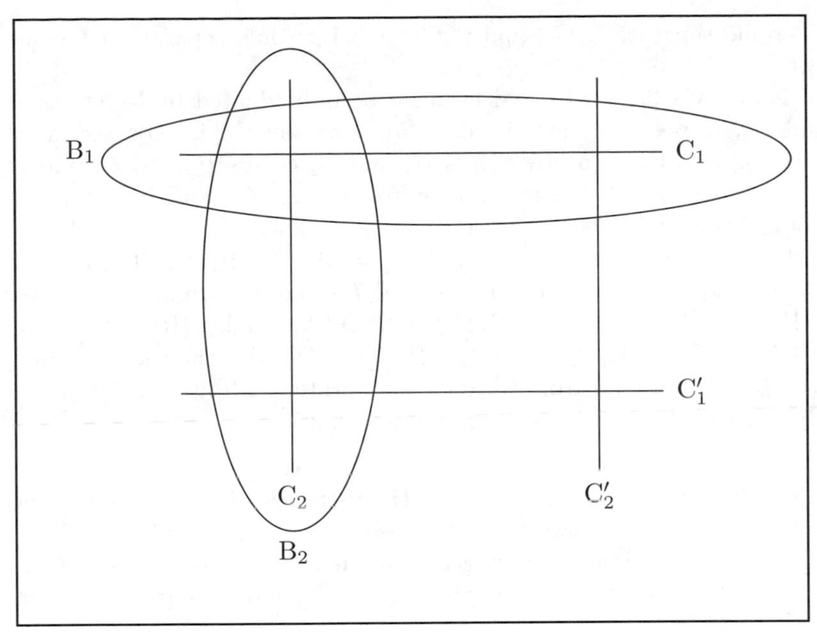

Figure 3

The significant thing about ovals B_1 and B_2 is that they *intersect.* This suggests the following testable hypothesis, which is formulated somewhat informally with particular reference to Figure 3:

9.3.5 Hypothesis. It is not possible to draw non-intersecting closed curves B_1 and B_2, such that B_1 separates C_1 and C_1' and B_2 separates C_2 and C_2'.

The plausibility of this hypothesis, and of a generalization that applies to the surfaces of bodies in general, will returned to below, but first its most important consequence will be cited, which again is somewhat informally stated:

9.3.6 Theorem. The surface on which Figure 3 is drawn is at least two-dimensional.

Proof. Suppose that the surface is less than two-dimensional, and assume that C_1 and C_1' are separate U-systems and C_2 and C_2' are also separate U-systems. Then $\Pi(C_1)$ and $\Pi(C_1')$ are closed, disjoint point sets, and $\Pi(C_2)$ and $\Pi(C_2')$ are also closed, disjoint point sets. Given that the surface is not at least two-dimensional, it would follow from 9.2.2 that there existed disjoint closed sets, Γ_1 and Γ_2 and disjoint open sets A_1 and A_1' such that $\sim \Gamma_1 = A_1 \cup A_1'$ and $\Pi(C_1) \subseteq A_1$ and $\Pi(C_1') \subseteq A_1'$, and such that $\sim \Gamma_2 = A_2 \cup A_2'$ and $\Pi(C_2) \subseteq A_2$ and $\Pi(C_2') \subseteq A_2'$. Hence

Γ_1 would separate $\Pi(C_1)$ and $\Pi(C_1')$ and Γ_2 would separate $\Pi(C_2)$ and $\Pi(C_2')$.

Now, by 7.2.2(3), Γ_1 and Γ_2 must be individuated by I-systems B_1 and B_2, respectively, and by 6.4.7(5) there must exist U-systems, B_1 and B_2, such that: (a) B_1 covers $\underline{B}_1$ and B_2 covers $\underline{B}_2$, (b) B_1 and B_2 are separate, and (c) B_1 is separate from C_1 and C_1', and B_2 is separate from C_2 and C_2'. Let $D_1 = A_1 \cap \sim \Pi(B_1)$ and $D_1' = A_1' \cap \sim \Pi(B_1)$, and let $D_2 = A_2 \cap \sim \Pi(B_2)$ and $D_2' = A_2' \cap \sim \Pi(B_2)$. Then clearly D_1 and D_1' are disjoint open sets, and D_2 and D_2' are also. Moreover, $\sim \Pi(B_1) = D_1 \cup D_1'$ and $\sim \Pi(B_2) = D_2 \cup D_2'$. Finally, $\Pi(C_1) \subseteq D_1$ and $\Pi(C_1') \subseteq D_1'$ and $\Pi(C_2) \subseteq D_2$ and $\Pi(C_2') \subseteq D_2'$. Together, (a)–(c) imply that B_1 separates C_1 from C_1' and B_2 separates C_2 from C_2'. But B_1 and B_2 being separate contradicts hypothesis 9.3.1.

Now we will generalize.

9.3.7 Second General Flatness Hypothesis. In any surface space it is possible to draw two pairs of closed curves, C_1 and C_1' and C_2 and C_2', for which it is not possible to draw non-intersecting closed curves B_1 and B_2, such that B_1 separates C_1 and C_1' and B_2 separates C_2 and C_2'.

This and Theorem 9.3.4 imply immediately that:

9.3.8 Theorem. Every surface space is 2-dimensional.

Thus, given flatness hypotheses, and assuming the general principles of the theory of surface topologies, it is possible to 'vindicate' the intuition that the surfaces of physical bodies are 2-dimensional, by reference to operational tests both for being at least 2-dimensional and for not being at least 3-dimensional. Although this is much more complex, it is analogous to vindicating the intuition that the lines composing the figure '$-|$' are not parallel, by showing that when the horizontal line is extended it intersects the vertical one. Of course everything depends on the validity of the hypotheses presupposed by the tests, just as Euclid's test for parallelism (Definition 23 of Book I, Heath, Vol. I, pp. 190–194) depends on Postulate 2, that straight lines can be extended or 'produced' indefinitely.

We end this section with brief comments on the flatness hypotheses, which, as said, are obviously empirical although they are so complicated that it is out of the question to examine them thoroughly in the present context—hence the term 'hypotheses'. This and the following section will confine their comments to Hypotheses 9.3.5 and 9.3.7, that are presupposed in arguing that bodies' surfaces are at least 2-dimensional.

The hypotheses are about what can be drawn on the surfaces of bodies and the properties and relations of the things drawn,[7] and only experience can confirm them. Given horizontal and vertical lines like C_1, C'_1, C_2, and C'_2 in Figure 3, perhaps Hypothesis 9.3.5, that it is not possible to draw 'separators' like B_1 and B_2 that are separate from each other is obvious, but it should not be taken for granted. B_1 and B_2 are simple closed curves that have the property of separating the entire diagram into an 'inside' and an 'outside', in such a way that C_1 lies inside B_1 and C'_1 lies outside it, and C_2 lies insides B_2 and C'_2 lies outside it. But modern topology teaches us that the class of curves that 'separate' a space in this way includes some very complicated ones (cf. the Jordan Curve Theorem, Courant and Robbins, pp. 244–246). Since we have by no means considered all possible separators of this kind, our example merely 'supports' the general hypothesis that no one of them that separates C_1 from C'_1 could be separate from one that separated C_2 from C'_2.

As to the *general* flatness hypothesis, 9.3.7, it affirms that what is true of the surface of the page is true of the surface of any body, namely that it is possible to draw a figure like Figure 3 on it. Since this involves the general concept of a body, it is obviously more questionable than the special hypothesis. In a sense it assumes that any body has a surface which, if 'irregular' at certain points, e.g., at the tip of a pin, it at least has small flattish or smoothly rounded spaces like the head of the pin, on which figures like Figure 3 can be traced. But fractal theory alerts us to the possibility of 'infinite irregularity', and we turn to that in concluding this chapter with particular reference to Hypothesis 9.3.5.

9.4 Fractal Possibilities: Methodological Remarks

If it is true that 'separators' like B_1 and B_2 in Figure 3 always intersect, as Hypothesis 9.3.5 affirms, there cannot be lacunas or 'holes' in the surface at the points where the separators would intersect, e.g., as pictured by the small circles in Figure 4. If there were such holes in the surface space, and the points in the space were the ones lying *outside* the holes, then segments C_1, C'_1, C_2, and C'_2 would be unaltered from Figure 3, and B_1 and B_2 would also be as they were in Figure 3, *except for the parts in the holes*, as in Figure 4:

[7]Note, though, that only in the indirect sense that drawing may involve *motion* are the hypotheses even remotely related to possible degrees of freedom of motions, which Poincaré and others occasionally argued are essential to the idea of spatial dimensionality. Cf. "Why Space has Three Dimensions," in *Last Essays*, especially pp. 37,38. Leibniz in is essay "The Metaphysical Foundations of Mathematics," Wiener collection, 1951, pp. 201–216 also linked the ideas of point, line, and plane to motion.

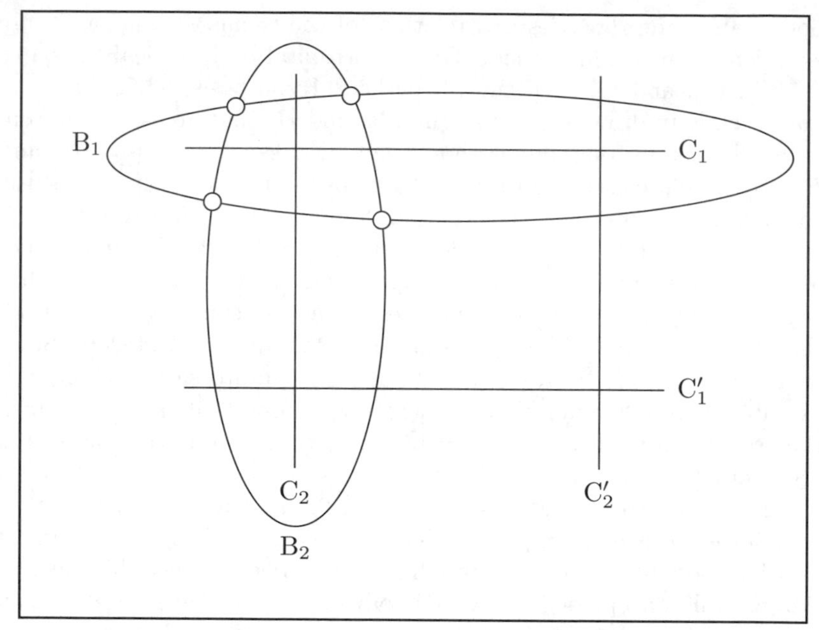

Figure 4

But this would mean that B_1 and B_2 no longer intersected, although B_1 would still separate C_l from C_1', since any continuous line drawn from C_1 to C_1' would still have to cross B_1, and B_2 would still separate C_2 and C_2' in this sense. Of course this wouldn't mean that B_1 and B_2 couldn't be redrawn in such a way as to separate C_1 from C_2 from C_2' and intersect each other, which would still support the hypothesis. But what if there were holes 'everywhere', that couldn't be avoided? That is a real possibility.

If the holes consisted of single points and these points were *dense* in the surface space in the sense that every neighborhood in the space contained one of them, then, assuming that the space that included the single points was 2-dimensional, the space without these points would be 1-dimensional because every point in it would have arbitrarily small neighborhoods containing points outside of it. It is true that spaces formed in this way, with single-point holes in them, are not compact, and therefore they do not satisfy the basic postulates of our theory of surface topologies (cf. Theorem 7.2.2(4)). But certain kinds of fractal spaces provide examples of compact surface-like spaces with holes 'everywhere'.

Simple examples are 'fractal carpets' (cf. Mandelbrot, 1977, pp. 133, 318), of which the 'triadic Sierpinski carpet' depicted in Figure 5 is the simplest.

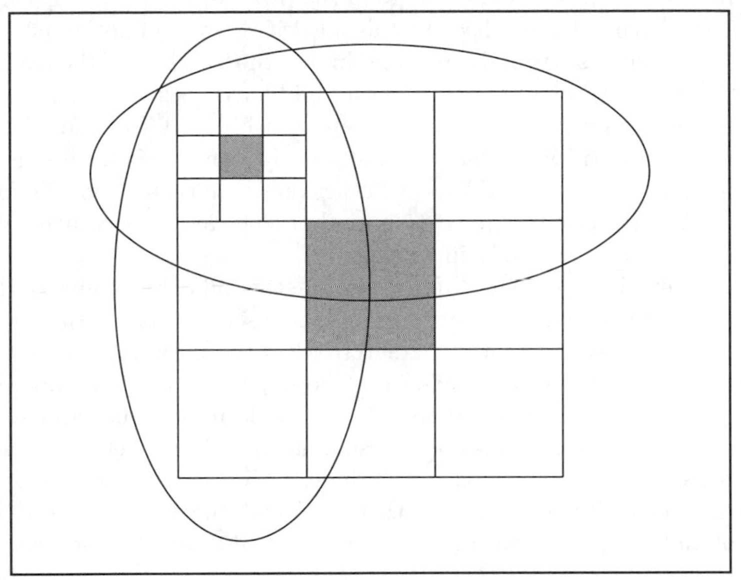

Figure 5

This is a direct generalization of the Cantor Discontinuum (Kelley, 1955, p. 165) to two dimensions. It is generated by first dividing a unit square into 9 equal parts, as shown, and then removing the middle part. Then each remaining part is itself divided into 9 parts and the middle one removed, and 'so on to infinity'. The resulting point-set is in fact a closed, compact subset of the unit square. Moreover, it is at least 1-dimensional since it contains all of the points on the boundary of the original square, but since every point of it has arbitrarily small neighborhoods that contain points outside of it, it is less than 2-dimensional.[8] Therefore it does not satisfy the general flatness Hypothesis, 9.3.7.

The Sierpinski carpet doesn't by itself invalidate our hypotheses, since the topology of this carpet is almost self-evidently not that of the surface of any ordinary body. We simply don't find bodies having holes in them exactly as described in the Sierpinski case. For one thing, this carpet has 0 area, and normal bodies have positive surface area. More general Sierpinski carpets can be generated starting by removing

[8]Its Hausdorff, fractional dimension is $log8/log3 = 1.8927$ (Mandelbrot, 1977, p. 133)

'bits' of the original square that are smaller than 1/9th of it, in fact ones of arbitrarily small size (Mandelbrot, 1977, p. 318), which are also topologically 1-dimensional, but so long as the same proportion of the remainder is removed at each stage in the process, the resulting 'carpet' also has 0 area. This result can be avoided if the proportion removed at each stage decreases rapidly to 0, and in fact the 'solidity' of the resulting figure, i.e., the ratio of its area after all bits have been removed to what it was before any were taken away, can be arbitrarily close to 1.[9] Such carpets would still invalidate the flatness hypothesis under discussion, and it only has to be considered whether ones *like* them might be found in Nature. This leads to quite deep methodological issues, which will be commented on very briefly in closing.

In 'ascertaining the facts', including whether all parts of surfaces have holes in them, we might seem to be best advised to consult the ways in which surfaces are currently characterized in the natural sciences—and the sciences that are most relevant to our present concerns are those that pertain to the nature of matter. But modern molecular and atomic theories teach us that what appears solid is really largely empty space,[10] which seems to disconfirm the hypotheses on which our account of surface dimensionality is based. But then, although the descriptions of these bits of matter are generally given in geometrical terms, Physics does not

[9]This may be related to results on density cited in Mandelbrot, 1977, p. 85a, but Adams, 1988, discusses a very simple form. Of course, if the sizes of the bits were reduced to 0 and they shrank to single points then removing them would diminish its dimension to 1, but it wouldn't diminish the figure's area at all and it would be visually indistinguishable from the original (but, as said, it would no longer be compact). The figure resulting from removing proportionately smaller bits *ad infinitum* from the original square would be visually distinguishable from the square, and it would also have the 'theoretical defect' that it wouldn't satisfy an intuitively desirable 'cosmological principle', namely that the smaller parts of the resulting figure should be geometrically similar to the larger ones. In this case the smaller parts would be more solid than the larger ones on the average.

[10]But modern criticism teaches us that this too is overly simple, though at least the early planetary models of the atom depicted it as being like the solar system in the way the 'electron planets' circle the 'nuclear sun', with mostly empty space between. However, our flatness hypotheses may be questioned at the more superficial level of naive observation and current metallurgical technique. The eye almost unaided perceives that most bodies have small cavities in their surfaces, and H.W. Gillett has commented that "Attempts to produce objects quite free of porosity by the use of extremely high pressures, to fill pores with some other metal such as copper, and improve properties by heat treatment of carburized objects have not been notably successful" (in the entry "Iron and Steel" in C. Carmichael (ed) *Kent's Mechanical Engineer's Handbook, 12th Ed.*).

The extent to which fractal theory accommodates modern physical pictures of the structure of matter is unclear to me, but that is beside the question of how well the theory developed in this work relates to physical theory.

apply to surface features like marks, which are the things that our theory applies to.[11] This leads to the final remarks of this chapter.

We are now concerned with what might be called the Problem of the Microsurfacial, but although our theory is of surface topology, its basic data are superficial 'appearances'. For example, we take the appearance of continuity in a rectangular mark like ■ as a 'datum', and we do not delve below the level of the appearance to a 'reality' that underlies it. But if we criticize this naive way of proceeding, e.g., by following well-worn arguments from the relativity of sense perception that lead to a subjectivist view of sense data, we find that there are no sciences other than the physical sciences that can provide us with an account of this reality, and none of those deal with the objects of the present theory, i.e., with features of the surfaces of bodies. Therefore it could seem that there are no theories akin to atomic and subatomic theories of matter, that might tell us that we should refrain from applying to the microsurfacial realm postulates that are only valid for macrosurfacial data. But that would be too simple.

Our theory has microsurfacial implications from the very outset. For example, along with undeniable facts of sense perception, it implies that points on surfaces are too small to be seen. This follows from the First Abstraction Principle, 6.6.1, together with fact that any visible mark like the small dot '•' can be flanked by separate marks, both of which touch the dot. Of course infinitely small points belong to the geometrical realm and not to the physical one, but in a sense they are the *measure* of the physical realm. And, speculating, if this is true then questioning our theory at the microphysical level implicitly questions its principles of abstraction at that level, and these are essentially principles of physical measurcment. Of course, that is the way of modern physics, and here we can only remind the reader that the principles of measurement of modern microphysics are currently very much in dispute. Part III of the present work will comment on *classical* metrology in more detail, and

[11]That is not to say that fractal considerations don't apply to surface features. Assuming that our surface spaces are compact, metrizable, 2-dimensional, and so on, it follows from the postulates of our theory that these spaces contain I-systems with 'tremas' like Sierpinski carpets. Thus, in a sense our theory is consistent with fractal theory (though having a subspace like a Sierpinski carpet doesn't imply that the whole space has that structure). On the other hand it is doubtful that the modern physical picture of 'Nature' is entirely adequately describable in fractal terms. The fractal picture seems to presuppose that quantities of matter occupy precisely delimited regions of space (e.g., the discussion of the question "How long is the coast of Britain?" in Chapter 5 of Mandelbrot, 1977, seems to suggest that "Britain" denotes a precisely defined spatial region), but modern quantum theory can be interpreted as denying exact locations to the distribution of matter in space.

here we will only conclude with a modest claim in support of our theory, independent of microphysical perplexities.

Although the properties that we have postulated of our surfaces features, e.g., the Separation Postulate, may at best be *a priori* plausibilities that rely on the visual inspection of a few special cases, following up their consequences does allow us to show how modern abstract topological theory can be operationalized in such a way as to explain a variety of common topological and proto-geometrical intuitions about the surfaces of bodies. These include such things as why we think that small dots are too large to be points, why we don't think that the segments of the 'tepee', X̲, meet at a point but we think that those forming the star, Ӿ, do, why we think that thin streaks like '————' aren't perfectly thin, and moreover anything perfectly thin couldn't be seen, although the boundaries of things that can be seen may be perfectly thin, and why we should say that the surfaces of bodies are 2-dimensional. These 'findings' are parts of an incomplete but very difficult program of operational 'explication', but it is hoped that enough of it has been carried out to suggest that it has some promise, though only the reader can judge this. The next chapter sketches one more 'pre-superpositional' stage in the program, which focuses on the concept of linearity. In some ways this is even more difficult to analyze than that of dimensionality, and the sketch will end by outlining a number problems with it that are unsolved as yet.

10

Aspects of a Platonic Account of Linearity[1]

10.1 Introduction

Trivially, while the boundaries of typical surface features like the circle in Figure 1 of Chapter 8 are 1-dimensional and qualify as lines in the vague Euclidean sense of having breadthless length, modern topology teaches us that not all 1-dimensional point-sets ought to qualify as lines in this sense. For example, the class of points in the unit square with one rational coordinate is 1-dimensional (since it is a countable union of 1-dimensional line-segments, cf. Theorems II.2 and IV.1 of Hurewicz and Wallman, 1948), but we would hardly regard this as the locus of a line. 1-dimensionality is clearly a necessary condition for linearity, but we want to consider additional conditions that may be sufficient for it. Reasons for this inquiry include the fact that it is a prolegomenon to the analysis of the specifically geometrical concept of a *straight* line, and it might seem to be presupposed in applications of certain aspects of algebraic topology—though this will be reconsidered in section 10.5.

Our inquiry will focus on *line segments* as, arguably, the basic components from which lines in general are formed, which also include circles, and branching and infinite lines.[2] Our approach to the problem of characterizing these segments may be regarded as Platonic in the generalized sense already hinted at in section 8.6. Making this precise, we postulate that they are characterized by their *form*, which is that of a particular abstract mathematical object.[3] Specifically, we postulate that a line seg-

[1] This chapter develops ideas originally discussed in section 8 of Adams, 1996.

[2] What we are here calling line segments are called *arcs* in topological theory (cf., Steen and Zeebach, 1970, p. 29). To avoid the suggestion that they are necessarily straight they might better be called *segments* of lines, where 'line' is construed in the classical geometrical sense that includes circles and infinite and broken lines.

[3] According to Wedberg, 1955, pp. 53 *et. seq.*, Plato conflates ideas with the things

ment must have the 'form' of the *unit interval of real numbers*, **I** = [0,1], and will say that a set of surface points constitutes a line segment if it has the topological form of this interval, in the sense that it is *homeomorphic* to it. Other figures like circles and infinite lines can be regarded as being composed of such segments in more or less complicated ways, though we will not attempt to delimit the class of these composite lines precisely.[4]

Our focus will be on the question: what *operational* conditions must be satisfied by a 'surface entity', say some part of a feature-boundary, for it to be homeomorphic to **I**? Of course we might say that a set of surface points is a line segment if it is homeomorphic to **I**, but while this characterization is precise, it is not operational. What we would like to do is to give an *intrinsic* characterization of the form, which lists empirical attributes, implying 1-dimensionality, possession of which is necessary and sufficient for the set to be homeomorphic to **I**.[5] In fact, intrinsic topological attributes of **I** are well known and will be cited below, but an operational analysis should translate them into criteria that can be applied to our surface topologies. However, we will completely succeed in this only in the case of parts of boundaries of observable surface features, and the more general problem of characterizing linearity in arbitrary point sets will only be partly resolved.

We will begin by giving the abstract characterization of a line seg-

---

to which they apply. For instance, the *idea* or *form* of the circle is conflated with the ideal circle. The latter is eternal and not accessible to sensory observation, though it can be 'grasped by reason' and it is the subject of geometrical theory (although this is complicated by fact that geometry treats of plural ideal circles). To transpose this to our case we should have to conflate the interval **I** with the form that anything would have to share with it for it to be an ideal line segment. We shall avoid this confusion, but there are still enough points of similarity to justify our regarding our approach as Platonic. Thus, the ideal interval **I**, which is a timeless mathematical abstraction, has to be 'grasped by the mind', although 'sensible' but more or less ephemeral objects may bear a likeness to it. We go beyond Plato in specifying in what this likeness consists, namely homeomorphism—although the *most* sensible objects of our theory are independent features that can be drawn on surfaces, and since we have just argued that they are 2-dimensional, they cannot be homeomorphic to **I**. However, we will try to show that dependent ones like parts of feature-boundaries can be 1-dimensional and homeomorphic to **I**, even if they may not be perfectly straight.

[4]Heath's discussion of lines in connection with Euclid's Definition 2 of Book I (Heath, Vol. 1, pp. 158–165) suggests that the ancient Greeks did not attempt this either.

[5]The comment on p. 93 of Newman, 1995, is significant in this regard:

> Among all particular topological spaces the most fundamental are the open line R1 and its closed subset [0,1]. It is therefore not very satisfactory that these spaces should be defined by means of the theory of real numbers, instead of by simple topological properties.

ment, and then state an intrinsic topological characterization of it, together with a corollary which, together with certain additional empirical assumptions, will be operationalized in the following section.

10.2 Abstract Characterization and its Application to Surface Spaces

10.2.1 Definition. A point set Γ in a surface space is a *line segment* if and only if it is homeomorphic to **I**.

The following result of Janiszewski (1912) gives an intrinsic condition that is equivalent to this abstract definition:

10.2.2 Theorem (Janisewski). A continuum, Γ, (i.e., a closed, compact and connected subset of a compact and metrizable space) is homeomorphic to **I** if and only if all but at most two points, π, of Γ are *cut points* in the sense that $\Gamma \sim \{\pi\}$ is not connected.[6]

The next theorem which depends now on the postulates of our theory of surface spaces, is easy to operationalize. It presupposes a concept of *betweenness* that holds between closed, connected 'parts' of segments:

10.2.3 Definitions

(1) Given three mutually disjoint, closed connected subsets Γ_1, Γ_2, and Γ_3 of Γ, Γ_1 is *between* Γ_2 and Γ_3 if any other closed, connected subset of Γ that intersects both Γ_2 and Γ_3 necessarily covers Γ_1.

(2) A point of Γ, π_1, is between two other points, π_2 and π_3 if they belong, respectively, to disjoint *parts*, i.e., closed connected subsets with more than one member, Γ_1, Γ_2, and Γ_3 of Γ, the first of which is between the other two.

(3) The betweenness relation on points is *Hilbertian* if for all π_1, π_2, π_3, and π_4,

(α) If π_2 is between π_1 and π_3 then it lies between π_3 and π_1:

(β) If π_1 and π_3 are distinct then there exists at least one point π_2 lying between π_1 and π_3;[7]

[6] This is a direct restatement of the version of the Jenisewski theorem that appears on p.93 of Newman, (1995); a similar theorem, Theorem 12.1 on p. 99, gives an intrinsic topological characterization of the conditions for a point set to be homeomorphic to a circle. However, this version seems dubious because, intuitively, it seems that these conditions should be satisfied by the open and half-open intervals (0,1), [0,1), and (0,1], as well as by **I** = [0,1]. The intuitively right conditions are that Γ is homeomorphic to [0,1] or (0,1) according as either no points or two points of it are not gaps, and it is homeomorphic to (0,1] or [0,1) of exactly one point if it is not a gap. If this is right then Corollary 1.2.2 should be modified accordingly, as will be commented on in footnote 9.

[7] Hilbert's second axiom added "and at least one point π_4 such that π_3 lies between π_1 and π_4," which is characteristic of the open interval (0,1), but which must be omitted in application to **I**.

(γ) Of any three distinct points there is always one and only one that lies between the other two; and

(δ) Any four distinct points π_1, π_2, π_3, and π_4 can always be so arranged that π_2 shall lie between π_1 and π_3 and also between π_1 and π_4, and furthermore π_3 shall lie between π_1 and π_4 and also between π_2 and π_4.[8]

10.2.4 Theorem. A continuum, Γ, is homeomorphic to **I** if the betweenness relation on it is Hilbertian, and it has at most two points that are not between other points, and if, given any part Γ_1 of Γ, if Γ_1 is between any other such parts of Γ then $\Gamma \sim \Gamma_1$ is the union of two non-empty disjoint open connected subsets, and if Γ_1 is not between any such subsets then $\Gamma \sim \Gamma_1$ is an open connected subset of Γ.[9]

Sketch of Proof. This uses the well known fact, which will be used again later, that the only non-trivial connected subsets of **I** (non-triviality will be assumed throughout in the rest of this argument) are *subintervals* of it, and these are open or closed in the topology of **I** according as they are open or closed intervals. Simple algebra applied to the their endpoints then establishes that the closed subintervals of **I** have the properties stipulated in the corollary, and therefore if Γ is homeomorphic to **I** its closed connected subsets must also have these properties.

Suppose conversely that the parts of Γ have the properties stipulated in the corollary. Then the following argument shows that Γ also has the properties stipulated in Janisewski's theorem, and therefore it is homeomorphic to **I**. First suppose that π_1 is a point of Γ that is between two other points of Γ, π_2 and π_3. By Definition 10.2.3, there exist closed, connected subsets Γ_1, Γ_2, and Γ_3 containing π_1, π_2, and π_3, respectively, such that Γ_1 is between Γ_2 and Γ_3. By the conditions of the Corollary,

[8]Except for axiom (β), which modifies Hilbert's axiom II,2, as noted in the previous footnote, these are essentially Axioms II,1–II,4 on p. 6 of Hilbert, 1902. The latter axioms define *strict* or *exclusive* betweenness, but betweenness relations are not always characterized in exactly this way. Thus, DEFINITION 1 on p. 84 of Suppes, *et al.*, (1989) defines *inclusive* betweenness in such a way that an element is always between itself and any other element.

[9]This presupposes the version of Janisewski's Theorem stated as Theorem 10.2.2. Querying this as in footnote 6 would require modifying the present Corollary, and a plausible modification would be that for Π to be homeomorphic to **I** requires it to have two disjoint parts that are not between other parts. This in turn would require a modification in the proof given below, most plausibly along the following lines: The modified Janisewski Theorem implies that the class of all points of Π that belong to parts of it that are between other parts must be an open subset including all but two points of Π, and it must be homeomorphic to the open interval (0,1). Then its closure will equal Π, and that must be homeomorphic to the closure of (0.1), i.e., it must be homeomorphic to **I**. But the details of this argument remain to be filled in, and it will not be pursued here.

$\Gamma \sim \Gamma_1$ is the union of disjoint open, connected subsets Θ and Ψ of Γ. Trivially, $\Theta \cup \Psi$ cannot be connected, since the union of two disjoint open sets cannot be connected.[10]

Now we use Theorem 8.3.2, that the classes of interior points of U-systems are bases for surface topologies. Therefore the classes of interior points of connected unions of the marks that compose these U-systems are also bases for these topologies. Moreover, since the marks themselves are connected, connected unions of them are also connected. Then, if π is a point of Γ that is between other points of Γ, and Γ_1 is the closure of a basic neighborhood of π, which is a union of connected marks, hence by the stipulations of the Corollary, $\Gamma \sim \Gamma_1$ is the union of disjoint open connected subsets Θ_1 and Ψ_1 of Γ. Moreover, the closures of Θ_1 and Ψ_1 cannot contain π, since π is an interior point of Γ_1, which is separate from Θ_1 and Ψ_1.

Now let Γ_2 be another closure of a connected basic neighborhood of π, which must also be such that $\Gamma \sim \Gamma_2$ is the union of disjoint open connected subsets, Θ_2 and Ψ_2, of Γ. Clearly neither Θ_2 nor Ψ_2 can intersect both Θ_1 and Ψ_1, for if, say, Θ_2 intersected both Θ_1 and Ψ_1 all three would be connected and so would their union (Corollary 1, p. 74 of Newman, 1995), and therefore so would their closure (Theorem 20, Kelley, 1955). But by the definition of betweenness this would imply that the closure of Θ_2 covered Γ_1 and therefore π, contradicting the fact that the closures of neither Θ_2 nor Ψ_2 can contain π. Therefore, we can assume without loss of generality that Θ_2 is separate from Ψ_1 and Ψ_2 is separate from Θ_1.

Generalizing, it is clear that every closure of a connected basic neighborhood, Γ_ι, of π, for ι in some index set I, must be such that $\Gamma \sim \Gamma_\iota$ is the union of disjoint open connected subsets, Θ_ι and Ψ_ι, of Γ, and moreover Θ_ι must be disjoint from all Ψ_κ and Π_ι must be disjoint from Θ_κ, for all ι and κ in I. Moreover, clearly the unions $\bigcup_{\iota\epsilon\iota} \Theta_\iota$ and $\bigcup_{\iota\epsilon\iota} \Psi_\iota$ must be disjoint connected open sets whose union is $\Gamma \sim \{\pi\}$. This proves that all points of Γ between other points of it are cut points.

Completing the argument, Γ cannot have three cut points, since of any three points one must be between the other two (Hilbert Axiom γ—note that this is the only one of Hilbert's four axioms that is used in

[10]In spite of its obviousness, in browsing through several elementary topology texts the author has not been able to find an explicit statement of this fact. A simple argument for it goes as follows. Suppose that Θ and Ψ are disjoint open sets, but, say, Ψ contained a boundary point of Θ, say π—hence Θ and Ψ are not separated, as required if their union is to be connected. Then, since Ψ is an open set it contains a neighborhood, N, of π, and since π is a boundary point of Θ all of its neighborhoods contain points of Θ. Hence a subset of Ψ contains a point of Θ, and therefore Θ and Ψ are not disjoint, contrary to hypothesis.

the proof).

We will return later to the way in which the betweenness axioms generate the orderings that are characteristic of line segments, but first let us operationalize the conditions for linearity that are given by Corollary 10.2.4.

10.3 Operational Characterization of Linearity in the Case of Boundary Segments

Now consider Figures 1 and 1R, below:

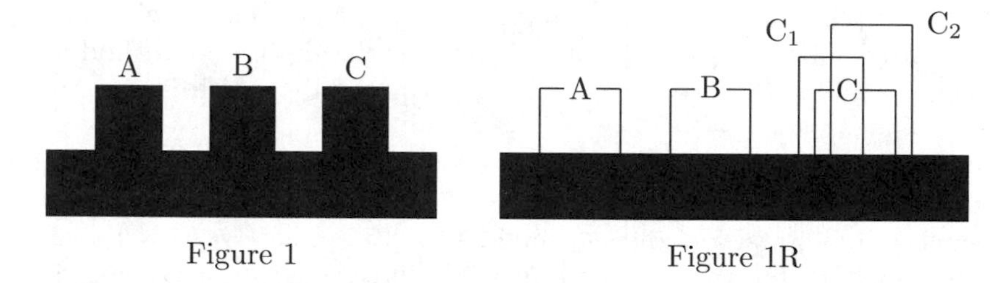

Figure 1 Figure 1R

We are concerned with the upper edge of the long rectangle in Figure 1, which three small squares A, B, and C rest on. For ease in conceptualization, Figure 1R represents the squares by outlines, although, since the representing rectangles also have interior spaces, this figure is clearly more complicated than the one it represents.[11] In any event, our primary concern is with Figure 1, where we hope to be able to show that the upper edge of the long rectangle not only looks linear but it has the properties stipulated in Corollary 10.2.4, and therefore it is a line segment in the sense of 10.2.1; i.e., it is homeomorphic to **I**.

First, note that it seems plausible that the intersections of squares A, B, and C with the upper edge of the long rectangle in Figure 1 should satisfy the conditions for being closed, connected parts of the edge, according to Theorem 8.4.2(3). This is suggested by the representation of rectangle C in Figure 1R, which pictures it as being covered by a union of rectangles C_1 and C_2, which could not cover it if their union were not connected.

[11]That Figure 1 only *represents* a long rectangle with squares resting on it must again be insisted upon, since what is actually before the reader is a single connected figure with rectangular 'bumps' on its upper edge. Recalling the genetic characterization of surface features and their incidence relations, the figure can only be regarded as being composed of the rectangle and four squares if these 'components' are conceived as having been drawn in a particular order. As discussed in section 2.3, the incidence relations among the components is defined by the manner in which they are drawn.

Second, of the three squares A, B, and C, clearly B is between A and C according to 10.3.2(1). That is, it is clear 'by experiment' that any attempt to produce a U-system, D, intersecting the intersections both of A and of B with the upper edge of the long rectangle, and whose intersection with the upper edge of that rectangle is connected, would have to cover the intersection of the edge with rectangle B.[12]

Finally, because the intersection of square B with the upper edge of the long rectangle lies between the intersections of A and C with this edge, the intersection of B with the edges *divides* it into separate regions to left and right of the intersection, whose union, being the complement of the intersection of B with the edge, which is closed, is an open set in the topology of the edge, as stipulated in Corollary 10.2.4. This suggests if it does not prove that *any* closed connected part of the upper edge of the rectangle that is between two other such parts divides the edge in the same way. Hence, we posit:

10.3.1 Hypothesis. The upper edge of the boundary of the large rectangle in Figures 1 and 1R is a line segment.

And, by a further generalization:

10.3.2 Hypothesis. 'Sufficiently small parts' of the boundaries of U-systems are line segments.

Clearly these hypotheses are empirical; moreover, they are independent of the previously formulated assumptions of our theory. But to the extent that they are true, they make clear what is presupposed in holding that no matter how thin and elongated they are, surface features are not ideal Platonic lines, although parts of their boundaries may be. Note how complex this is, as well as how much this explication depends on the way that boundary parts are represented. This is necessarily a matter of concern, because of the possibility of mistaking properties of the representation for properties of the things represented.[13] This will be returned to in section 10.5, but given the almost universal reliance on representations, it is no wonder that Plato should have come to think that everything sensory is representation! First, however, we will comment on linear orderings.

---

[12]Note that this would not be true of the *whole* boundary of the large rectangle, since by going around its ends and lower edge, A and C could be connected without intersecting rectangle B.

[13]Recall the incident in Mark Twain's *Tom Sawyer Abroad*, in which Huck, Jim, and Tom mistakenly suppose that the colors of the states they pass over in a balloon must be same as those of the regions that represent them on a map.

10.4 Linear Ordering

It is part of our conception of a line segment that the points on it should be ordered, as is obviously the case with the ideal line **I**. Also, this has usually been assumed to be true of the order in which points are traversed when lines are traced in the way that Leibniz and others assumed that they are generated.[14] Continuity itself is associated with the 'absence of gaps' in orderings,[15] and in 1895, prior to the development of theories in terms of which lines and their topologies could be characterized, Cantor characterized *continuous orderings* thus:

10.4.1 Definition. (Cantor, 1895) A binary relation $\prec$ on Π is an *ordering of type* Θ if and only if:

(1) it is a simple ordering of Π, i.e., for any α, β, and γ in Π,

(a) if $\alpha \neq \beta$ then $\alpha \prec \beta$ if and only if $\beta \not\prec \alpha$, and

(b) if $\alpha \prec \beta$ and $\beta \prec \gamma$ then $\alpha \prec \gamma$;

(2) there is a countable subset, Π', of Π that is dense in Π in the sense that for any $\alpha \prec \beta$ in Π there exists γ in Π' such that $\alpha \prec \gamma$ and $\gamma \prec \beta$;

and

(3) (Dedekind's Postulate) if Π_1 and Π_2 are disjoint non-empty subsets of Π, such that $\Pi_1 \cup \Pi_2 = \Pi$, and $\pi_1 \prec \pi_2$ for all $\pi_1 \in \Pi_1$ and $\pi_2 \in \Pi_2$, then either there is some $\pi_1' \in \Pi_1$ such that $\pi_1 \not\prec \pi_1'$ for all $\pi_1 \in \Pi_1$, or there is some $\pi_2' \in \Pi_2$ such that $\pi_2 \not\prec \pi_2'$ for all $\pi_2 \in \Pi_2$.[16]

Adding that Π must have a first and a last element, we can then say:

10.4.2 Theorem. Given an ordering of type Θ on a point set Π, if it has elements, α and β, that are *first* and *last* in the sense that for all γ in Π, if $\gamma \neq \alpha$ then $\alpha \prec \gamma$, and if $\gamma \not\prec \beta$ then $\gamma \prec \beta$, then the order topology of Π is homeomorphic to **I**.

But this alone doesn't imply that a subspace of a surface space, for instance the space of points on the upper edge of the large rectangle in Figure 1, is homeomorphic to **I**. It only says that if the subspace has *some*

---

[14] Leibniz, 1715, "The Metaphysical Foundations of Mathematics," in the Wiener collection, 1951, p. 206.

[15] Leibniz, 1715, "The Metaphysical Foundations of Mathematics," in the Wiener collection, 1951, p. 206. Similarly, "... You ask me for some elucidation of my Principle of Continuity. I certainly think that this Principle is a general one and holds not only in Geometry but also in Physics" and "... the Laws of Collision of Bodies left to us by Descartes are false; but I can show that they are false only because they would allow *hiatuses* in events by violating the Law of Continuity ... " (letter to Varignon, 1702, pp. 184–6 in Wiener collection, my italics). Thus, Leibniz associated continuity with absence of hiatuses or 'jumps', and he generalized the Principle of Continuity from Geometry to all of Nature (cf. Russell, 1900, p. 111).

[16] The present characterization is essentially equivalent to the one given on p. 49 of Huntington, 1917.

ordering of type Θ with a first and a last element, then the topology generated by that ordering is homeomorphic to **I**. But the topology generated by *some* ordering of the subspace (cf. the definition of an order topology in Problem I, pp. 57–8, Kelley, 1955) is not necessarily the same as the one generated by the surface topology as characterized in Definitions 7.2.1, and proving this requires showing that the betweenness relation characterized in Definitions 10.2.3 generates an ordering of type Θ, with first and last points. If this is true, say of the points in the set Π on the upper edge of the long rectangle Figure 1, it depends on empirical facts about this set, as follows.

Empirical observation establishes that this holds of the betweenness relation among the parts of that is characterized in Definition 10.2.3(1), i.e., observation shows that disjoint closed connected subsets of Π, satisfy all four of Hilbert's betweenness axioms. This implies:

10.4.3 Theorem (Hilbert).[17] If we have given any finite number of disjoint parts of Π we can always arrange them in a sequence A, B, C, D, E,..., K so that B shall lie between A and C, D, E,..., K, C between A, B and D, E,..., K, etc. Aside from this order of sequence, there exists but one other possessing this property namely, the reverse order, K,..., E, D, C, B, A.

Given 10.2.3(2), this extends to the *points* of Π, hence to any finite number of members $\pi_1, \ldots, \pi_n$ of Π: i.e., they can be ranged in a sequence in such a way that for all $i, j, k = 1, \ldots, n$, π_i is between π_j and π_k if and only if either $j < i < k$ or $k < i < j$. Calling a sequence arranged in this way a *betweenness sequence*, we can say that for any π_h, π_i, π_j, and π_k, not necessarily distinct, π_h and π_i *occur in the same order in the sequence* as π_j and π_k if $h < i$ if and only if $j < k$. It also follows from the axioms of connection that if π_h and π_i occur in the same order as π_j and π_k in some betweenness sequence, then they occur in the same order in all such sequences in which they occur at all. Consequently we can define π_h and π_i to occur in the same order as π_j and π_k, if they occur in the same order in some betweenness sequence.

Given the above, and given any distinct π_h and π_i, we can define a binary relation, $\beta \prec \gamma$, on Π to hold if and only if β and γ occur in the same order as π_h and π_i. Then $\beta \prec \gamma$ is itself a simple ordering of Π; moreover there only two such orderings of Π, and they satisfy Hilbert's betweenness axioms. But this still does not prove that the ordering $\beta \prec \gamma$ is of type Θ with first and last elements. All that has been done so far is to provide an observational foundation for *an* ordering of Π, and it must still be shown that this ordering: (a) has a countable dense

[17] Hilbert, 1902, p. 6.

subset, and (b) satisfies Dedekind's Postulate. The former follows fairly routinely from the fact that the basic surface topology is metrizable and therefore so is the topology of its subspace Π. But the author has found establishing that Dedekind's Postulate is satisfied very difficult. In fact, Janisewski's Theorem came to the author's rescue in this regard, which makes it possible to go directly to isomorphism to **I** without going by way of an empirical ordering of Π. But we have now established a connection between linearity and order, which implies that the empirical ordering $\beta \prec \gamma$ just described is in fact an ordering of type Θ, which in turn explains why ordering should be regarded as essential and even definitive of linearity.

The foregoing finishes our inquiry into the empirical meanings of topological concepts in their application to physical bodies' surfaces. Clearly so far our investigation has been of very limited extent, and this chapter's concluding sections will point out certain directions and problems for further inquiry, beginning with some further remarks on topological representation, focusing now on the representation of lines.

10.5 Representing Lines

Introductory geometry texts are replete with diagrams in which ideal lines are represented by thin streaks, as in Figure 2 below, which represents a circle with a diameter drawn horizontally through it:

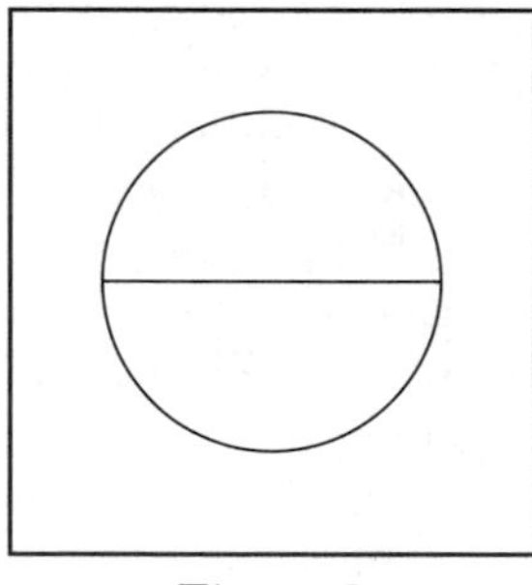

Figure 2

Figure 3

But we may ask how the horizontal streak, which has two separate sides, can represent an ideal line whose two 'sides', if it can be said to have them at all, are coincident. Plato's Cave Analogy seems very apt here, since 'reason' somehow grasps what the diagram only 'represents', though no more than the shadows on the walls of the Cave is the diagram exactly like the thing it supposedly corresponds to 'in the real world'.[18]

[18] Plato's views on Geometry are too complicated to be summarized briefly, but the following quotation from Wedberg, 1955, illustrates this point:

But we should ask: what does the reason really grasp in this case?

Consider *Euler's Formula*, which is supposed to characterize the relation between the number of vertices, V, the number of line segments, L, and the number of undivided regions, R, into which the lines in a diagram divide the space in which it appears. The general formula is $V - L + R = 2$,[19] and it seems to be born out in Figure 2, since the upper and the lower semicircles as well as the diameter count as distinct line segments, there are two vertices in which these segments meet, and if the space outside the diagram is counted as an undivided region, then $V = 2, L = 3, R = 3$, and $V - L + R = 2 - 3 + 3 = 2$.

But the foregoing is inexact, since if its 'line segments' really have *two* edges that 'count as lines', then Euler's Formula is not satisfied. That is most easily seen from the fact that, assuming that the 'segments' actually have some thickness, the whole diagram is homeomorphic to a circle with two small interior ovals, as shown in Figure 3. But this diagram has *no* vertices, it has *three* lines, each of which is a circle or oval, and it divides the plane into *four* undivided regions, hence $V - L + R = 0 - 3 + 4 = 1$. Hence if Figure 2 'illustrates' Euler's formula, what *topological lines* it contains, namely the *edges* of the thin streaks, 'don't count'— they don't 'represent' anything. A little reflection on this and similar examples suggests that what 'ideas' diagrams are intended to illustrate or represent depend on the context in which they are 'presented', much as signs like the north direction arrows that appear on maps must be read in a special 'map representing way' in order to be properly interpreted.[20]

> *The concepts of geometry are ideas (in the sense of the general theory of ideas).*
>
> A spatial object that exemplifies a Euclidean concept or that, in this sense, "partakes" of a Euclidean Idea, we shall here designate a "Euclidean object." *There are no truly Euclidean objects in the sensible world.* (p. 49).

[19]This formula appears to have been discovered by Descartes in 1640, cf. Courant and Robbins 1941, pp. 236–240. Actually, Courant and Robbins discuss the formula as it applies to the surface of a polyhedron, where V, L and R are, respectively, the numbers of vertices, lines formed by edges, and regions (faces) of the polyhedron's surface. The formula is 'transposed' to diagrams by treating the region outside of the diagram as another 'face' of a polyhedron.

Oddly, however, what is diagrammed in Figure 3 cannot represent the surface of any finite polyhedron. That is because every such polyhedron has a positive number of vertices (of course, infinite polyhedra like an infinite triangular prism can have no vertices). Looking at Courant and Robbins' proof of Euler's Formula for polyhedra, we see immediately what assumptions of the proof are violated. This observation parallels points made by Lakatos in his famous paper "Proofs and Refutations."

[20]Very similar observations apply to the use of lines in diagrams intended to illustrate map-coloring ideas. Thus, Courant and Robbins illustrate the Four Color

But these are only the most superficial comments on a subject that demands far more extensive study, and the final section comments on still another matter for further study.

10.6 Open Problems

First, a point of general methodology. All but the most elementary parts of topological theory make extensive use of ideas that are far from empirical. One that is very far has just been encountered: namely that of a homeomorphism. For example, though we may see a circle and a square clearly, we cannot 'see' in any direct way that they are homeomorphic; i.e., that there is a one–one function mapping the points and open sets of one onto points and open sets of the other. Perhaps we sense this intuitively in a circle and an ellipse, given the idea that topology is 'rubber sheet geometry', but it stretches things somewhat to say that to see it in the case of the circle and the square.[21] This must be proven, and, as in the case of the abstract, Platonic interval **I** and the boundary segment in Figure 1 discussed in section 10.3, this may require painstaking enumeration of the attributes of the entities whose identity of topological form—their homeomorphism—is to be established.

A further point concerning the difficulty of relating topological concepts to empirical application is that, while topological theory may originally have been inspired by empirical examples, e.g., the Königsberg Bridges Problem, Möbius bands, map-coloring problems, and so on,[22]

Theorem by Figure 129 on p. 246, which supposedly represents regions that must be colored with four colors if each region is to be colored with a color different from a region with which it has a common border. Actually, however, this figure is homeomorphic to one in which there are four separate circles colored white 'embedded in a black sea'. And, once again, we seem to be driven back to a Platonic account, in which the reason somehow grasps what the senses only imperfectly present. Again it appears that the reason can grasp the same figure in different ways, depending on what it is meant to 'show'. Still other figures use lines in much more complicated ways to illustrate other topological ideas. Thus, Figure 139 on p. 260 pictures a Möbius band, which is supposed to be one-sided surface, but this requires the reason to grasp the idea that the 'real' Möbius band has zero thickness.

[21]The intuitive 'rubber sheet geometry' characterization is ubiquitous (cf. Firby and Gardiner, 1982, p. 11), but the circle and the square show the need to transcend it in other cases.

[22]The intuitive geometrical origin of many concepts of elementary topology is evident in terms like 'closed', 'open', 'boundary', 'intersecting', 'connected', 'path', etc. Courant and Robbins, 1941, make a significant comment in this connection:

> ... the pioneers, such as Poincaré, were forced to rely largely on geometrical intuition. Even today a student of topology will find that by too much insistence on a rigorous form of presentation he may easily lose sight of the essential geometrical content of a mass of formal detail. (*What is Mathematics?*, p. 235).

the primary applications of modern mathematical topology are within the domain of mathematics itself. The 'space' of real numbers in the interval $\mathbf{I} = [0,1]$, whose 'points' are clearly defined mathematically, is an elementary example. In this and like cases mathematicians are at liberty to formulate their theories in terms of concepts of arbitrary levels of abstraction, and they are under no compulsion to relate them to observation, or indeed to anything in the physical world. But what is simple in abstract theory, like the idea of a one-one mapping, may be anything but simple in application. Here are two examples, the second of which is extremely important to applied topology and geometry.

The first still concerns lines: How do we know, if it is true at all, that there are lines on surfaces connecting any two points on them? Euclid postulates that a straight line can be drawn between any two points (Postulate 1 of Book I, Heath, 1925, p. 195), but clearly even the drawability of curved lines between any two points cannot follow from the anything that our theory has postulated so far. That is because nothing in it rules out the possibility that bodies' surfaces might themselves be disconnected, e.g., as the 'composite surface' of two separate billiard balls would be. In fact, it is not self-evident that lines connect any two points on the surface of a single billiard ball; i.e., that given any two points on it there exists a point set Π homeomorphic to $\mathbf{I}$ and containing these points (technically, that the surface is *connected by arcs*, Steen and Zeebach, 1970, p. 29). It would be even more apposite so far as concerns application to generalize to lines the kind of 'approximation' already considered in Chapter 6 in which small dots approximate ideal points, and then prove that given any two small dots it is possible to draw marks or U-systems connecting them that arbitrarily closely approximate ideal lines. This brings us to our final point.

What about bodies' surfaces themselves? We have given reasons for thinking that they are 2-dimensional, but no more than the fact that a point set is 1-dimensional entitles us to conclude that it is linear, does the fact that a surface is 2-dimensional entitle us to conclude that it is planar, or even 'smooth'. The natural generalization of our approach to linearity would be to characterize this kind of smoothness in terms of topological sameness to other Platonic ideals e.g., homeomorphism to the Euclidean plane or some suitable part thereof. In fact, this is the standard approach used in differential topology. Thus, Munkres defines a *manifold*:

We might add that even in 1998 persons applying topology to the physical world, and to bodies' surfaces in particular, still rely largely on the intuitions that the present work seeks to clarify.

> A (topological) manifold M is a Hausdorff space with a countable basis, satisfying the following condition: There is an integer m such that each point of M has a neighborhood homeomorphic with an open subset of H^m or R^m. (Munkres, 1966: setting m = 2 yields Euclidean 2-space, which is here symbolized R^2).

This is precise enough, but, as with linearity, which can be regarded as the special case in which m = 1, it is not operational. How do we know that the surface of a billiard ball is a 2-dimensional manifold or even approximates one? The author has not been able to locate a two-dimensional analogue of Janisewski's Theorem in the topological literature, that would provide an intrinsic characterization of this kind of manifold, i.e., one which enumerates the topological properties that a surface space or neighborhood in it must possess in order to be homeomorphic to the Euclidean plane or a subspace of it. In fact, modern geometrical theory tends in the direction of more rather than less abstractness when it characterizes the Euclidean plane by reference to 2-dimensional Cartesian coordinate representations of points in it,[23] and these are very far from being 'operationally transparent'. This approach derives ultimately from Descartes' *Geometrie*, but as Manders has shown, while that work is surprisingly 'synthetic' in the traditional Euclidean sense, the path from it to modern Cartesian coordinate representations is by no means straightforward.

But, to revert to the theme of 'sameness', there is one way of establishing it that was central to traditional geometry, but which has dropped out of sight almost entirely in modern topology and even geometry. That involves *superimposing* one surface on another, in such a way as to bring figures on one surface into coincidence figures in the other. Now, to the extent that the operations presupposed in our basic theory, like drawing figures on surfaces, involve bringing drawing instruments into contact with the surfaces on which the drawings are made, they involve something akin to superposition, but so far we have not considered interconnections between *the marks* that may appear on the superimposed surfaces. It is now time that we examined that.

[23]Thus, R^2 is characterized this is the way in the quotation from Munkres, 1966, cited above.

Part III

Superposition

11

The Method of Superposition and Its Problems[1]

11.1 Historical Background

Euclid gave the method of superposition pride of place as *Common Notion 4*, "Things that coincide with one another are equal to one another," (Heath, 1956, p. 155), which is used in Book I, e.g. in the proof of Proposition 4, in which one triangle is 'applied' to another in order to prove the congruence of the triangles (*ibid*, p. 247). However, it has become customary to deride this use, as Russell did when he wrote:

> The fourth proposition is the first in which Euclid employs the method of superposition—a method which, since he will make any detour to avoid it, he evidently dislikes, and rightly, since it has no logical validity, and strikes every intelligent child as a juggle. (*The Principles of Mathematics, Second Edition*, p. 405.)

And Heath comments:

> In the note on *Common Notion 4* I have already mentioned that Euclid obviously used the method of superposition with reluctance, and I have given, after Veronese for the most part, the reason for holding that the method is not admissible as a *theoretical* means of proving inequality, although it may be of use as a *practical* test, and thus may furnish an empirical basis on which to found a postulate. (Heath, 1956, Vol. I, p. 249).

Indeed, the dubiousness of the method in deductive proofs is highlighted in a curious work on practical geometry, *Artis Cuislibet Con-*

[1]Much of this chapter and parts of succeeding chapters are drawn from Adams, 1994.

summatio by an anonymous 12th century author, that 'proves' that the circumference of a circle is 22/7 of its diameter, as follows:

> If the diameter be tripled and its seventh part be added to it, the circumference of the circle will be had. If a straight line could be curved and if it could be superposed to [the curve], it would neither exceed [it] nor be exceeded [by it].[2] (S. K. Victor's translation, 1979, p. 173)

Given the questionableness of superposition arguments in deductive geometry, modern writers on that subject have eschewed them. Hilbert, for instance, broke down the different aspects of Proposition 4 and stated them without proof as *axioms of congruence*, namely as Axioms IV,1–IV,6, pp. 12–15 in *The Foundations of Geometry*.[3] But that isn't quite the end of the story. First, as Heath suggests, superposition methods might be used as *practical tests*, which could furnish an empirical basis on which to found the postulates.[4] Indeed, whatever the theorists may say, meter sticks and the like are still superimposed on the objects they are used to measure, in order to determine their geometrical dimensions. Furthermore, superposition has not been entirely banished from geometrical theory. Thus, the fundamental "facts" of the approach to Geometry advanced in Helmholtz's celebrated paper "On the Facts Underlying Geometry" (Helmholtz, 1868) were relations of superposition, and in the hands of Sophus Lie and others this became the transformational approach to Geometry that Hilbert treated with the same seriousness as he did the more 'classical', Euclidean approach (cf. Appendix IV of the 8th edition of *The Foundations of Geometry*, 1971).

However, making relations of superposition the basis of the transformational approach to geometrical theory can also transform it into a non-empirical theory. In fact, the relation that Helmholtz took as fundamental to his theory holds between a physical body and the region of space occupied by the body at a given time. Thus, he wrote:

> The existence of mobile but rigid bodies, or point systems, is

[2]However, that *Artis Cuislibet Consummatio* was a work on practical geometry, which was a recognized if not highly valued specialty of its time, may only prove that this was a 'practical proof', which, as Heath suggests, it might well be.

[3]I have been unable to trace down the earliest treatises that dispense with superposition methods in geometrical theory. It is noteworthy that as late as 1848, the 15th edition of Legendre's celebrated text *Les Éléments de Géométrie* still proves essentially Proposition 4 using the method, saying "En effet, ces triangles peuvent être posés l'un sur l'autre de maniere qu'ils cöncident parfaitment" (Éléments de Géométrie, 15th ed., p. 10). Curiously, this text did not state the common notions explicitly.

[4]Though we will see in Chapter 12 that superposition is a much more limited *practical* means of establishing congruence than Heath seems to suggest.

> presupposed such as is needed for the comparison of spatial magnitudes of congruence (Helmholtz, 1868, Lowe translation, p. 43 in Cohen and Elkana, 1977.)

The spatial magnitudes referred to here are regions of an 'n-fold extended magnitude', and different regions in this space are compared by moving rigid bodies from one to another. But not only is one *relatum* of this relation, namely a region of space, not an empirical observable, but if its place is taken by a another physical body, then superposition becomes impossible because two bodies cannot be in the same place at the same time. Paul Hertz commented on this in the Hertz-Schlick Centenary edition of Helmholtz's epistemological writings:

> Yet three-dimensional rigid bodies in three-dimensional space are impenetrable. Thus, *that* stationary space is *only an object of thought... It is a different matter in the planimetric case—for one can let one lamina slide along another.* (ibid, p. 62, author's italics).

But, as the italicized part of the above quotation makes clear, this difficulty does not arise in the planimetric case.[5] Plane figures or 'lamina' are not impenetrable, and they can be superimposed on one another, as Euclid presupposed—in fact, in the author's opinion this is the underlying motivation for putting the planimetric books first in *The Elements*. Though superposition methods may be objectionable in deductive Geometry, nonetheless they are important to its practical application, and *The Elements* can be regarded as steering a course between theory and application.[6] *However*, problems arise even when attention is restricted to plane figures, not solely because such figures are two-dimensional and their existence may be questioned on metaphysical grounds, but also on certain 'logical' ones that occupied the ancients, and which are beginning to receive attention at the present time. These are commented on in the following section.

---

[5]Interestingly, Hilbert restricted himself to this case in his exposition of the transformational theory.

[6]This thesis is reinforced by noting the form in which typical geometrical propositions are stated in *The Elements*. Thus, 'problem propositions' are usually given as "To do such and such..", e.g., Proposition 9 of Book I, "To bisect a given rectilineal angle." The *demonstration* proceeds by giving a 'recipe' for bisecting the angle, followed by an argument showing that what is produced in accord with the recipe satisfies the requirement of bisecting the given angle. It is notable that in *Euclid and his Modern Rivals*, 1885, Charles Dodgson (Lewis Carroll) speaks regularly of elementary geometry texts, including Euclid's, as "manuals" (cf. p. xiv).

11.2 Logical Problems of Surface Superposition

From Ancient Greek times to the present philosophers of Geometry have had difficulty with superposition. If one body is superimposed on another, seemingly points on the first body's surface come into coincidence with points on the second body's surface, hence the bodies have points in common. And if so, in some sense they unite to form *one* body. The Hellenistic philosopher Sextus Empiricus found this paradoxical:

> ... if juxtaposed bodies touch one another, they contact one another with their boundaries—for instance with their surfaces. But the surfaces will not be unified as a whole with one another because of the touching, for otherwise the touching would be a fusion and separating the things touching would be a tearing apart, which we do not find to be the case (p. 179 of Mates', 1996, translation of *Outlines of Pyrrhonism*).

The debate concerning contact continued through medieval times, and was rekindled by Bolzano and Brentano just prior to the development of the modern topological account of boundaries, with its distinction between closed and open sets. Thus, Brentano objected to Bolzano's 'monstrous doctrine' that there "exist bodies without surfaces," which is required because contact is "possible only between a body with a surface and another without" (quoted from Zimmerman, 1996, pp. 156,7), which seems to anticipate modern views.[7] But even now certain concepts of 'ultramodern' differential topology can seem to reintroduce traditional problems. Thus, Firby and Gardiner describe the *connected sum* of two surfaces as being formed by 'gluing together' the surfaces' boundaries 'point by point', in such a way that when two points are glued together they coalesce into single points of another surface, which is the 'sum' of the original surfaces (*Surface Topology*, 1982, p. 25),[8] which sounds like

[7]Zimmerman's historical summary is extremely useful (e.g., on p. 149 it points out that Cantor was familiar with scholastic disputes about the infinite), especially in emphasizing the interconnections between the problem of physical contact and their relation to other problems of matter and space.

There are obvious similarities between the present considerations and considerations commented on in section 8.1, relating to the distinction between open and closed sets. However there is an important difference between these cases. Juxtaposing bodies involves *changing* their relation by bringing them into contact, but once they have come into existence, features of the surface of a single body are assumed either to touch or to be separate permanently.

[8]See also the operation of forming a *recollment des espaces*, discussed in section 5 of chapter 2 of Gramain, *Topologie des surfaces*, 1971. Of course terms like 'gluing' and 'fusing' do not figure in the formal theory of differential topology, and they only enter into informal descriptions of its applications to physical surfaces. But, as with the elementary topological theory with which the present work is concerned,

the 'fusion' of which Sextus speaks.

It seems to the author that these confusions, if they are such, are rooted in a tacitly assumed dualism of Space and Matter – often now fancified with mereological trimmings, but still a dualism. Intuitively, physical objects in contact have something physical in common, say physical points, and what they have in common must be material. And what can material objects have in common but matter? Something like this seems to be presupposed in one of Russell's animadversions against superposition:

> ... to speak of motions (e.g., of one triangle being superimposed on another) implies that our triangles are not spatial, but material. For a point in space *is* a position, and can no more change its position than a leopard can change its spots. (*Principles of Mathematics*, 2nd. ed., 1903, p. 405; author's parenthetical insertion).

Assuming that a tacit dualism underlies the foregoing arguments, readers familiar with the argument of Parts I and II of the present work may easily imagine our response to these arguments. The basic building blocks of the theory of the topologies of physical bodies' surfaces that was developed in Parts I and II were publicly observable, concrete *features*, typified by visible marks, that can be seen or otherwise 'sensed' on these surfaces. Moreover, Chapter 2 argued that these are concrete but not material, and therefore they have no place in the dualistic doctrines that are presupposed in paradoxical arguments like the ones outlined in this section. Part III will now put these building blocks to further use, among other things to resolve the paradoxes and build on them a hopefully improved positive theory of superposition and related notions.

11.3 Suggested Resolutions

The key idea of the theory to be outlined in what follows is formally quite simple, although its empirical foundation and mathematical details are less so. The idea is that when one body is superimposed on another, marks on the different bodies' surfaces come into temporary coincidence with one another and generate a 'composite surface topology', whose properties are essentially the same as those of the component bodies.[9]

---

it is difficult to describe in detail the application of differential topology to *non*-mathematical phenomena. Chapter 16 comments on *matter* and *substratum*, and their relation to surface features.

[9]'Essentially' should be stressed. Certain properties of the topologies of single bodies' surfaces are not 'preserved' in the topologies of the composite surfaces formed when they are superimposed on other bodies. The most notable of these is *orientability*, to be commented on in Chapter 14.

But the important thing is that when surfaces are superimposed their *features* become fused, and not their *matter*. *Ipso facto*, the *bodies* whose surfaces are superimposed are not fused, and they can be taken apart or separated (not *torn apart*, as Sextus says) later. Furthermore, when features that were temporarily fused are later separated they 'regain their identities'.

It is especially important to applied geometry that if features on separate surfaces exactly coincide when they were superimposed, they can be said to be *geometrically equal*, or *congruent*, at least during the period of their superposition. Beyond that, if the surfaces are not deformed in the process the features remain congruent thereafter. This provides a seemingly empirical basis for judging the equality of features, by seeing whether they coincide exactly when they are superimposed—as happens when a measuring rod is placed against a surface and certain graduations of the former are found to coincide with certain features of the latter.[10] With an empirical account of superposition in hand we can turn to more properly geometrical matters, which will be discussed at some length in the following chapters—although, unlike the matters discussed in Part II, what follows will be more informal.

11.4 Looking Ahead

As said, we are now in a position to consider properly geometrical topics, and the remaining chapters of Part III are concerned with the following ones. Chapter 12 considers both the empirical basis of 'composite surface topologies' formed when two or more surfaces are superimposed, either partially or wholly, and the most elementary properties of these topologies, including their dimensionality. Among the matters discussed are: (1) what can be *observed* when one surface is 'fitted' against another; (2) the empirical conditions that must be satisfied by the topologies of

[10] Reichenbach's comment, "If the measuring rod is laid down, its length is compared only to that part of a body, say a wall, which it covers at the moment," (*The Philosophy of Space and Time*, p. 16) hints at something like the above, though it manifests some of the confusions that present work aims to correct. The most obvious is the mereological reference to a *part* of a body, as though something could be compared to the part if it were not observable—say by means of some feature or figure that coincides with the part. if indeed the part is itself 2-dimensional and can coincide with a figure. The same difficulty is implicit in the phrase 'its length', as though *that* can be observed independent of things marking out the length. Of course Reichenbach, following Poincaré and anticipating Grunbaum, was concerned primarily with the 'non-deformation' requirement that must be satisfied if things that coincide exactly when they are superimposed are to be held to be equal after they are separated, which will be returned to in Chapter 14. However, even the seeming empiricalness of the 'superposition test' for *non*-separate equality will be reexamined in section 12.1.

individual surfaces, if they are to be 'compatible' in the sense they form a unified composite topology when the surfaces are superimposed; and (3) the properties of these composite topologies, including compactness, metrizability, dimensionality, and so on.

Side issues that will be commented on briefly include: (1) 'infinite fittings together', that make it possible to construct the topological plane; (2) impenetrability, i.e., the significance of the fact that only *two* surfaces can be superimposed face to face in our physical space; (3) nonorientability, which, seemingly, is a property that only composite surfaces can have, and (4) the 'empirical meaning' of *differential topology,* as it might apply to physical surfaces, which is much more difficult to apply than elementary topology is.

Chapters 13 and 14 are concerned with possible ways in which bodies can be fitted together or superimposed on other bodies. These are empirical matters, and investigating them is the starting point for more properly geometrical inquiry. That is because *rigid* bodies can be roughly characterized as ones that can fitted together with or *superimposed on other bodies in a minimum number of ways.* Chapter 14 is concerned with this minimality, since we want to allow that rigid bodies can 'move freely', but not so freely that they become deformed, grow, or shrink in the process. Although we will not definitively solve that problem, it is suggested that solving it would put us in a position to inquire into the empirical 'meanings' the concepts of *congruence*, *length*, and *distance*, e.g., between points on the surfaces of rigid bodies. These topics will be discussed in sections 14.2 and 14.3, where they will be approached along lines that were already commented on briefly in section 4.7 in connection with *standards*, e.g., standards of color and changes in color, especially at times when colored things are not being compared directly with standards.

In addition to congruence and distance, we will begin discussion of the ideas of *straightness* and *flatness*, both of which are very difficult, and in the case of the former the many competing views of its empirical meaning have been advocated, which will be commented on extremely inconclusively in section 14.4.[11]

Chapter 15 concludes Part III, with a discussion of probably the most difficult and fundamental concept of all, namely that of *space.* Everything through Chapter 14 is concerned exclusively with infinitely

[11] This might also serve as a bridge to comments on *light* and *lines of sight*, which would in turn connect the subject to 'modern' physical geodesy to be discussed in Chapter 16. However, this work will not enter into the special role that light and sight, in contrast to purely tactile exploration, can play in establishing the properties of physical spaces.

thin and perhaps infinitely extended surfaces, which are 2-dimensional, but at least since Kant the literature on the foundations of Geometry has taken 3-dimensional Space to be the fundamental object of inquiry. In our view this is a mistake, perhaps as radical as it is to assume that *Anschauungen* pertaining to natural numbers should be the starting point of inquiries into the concept of number. Some comments on this will be made in section 15.1, but our positive approach to space, which is Leibnizian, is outlined in the remaining sections of the chapter. The key idea is that of a *rigid frame*, characterized by a system of rigid bodies rigidly fitted together, and arbitrarily extensible. A *position* in such a frame can be described *constructively*, by reference possible extensions of the frame that would bring a feature of it into coincidence with the position. E.g., to verify that an object is located at what is described as "the point six feet above the center of a floor," we first locate the floor's center, then erect a perpendicular to the floor at that point, then locate a mark on the perpendicular six feet from the point at which it touches the floor, and finally verify that this mark is coincident with the object in question. The *space* of this frame can then be characterized as the set of locations in it; however the characterization of its *topology*, and especially that of its dimensionality, is difficult. Moreover, because this difficulty increases when it comes to incorporating different 'relative spaces' into single 'universal spaces' of the kind that physicists tend to theorize about, and locations in them, we shall eschew this discussion entirely.

The following chapters discuss the matters touched on above in greater detail. However, they are much more informal than the chapters in Part II were. Deeper inquiries into many of the topics to be discussed lead to mathematical difficulties that are beyond the author's powers to resolve.

12

Phenomena and Topology of Superposition

12.1 Introduction: Empirical Difficulties

As said in the previous chapter, we conceive the superposition of one surface on another as bringing certain markings on the first into coincidence with other markings on the second, somewhat as characters on a sheet of paper are printed from raised type. This is how superposition will be conceived theoretically in what follows, but we should be clear that what is conceived in this way is not entirely empirical. To see this, return for the moment to Hertz's comment on Helmholtz's theory, quoted in section 1.1:

Yet three-dimensional rigid bodies in three-dimensional space are impenetrable. Thus, that stationary space is only an object of thought... It is a different matter in the planimetric case—for one can let one lamina slide along another.

Figures 1a and 1b picture two laminae with congruent rectangles drawn on them,[1] first as directly superimposed in Figure 1a on the left, and then with the upper lamina slid to the right, in Figure 1b:

[1]Of course the laminae need not be thin; all that is required is that the lower surface of the upper lamina be superimposed on the upper surface of the lower one.

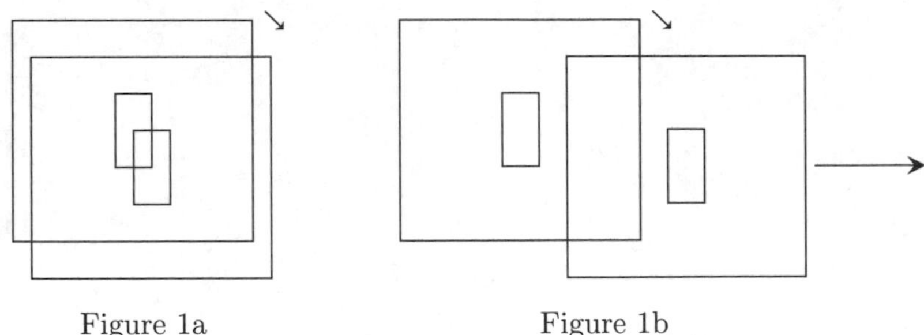

Figure 1a   Figure 1b

This picture might closely approximate what actually happens when a flat surface with a small rectangle on it is slid over another one, but the point is that it is not possible to *observe* the small rectangles when they are superimposed in this way, much less determine by observation that one is exactly superimposed on the other. That is most evident if the laminae are opaque, but even if they should be transparent a problem would arise that is hinted at by a most interesting entry in Berkeley's *Commonplace Book*, which is worth quoting at length:

> The Mathematicians should look to their axiom Quae congruent sunt aequalia (things that coincide are equal). I know not what they mean by bidding me to put one triangle on another. The under Triangle is no Triangle, nothing at all, it not being perceiv'd. I ask must sight be the judge of this Congruentia or not. if it must be then all Lines seen under the same angle are equal wch they will not acknowledge. Must the Touch be the judge? But we cannot touch or feel Lines & Surfaces, such as Triangles etc according to the Mathematicians themselves. Much less can we feel a line or Triangle that's cover'd by another Line or Triangle. (Comment 528 in Notebook A of the *Philosophical Commentaries*, Ayers, 1985, p. 301; author's parenthetical insertion).[2]

An essay would be required to comment on all of the themes hinted at here, but the key one for our purpose is that exactly superimposed triangles, or rectangles cannot be distinguished by sight or touch (hence according to *esse* is *percipi* they are not distinct things).

In view of the foregoing reflections, we shall not assume that exact

[2]Reflections of this sort may explain the absence of considerations of superposition not only in Berkeley's own later work, but in psychological theories of spatial perception from Mach to Marr, and the consequent difficulty that these theories have with 'exteriorness', 'space', and its dimensionality. Marr's theory will be commented on in section 17.7.

superposition 'face to face' can be established by immediate sensory observation. If the surfaces are of opaque substances then superimposing them makes the parts of them that are superimposed inaccessible to observation, and the quotation from the *Commonplace Book* brings out difficulties that arise even when they are not opaque.[3] Thus, as already said, we may not assume that superposition is a *practical* test of geometrical equality, as the proof of Proposition 4 of Book I of *The Elements* might suggest. That requires us to 'apply' one triangle, ABC, to another triangle, DEF, in such a way that A comes to coincide with D and the straight line AB falls on DE, as below:[4]

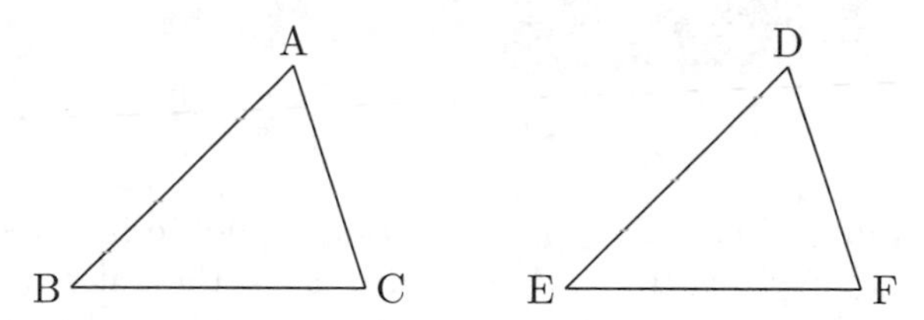

When the triangles are superimposed in the way that Euclid's demonstration requires, we are supposed to conclude that B must coincide with E because AB and DE are assumed to be equal, and moreover, given that AC and DF are supposed to be equal and the angle BAC is supposed to equal EDF, the third side, BC must coincide with EF (Heath, 1925, Vol. 1, pp. 247, 8). But it has just been argued that if ABC and DEF actually coincide in this way when one is applied to the other, it must be impossible to see all of either of them, since superimposing them face to face would make them inaccessible to observation. So what are we left with? Does it follow that superposition is entirely useless as a practical means of establishing equality? We suggest that that would be going too far.

Measuring rods and graduated rulers are superimposed on other objects as a practical means of establishing a kind of geometrical equality—congruence. Thus, when a graduated ruler is used to measure the distance between two features on a flat surface, certain of its graduations are directly, visibly juxtaposed to the features, the distance between which it is desired to determine. Moreover, it is possible that suitably

[3]As in footnote 11 of the previous chapter, we will put aside in this work consideration of purely optical methods that cannot be substantiated by tactile ones—that is, by methods that are accessible to blind persons. Thus, we will not consider the possibility of verifying that features that become 'tactually inaccessible' when they are on opposed faces of superimposed bodies, as in Figure 1a, might, if the bodies are transparent, be observed by optical means.

[4]This test will be returned to in section 13.2.

ingenious uses of this kind of 'edge juxtaposition test' might establish in a roundabout way what Euclid's argument suggests can be established by direct inspection.

To illustrate, the ruler might be placed on a horizontal slab, as illustrated below:

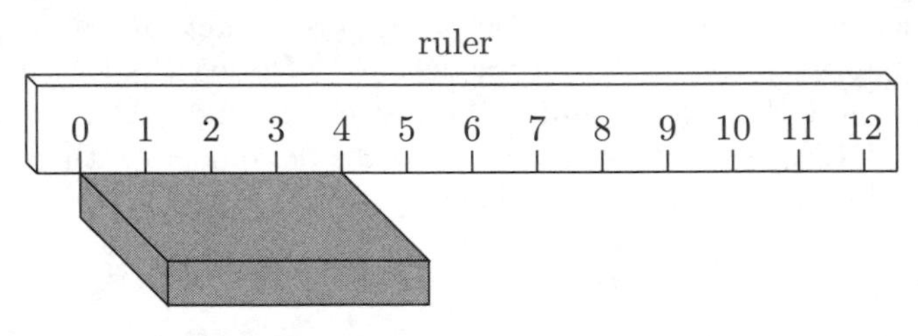

Figure 2

Clearly, nothing inaccessible to observation would be required to determine that graduations 0 and 4 on the edge of the ruler in the figure above were coincident with the upper rear corners of the slab. This is practical, and, it could show that the corners were 4 inches apart, as in the figure. Face-to-face superposition is not involved, and all that is required is to juxtapose the edges of the ruler and slab in such a way that, while it may obscure parts of the two bodies, it does not obscure parts that are observed in making the measurement.

It is noteworthy that Hilbert's reformulation of the proof of Proposition I.4 only presupposes direct congruence comparisons of line segments and angles (in fact that is all that Euclid's own proof requires), and Borsuk and Smielew in turn define angle congruence in terms of segment congruence.[5] Since I.4 and I.8 are the only Propositions in *The Elements* whose proofs involve superposition, all that is really required are line segment comparisons, which we have argued are practical. We also see the importance in testing or proving the propositions of plane geometry of *straightedge juxtaposition*, as against superposition in general. This also highlights the significance of a special feature that distinguishes straight lines from other curves: namely that they are the only ones that can be rotated in space without being displaced.[6] The possibility of arbitrary rotation in space is required if a straightedge is used as a planar measuring device—if it has to be set up at right angles to a plane surface and

[5]See the propositions of Group IV, pp. 12–15 of Hilbert's *Foundations of Geometry*, and p. 84 of Borsuk and Smielew, 1960.

[6]Heath (1925, pp. 168,9), comments at some length on this property as possibly definitive of straightness, and cites Heron of Alexandria (1st century BC) as the first proponent of it that he can find. Leibniz, Sacceri, and Gauss are cited in this connection.

made to fit 'snugly' to the surface when it is used to measure distances between features of it. But why then consider face-to-face superposition at all, as the proof of Euclid's Proposition I.4 seems to assume?

As said at the outset, whether or not they are accessible to direct sensory inspection, we will assume here that face-to-face superpositions are 'real facts' of the kind required by Helmholtz's theory. Moreover, we will assume that coincidences established by face-to-face superpositions have many of the properties that coincidence relations among features of separate, accessible surfaces have. But if the 'inaccessible coincidences' that come about as a result of face-to-face superpositions cannot be observed, what justifies the assumption that they satisfy the same principles that accessible coincidences have been assumed to satisfy—that the coincidence relation is symmetric and satisfies the Separation Postulate? Part of the answer to this question is simply the weight of tradition.

Clearly Euclid assumed that one figure could be 'applied' to another face-to-face, and even writers like Russell who objected to appealing to the result of such an operation in a deductive argument did not deny that it was physically possible to carry it out, or that its result could be much as Euclid described it. Even Berkeley, in arguing that exactly superimposed figures could not be distinguished by the senses, and therefore according to *esse* is *percipi* they must be 'metaphysically' one thing, did not deny that the two things could be brought into coincidence by superposition, whatever happens to them as a result. Coming closer to our time, combinatorial or algebraic topology deals with *polyhedral complexes*, which can be 'decomposed' into 'properly situated *simplexes*' such that two simplexes in the decomposition have at most two faces in common.[7] Thus, when two identical cubes are fitted together face-to-face, a 'solid' is formed that is a polyhedral complex in the topologist's sense, and this kind of fitting involves superposition. Similarly, *orientation* and *orientability*, which are now treated within the framework differential topology, can be considered as phenomena of superposition, since simplexes are assumed to be orientable,[8] but superimposing them may or may not generate complexes with coherent orientations.

But weight of tradition and current usage do not by themselves justify assuming that bodies and their surfaces can be superimposed face

[7]Cf. Pontryagin, 1952, especially pp. 12, 13. The point made in the Introduction, that the author has only an amateur's acquaintance with topological theory, must be even more strongly emphasized here. My only acquaintance with these subjects derives from my student days in the late 1940s.

[8]For example, on p. 24 of Firby and Gardiner, 1982, the authors say "Thus, the surface of every solid object we see about us is a bounded orientable surface. . . " This will be returned to in section 12.7.

to face, and when they are, features of the superimposed surfaces have topologies and geometries that conform to the same principles as those of the surfaces that are superimposed. That these things are taken for granted should itself be explicable by reference to data derived from sensory observation. There are two points to make in this connection.

The first is that intuitive realistic philosophy assumes that objects or figures retain their forms, topological or geometrical, when they are not being observed, and unobserved figures, such as ones on hidden surfaces, stand in the same relations of coincidence to other figures that they stand in when they are observed. For example, supposing that the small rectangles on the superimposed blocks in Figure 1b can be observed and measured when the blocks are slid apart, as they are in that figure, then, if the blocks are rigid, they would be assumed to retain their geometrical forms when the blocks are slid together, as pictured in Figure 1a. Similarly, if the blocks were made of rubber they would be assumed to retain their topological forms in these circumstances, and one of the rectangles would be assumed to remain in the same relation of contact or separation from another figure on the same surface when the blocks are slid together.[9]

The other point also involves continuity considerations. When the surfaces are superimposed, how do we know that figures on one surface even come in contact with figures on the other? If the figures are only partially obscured we can *see* that one is in contact with the other, as are A and B in the diagram below[10]:

[9]Exceptions must be made when surfaces of single flexible bodies are folded on themselves, and figures on them that were separate prior to folding become coincident afterwards. This is impossible when the bodies are rigid, and in what follows we will ignore the possibility even in non-rigid cases—though the reader may wish to consider this further.

[10]If it is objected that an edge is too thin to be seen, it may be replied that the laminae have some thickness, and the features in question, which we have been calling rectangles, may actually extend over the edges of the laminae, and when they are superimposed, as shown, we can see parts of marks on them that extend beyond their edges.

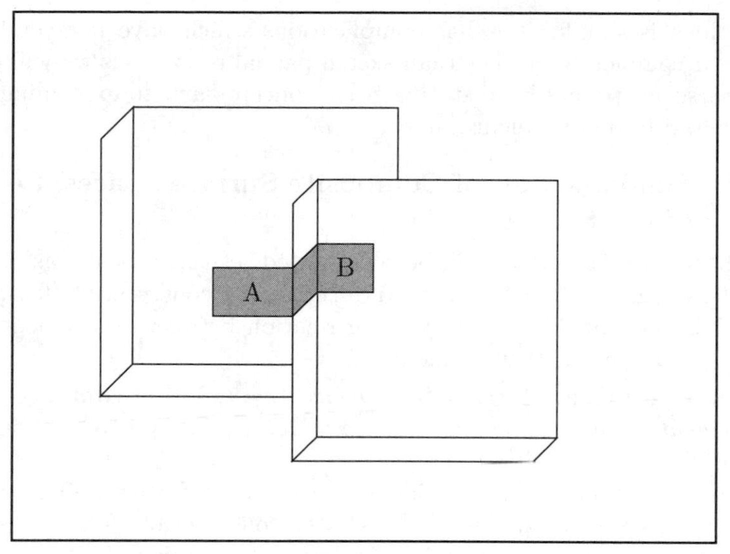

Figure 3

And, if the rectangles on the two surfaces are seen to be in contact at their edges, as above, why should they not remain in contact when the surfaces are slid together?

Obviously the congruence of whole figures cannot be established by direct observation, but section 12.3 will comment on observational data that might confirm the hypothesis that marks and other features on superimposed surfaces do stand in relations of contact and separation, and these satisfy at least some of the principles that coincidence relations between marks and other features of single surfaces do.

Returning to topology, as said in Chapter 11, the fundamental idea of the theory of composite surface topologies that will be outlined in this chapter is that the result of superimposing parts of the surfaces of two or more bodies, together with the marks and other features that are on them, is to create a *composite* surface topology. This is generated by coincidence relations among the features on the surfaces, which retain their relations of coincidence when they are on the same surface, but which add new relations of coincidence when features on one surface are brought in contact with the features on another. Assuming that the resulting system of coincidence relations satisfies the postulates already set forth in Part II, plus certain 'compatibility conditions' to be noted, the *composite topological space* generated by the augmented system of coincidences can be regarded as a 'union' of the spaces of the component surfaces, i.e., a union of its *component spaces*. However, while the basic idea is simple, it is has complications which have not yet been resolved completely. Rather than sketch partial resolutions we will confine ourselves primarily to stating basic concepts and to explaining the unresolved formal problems.[11]

12.2 Fundamentals of Composite Surface Spaces: Points of the Spaces

In what follows a surface will be represented set-theoretically as an ordered pair, $\Sigma =< \Im, \Re >$, where $\Im$ is the set its constructible features, and $\Re$ is the coincidence or touching relation between its elements. A *composite surface* is then defined:

12.2.1 Definition. A *composite surface* $\Sigma =< \Im, \Re >$ is *formed from component surfaces* $\Sigma^i =< \Im^i, \Re^i >, i = 1, \ldots, n$, if $\Im = \Im^1 \cup \ldots \cup \Im^n$ and $\Re^1 \cup \ldots \cup \Re^n \subseteq \Re$.[12]

Thus, we assume that the set of constructible features of the composite surface is the union of the sets of constructible features of its components,[13] and the coincidence relation $\Re$ contains all of the component coincidence relations, $\Re^i$.

Of course what makes superficial composition—superposition—interesting are the *intersurfacial* coincidences that arise when elements or features of different component surfaces are superimposed. In this case the union $\Re^1 \cup \ldots \cup \Re^n$ is a proper subset of $\Re$, and the composite surface is more than a simple 'union' of its components. How then is it related to them?

To begin with, it does not follow logically from the fact that component surfaces satisfy the postulates of Part II that a composite surface formed from them also satisfies these postulates. For instance, it is logically possible that the intrasurfacial coincidences that arise when one surface, $\Sigma^1 =< \Im^1, \Re^1 >$ is superimposed on a second surface, $\Sigma^2 =< \Im^2, \Re^2 >$ could include all pairs in the product set $\Im^1 \times \Im^2$, in which case the Separation Postulate would fail to hold. Formalizations of the theory of superposition would be expected to formulate postulates that imply that this sort of thing does not happen: i.e., that imply

[11] The topics discussed in this and immediately following sections are touched on briefly in Part 3 of Adams, 1986, and the conclusions reached are presupposed in Part 6 of Adams, 1994.

[12] Later we shall have to consider *countable* superpositions, such as could arise if arbitrarily many 'tiles' were juxtaposed to form an infinite plane. This will be postponed to Section 12.6.

[13] Note: If one solid block is laid across another, corners may be discerned at points where the blocks' surfaces cross, where nothing would have been discerned previously. As with the 'folding problem' mentioned in footnote 6, we will ignore this possibility in what follows.

that what holds for component surfaces holds for the composite surfaces formed from them, which would guarantee that their spaces have all the component space properties derived in Chapter 6. We will assume this here, and even inquire in section 12.3 into the empirical justification of the assumption that composite surfaces satisfy the Separation Postulate. But even this does not guarantee that their spaces are related to the spaces of the component surfaces that form them 'in the right way'. For instance, it does not yet follow that the points of composite surfaces can be identified with points on the surfaces of their components. To see this, consider the following.

Suppose that $\Sigma^1 =< \Im^1, \Re^1 >$ and $\Sigma^2 =< \Im^2, \Re^2 >$ are defined as follows. $\Im^1$ and $\Im^2$ are, respectively, the set of closed intervals of real numbers, $[\alpha, \beta]$, for $0 \leq \alpha < \beta \leq 1$, and the set of 'rectangles' in the unit square, which are cartesian products of these intervals. $\Re^1$ and $\Re^2$ are then taken to be the relations of non-empty intersection on the sets $\Im^1$ and $\Im^2$, respectively. These 'surfaces' are easily seen to satisfy the postulates of Part II, and the point sets generated by their PI-systems (Definitions 6.5.3(1) and 6.6.4) can be taken to be the elements of $\mathbf{I} = [0,1]$ and $\mathbf{I}^2 = [0,1] \times [0,1]$, respectively, i.e., they are equivalent to points in the unit interval, and to points in the unit square. But now let the composite surface $\Sigma =< \Im, \Re >$ be formed from Σ^1 and Σ^2, as follows.

$\Im$ is defined as above, i.e., $\Im = \Im^1 \cup \Im^2$ is the set of intervals $[\alpha, \beta]$ plus the set of rectangles $[\gamma, \delta] \times [\eta, \mu]$ in the 'universes', $\Im^1$ and $\Im^2$, of Σ^1 and Σ^2, respectively (note that the universes are disjoint). But if the relation $\Re$ is defined to hold *between* members $[\alpha, \beta]$ of $\Im^1$, and $[\gamma, \delta] \times [\eta, \mu]$ of $\Im^2$, if and only if $[\alpha, \beta] = [\gamma, \delta]$ (i.e., if the first is the 'first component' of the second), then clearly any element, $[\alpha, \beta]$, of Σ^1 corresponds to and is covered by the element, $[\alpha, \beta] \times [0,1]$, of Σ^2. Moreover, it is easily seen that the composite surface Σ satisfies at least the postulates of Chapter 6; that is, Σ satisfies the symmetry and separation postulates. But the class of *points* of Σ is not 'properly related' to those of Σ^1 and Σ^2, and in particular they are not properly related to those of Σ^1. This is seen as follows.

The I-systems of Σ^1 are *ipso facto* I-systems of Σ, but the PI-systems (the inseparable and indivisible I-systems) of Σ^1 are not PI-systems of Σ. In particular, the class of intervals $[0, \beta]$ is a PI-system of Σ^1; however, it is not one of Σ because this class is inseparable from both $[0, \beta] \times [0, \frac{1}{3}]$ and $[0, \beta] \times [\frac{2}{3}, 1]$, but these systems are separate from each other—i.e., the class is not indivisible in Σ, and therefore it does not individuate a point of it. It follows that the Σ^1-point individuated by the class of intervals $[0, \beta]$ does not correspond to a point of Σ, which it should if the points of Σ were properly related to those of Σ^1 and Σ^2.

The foregoing ought to provide some idea of what should be required if the points of composite surfaces are to be related properly to those of the component surfaces of which they are formed. It is not that the points of component surface should *be* points of the composite surfaces that are formed from them, but rather that:

12.2.2 Propositions. Let Σ be a composite surface formed from component surfaces $\Sigma^1, \ldots, \Sigma^n$.

(1) For $i = 1, \ldots, n$, a Σ^i-PI-system (a PI-system of Σ^i) that individuates a Σ^i-point (point of Σ^i) should also individuate a Σ-point (point of Σ); conversely, every Σ-point should be individuated by a Σ^i-PI-system for some $i = 1, \ldots, n$.

(2) Let $\Pi(\Sigma)$ and $\Pi^1(\Sigma), \ldots, \Pi^n(\Sigma)$ be, respectively, the sets of Σ-points and of Σ-points that are individuated by Σ^i-PI-systems for $i = 1, \ldots, n$, then $\Pi(\Sigma) = \Pi^1(\Sigma) \cup \ldots \cup \Pi^n(\Sigma)$.

We have seen that just satisfying the symmetry and Separation Postulates does not guarantee that condition 12.2.2 is satisfied, but we have not attempted to formulate postulates as elementary as symmetry and Separation, which would guarantee this. Doing this, and proving that these postulates would entail that condition 12.2.2 is satisfied, is part of what should be involved in developing a systematic *theory of superposition*, other aspects of which will be noted in succeeding sections and chapters. This will be begun in section 12.6 concerning the topologies of composite surfaces, but first we will comment on three less formal matters: (1) resolving the 'paradoxes of superposition' noted in section 12.1, (2) the rôle of superposition in a Euclidean argument, and (3) the empirical justification of the postulates of Chapter 6, in their application to composite surfaces.

12.3 The Paradoxes of Superposition

Consider first the argument of Sextus Empiricus that was quoted in section 11.2:

> ... if juxtaposed bodies touch one another, they contact one another with their boundaries—for instance with their surfaces. But the surfaces will not be unified as a whole with one another because of the touching, for otherwise the touching would be a fusion and separating the things touching would be a tearing apart, which we do not find to be the case.

According to the account of superposition sketched in the previous section, superimposing bodies involves bringing their surfaces into contact in such a way as to make marks on the surfaces that were separate prior to the superposition come into coincidence. But neither a surface nor a mark on it is 'unified' with another surface or mark, in the sense of becoming one 'thing'. Throughout the interval when the surfaces are in contact, they and the marks on them retain their identities—and separating them does not require 'tearing' them apart. Even a *point* on a surface remains distinct when it is superimposed on another point, but the PI-system that individuates it may come to individuate the other point too, as well as the point in the composite surface space formed as a result of superimposing them. But the composite space is ephemeral. It stays in existence only so long as the component surfaces remain superimposed—not only that, they must remain superimposed 'in the same way'.

It follows that we are not, as Bolzano apparently was, compelled to assume that when one body is superimposed on another, contact between them is "possible only between a body with a surface and another without." As just said, according to our theory, superimposing one body on another results in bringing their surfaces in contact, but it does not involve physical fusion, either of the bodies or of their features, which is what seems to have been repugnant to Bolzano.

As concerns Russell's remark, also cited in section 11.2,

> ... to speak of motions (e.g., of one triangle being superimposed on another) implies that our triangles are not spatial, but material. For a point in space *is* a position, and can no more change its position than a leopard can change its spots.

Clearly our theory does not deal with points in a single 'Space', which Russell seems to assume that all points are necessarily 'in'. Our theory does not even assume that a point on a particular surface changes its position *on that surface*. It only assumes that when it is superimposed on another surface, the PI-system that individuates it may come to individuate a point on the composite surface formed when the surfaces are superimposed. But the composite surface point comes into existence at that time, and it goes out of existence when the surfaces are separated.

Turning to positive matters, we will begin with some inconclusive comments on the empirical justification of superposition claims.

12.4 The Justification of Superposition Claims

Let us begin by recalling some points that were already made about superposition. We have assumed that what can be directly observed when one surface is superimposed on another are coincidences of marks or other features that are visible at the *edges* of the superimposed parts of these surfaces. This was illustrated in Figure 2, where the edge of the ruler comes in contact that corners of the block that it is being used to measure. The limitation to edge-comparisons seems partly to justify Berkeley's skepticism as to the empirical basis of coincidence judgments, as reflected in the comments cited earlier:

> The Mathematicians should look to their axiom Quae congruent sunt aequalia.... I ask must sight be the judge of this Congruentia or not. if it must be then all Lines seen under the same angle are equal wch they will not acknowledge. Must the Touch be the judge? ...Much less can we feel a line or Triangle that's cover'd by another Line or Triangle.

If what we said earlier was right, neither by sight nor touch can we directly sense a *whole* triangle superimposed on another—at best we may be able to see or touch their edges *if they are cut out*—and even then it is questionable whether we can see that there are *two* triangles. But let us examine this is more detail.

If we were to try to verify the Separation Postulate in its application to composite surfaces, we might be 'given' four marks, x, y, z, and w, arranged thus:

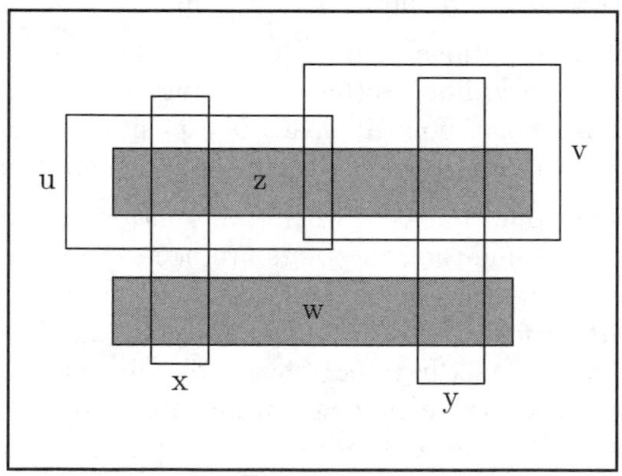

Figure 4a

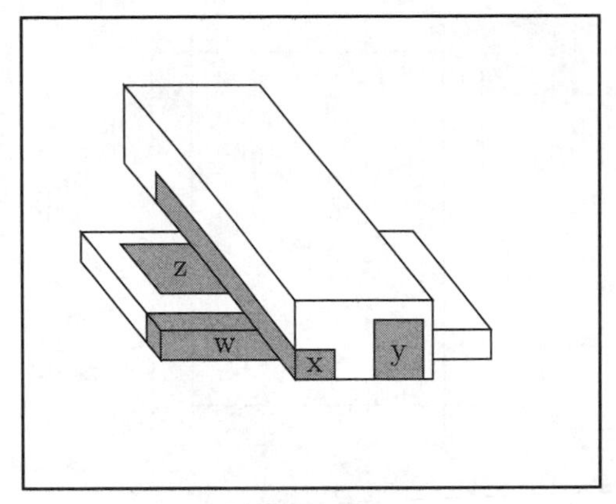

Figure 4b

Supposing that x and y are separate and z and w are separate, the Separation Postulate holds that it is possible to cover z with a U-system, say {u,v}, which is separate from w, no member of which touches both x and y—as represented by the outlined rectangles above. This was assumed to hold when x, y, z, and w are all on a single surface, as in Figure 4a, but now suppose that they lie on superimposed surfaces of different blocks, with x and y on the bottom surface of the upper block, and z and w on the upper surface of the lower block, as in Figure 4b. But now if we attempt to 'draw' rectangles u and v that cover z in such a way that (1) neither of them touches both x and y, and (2) both of them are separate from w, how are we to tell that they are separate from w when the upper block is superimposed on them?

Consider a simpler example. How might we determine whether the ace of clubs on a card that is placed face-down touches the ace of diamonds that is face-up on a card on which placed, as in Figure 5?

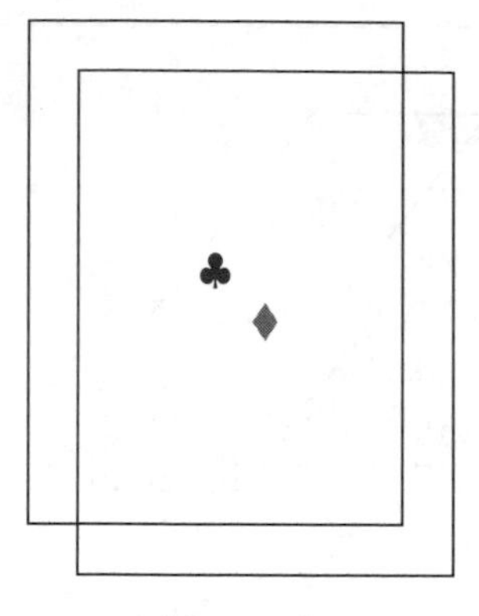

Figure 5

Lacking X-ray vision, but having previously ascertained that both are in the centers of their cards, we might infer that the club and the diamond must be separate if the club card is placed over the diamond card as they are depicted in Figure 5. But that would be an inference, and a geometrical one at that, since it would presuppose that the cards are not 'rubber sheets' on which the positions of the club and diamond might shift when the cards move. In fact, so far as the author can see, the relative positions of the diamond and club cannot be determined by direct observation when they are superimposed in this way, and for the same reason it seems that there is no direct observational way of verifying the Separation Postulate when it is applied to features like x, y, z, and w if they are on surfaces that are superimposed in the way illustrated in Figure 4b. Does this mean that the Separation Postulate is an untestable hypothesis in its application to composite surfaces? Actually there is some hope, in that it can be regarded as a plausible extrapolation from observable phenomena.

We may not be able to see enough of a feature to determine its relation of separation or contact with another feature when the features are on composite surfaces, but we can see that the exposed, accessible parts of these surfaces satisfy our postulates. Suppose again that x, y, z, and w on the surfaces of the blocks in Figure 4b. The figure suggests that the exposed parts of x, y, z, and w satisfy the Separation Postulate—in fact, even though the surface is composite, the ensemble of its exposed parts is phenomenologically indistinguishable from a single surface. And this in turn suggests that to the extent that the Separation Postulate can be tested observationally, namely in application exposed surfaces, simple or composite, it seems to be satisfied.[14]

The following sections generalize these considerations to the *topologies* of composite surfaces, beginning with the fundamental concept of an *open set*, and we will see that similar arguments apply to the topological postulates of Chapters 7–10 of Part II.

12.5 Composite Surface Topologies

We assume that basic open sets in the topology of a composite surface are defined in the same way as the basic topologies of its components:

12.5.1 Definition. The open sets of the topology of a composite surface, Σ, are arbitrary unions of Σ-point-sets that are complements of the point sets of Σ-U-systems.

Assuming that composite surfaces satisfy the other postulates and assumptions of Part II, e.g., the Finite Covering Postulate, 7.3.1, Countable Basis Hypothesis, 7.4.2, and so on, the following should be the case:

12.5.2 Propositions. Let Σ be a composite surface formed from component surfaces $\Sigma^1, \ldots, \Sigma^n$. Then:

(1) Σ is compact, Hausdorff, and metrizable.

(2) The sets of Σ-points that are individuated by Σ^i-PI-systems, for $i = 1, \ldots, n$, (PI-systems of Σ^i) are closed in the topology of Σ.

(3) If $\Pi(\Sigma)$ and $\Pi(\Sigma^1), \ldots, \Pi(\Sigma^n)$ are, respectively, the point-sets of Σ and the sets of points of $\Pi(\Sigma^i)$ that are individuated by PI-systems of $\Sigma^1, \ldots, \Sigma^n$ (cf. Proposition 12.2.3(2)), and if Π is a Σ-open set (i.e., open set in the topology of Σ) then $\Pi \cap \Pi(\Sigma^i)$ is an Σ^i-open set, and conversely, for $i = 1, \ldots, n$, the PI-systems that individuate points in every Σ^i-open set also individuate points in $\Pi \cap \Pi(\Sigma^i)$. The same is true of the closed sets of Σ and of $\Sigma^1, \ldots, \Sigma^n$.

We might also retrace the argument in Chapter 9, that component surfaces are 2-dimensional, to prove that composite surfaces are also 2-dimensional. However, given the foregoing propositions, there is an easier way to prove this. These imply that the topology of composite surface is essentially a finite union or 'sum' of the topologies of its component surfaces, and this, combined with the fact that the compo-

---

[14] Of course, less direct tests might be considered, such as involve the use of X-rays or other devices for sensing what lies under an opaque surface. But, as noted, here we are putting aside purely optical methods that cannot be verified by purely tactile means.

A very different kind of test could involve 'printing' figures on one surface onto another surface when the surfaces are superimposed. Thus if the club on the face-down card in Figure 5 were coated with ink in such a way as to leave a trace on the face-up card when it is superimposed on it, then the relative positions of the diamond and the newly printed club on the face-up card would show whether the diamond touched the club when the cards were superimposed.

nent surface topologies are essentially closed subspaces of the composite surface topology, implies that if the component topological spaces are 2-dimensional then the composite topological space must also be 2-dimensional (Hurewicz and Wallman, 1948, p. 30, Theorem III 2).

12.5.3 Propositions

(1) The topologies of composite surfaces are equivalent to sums of the topologies of the component surfaces forming them.

(2) The topologies of composite surfaces formed from finitely many 2-dimensional component surfaces are 2-dimensional.

These facts, which of course have not been established in detail, are not surprising, but we are more apt to be interested in the dimensionality of *infinite* composite surfaces like planes formed from countably many square 'tiles', to which the same dimensionality theorems apply. However, the topologies of infinite composites are less easy to characterize than those of finite ones, and this is one of two topics that will be commented on very unsystematically in the concluding sections of this chapter.

12.6 On Countable Composite Surfaces

The infinite plane, with which the planimetric books of Euclid's *Elements* are concerned, can be regarded as being composed of or 'tiled over' by a doubly infinitely extended array of congruent square tiles, a portion of which might be represented thus:

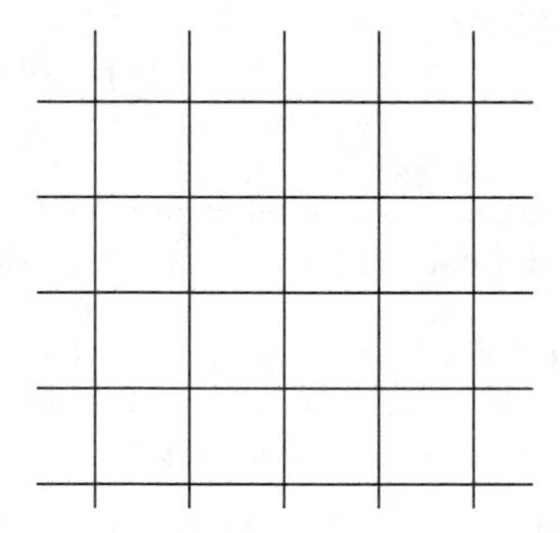

Constructed in this way, the plane can be regarded as being arrived at by successively adding tiles to infinity, until no further additions are possible. But, even restricted to the tops of the tiles, which are what form the plane, the topology of this composite surface cannot be characterized as simply as the topologies of its components can be. The reason is easy to see.

Letting $\Sigma^1 =< \Im^1, \Re^1 >$, $\Sigma^2 =< \Im^2, \Re^2 >, \ldots$ be an enumeration of the top surfaces of the tiles that form the plane, and letting $\Sigma =< \Im, \Re >$ be the surface of the entire plane, it is still the case that $\Im = \Im^1 \cup \Im^2 \cup \ldots$, and $\Re^1 \cup \Re^2 \cup \subseteq \Re$. Σ can still be assumed to satisfy the Separation Postulate, and, granted the Abstraction Principles of section 6.6, Σ-points can be assumed to be individuated by PI-systems of Σ. Moreover, if Π is the class of Σ-points, and for $i = 1, 2, \ldots, \Pi(\Sigma^i)$ is the class of Σ-points that are inseparable from elements of $\Im^i$, then $\Pi = \Pi(\Sigma^1) \bigcup \Pi(\Sigma^2) \cup \ldots$. But the open sets of Σ cannot be characterized as unions of classes of Σ-points that are exterior to Σ-U-systems.

Σ-U-systems are finite sets of features of the *component* surfaces, Σ^1, $\Sigma^2, \ldots$, each of which is itself finite in extent. Therefore each Σ-U-system is also finite in extent, hence its complement is not only infinite but occupies almost all of the infinite surface Σ (this was noted previously in section 7.6). It is true that unions of 'cofinite' complements of Σ-U-systems define a topology, but it is easy to see that it is not Hausdorff. Therefore it is not a 'natural topology' for the surface of the plane, and another way of characterizing that must be looked for. One such is outlined in section VIII of Adams, 1973, the intuitive idea of which is as follows.

The figure below is a hollow circle, which intuitively divides the space not on the rectangle itself into an inside and an outside:

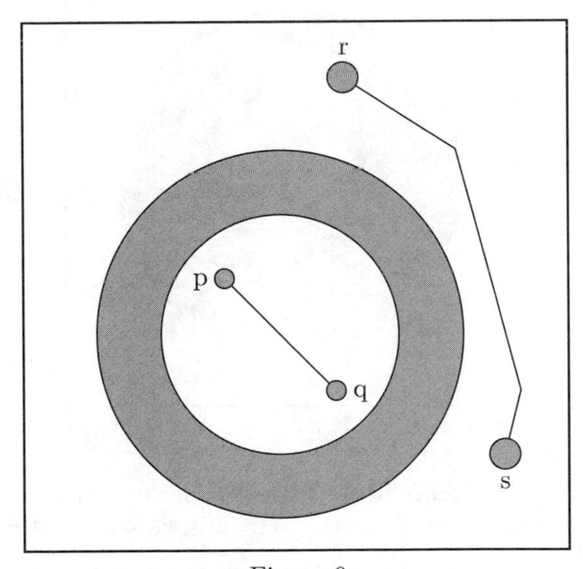

Figure 6

Of the four small circles p, q, r, and s, the first two are inside the large circle, and the last two are outside. What shows, not that the circles are inside and outside the large circle, but rather that p and q are on one 'side' of the large circle, and r and s are on the other of it, is that p and q are connected by a continuous line that never touches the large circle itself, and so are r and s, but it is not possible to connect, say q and r, by such a line without crossing the large circle. Now, the union of the inside and outside of the large circle is the totality of what is not on that circle itself, that is, of its topological 'exterior',[15] and if the whole plane on which the figure lies were infinite, the topological exterior would also be infinite.

But if that exterior is broken down into *components*, which may be called 'inside' and 'outside', the inside component is finite. In preceding chapters and previous sections of this chapter, we took the topological exterior, inside+outside, as a basic open set, but the approach sketched in Adams, 1973 took each component individually as an open set. However, to do this we are required to give mathematically precise characterizations of inside and outside, or, more generally, of the *connected components* of the topological exterior (Kelley, 1955, p. 54). This is essential in the figure below, which has two 'inside components':

Figure 7

The intuitive difference between what we have been calling the inside and outside components of the large circle is, as said, that we can draw

[15]For the moment we will continue to put single quotes around the word 'exterior', to emphasize the fact that it is the totality of all of the points not lying on the rectangle itself, and that includes the points *inside* it.

continuous paths that do not cross that circle between points or features of any one component, but we cannot do this between points or features that are in one component and ones that are in a different one (recall the 'boundary crossing feature' of our Basic Boundary Operationalization Theorem, 8.2.2, in section 8.2). However, there is now a problem: namely to characterize continuous paths, and this is difficult because the kind of continuity that we are concerned with is essentially topological connectedness, which is normally defined in terms of the more fundamental concept of an open set (cf. Kelley, 1955, p. 53). That is the way connectedness was defined in section 7.5 of Chapter 7, but we cannot do this now if open sets are defined as connected components of features like the large circles in Figures 6 and 7.

Our tentative solution to the new problem is to take the Connectivity Hypothesis, 7.5.1, as a fundamental postulate of our theory, but stating it now in terms of a new primitive concept, namely *connectivity*, and assuming this hypothesis, then to define the hitherto basic topological concept of an open set in terms of the now basic concept of connectivity. But this modification leaves a considerable amount of work to be done. It must be shown that if the Connectivity Hypothesis is postulated and open sets are defined to be connected components of U-systems, then the topology they generate is Hausdorff (but not necessarily compact!). It must be shown that the features that are *postulated* to be connected are also connected in the topology that is now generated from open sets that are themselves defined in terms of the new primitive concept. And so on. Most of this inquiry still remains to be carried out, and we can only point to what one would hope to be able to show.

The main point is that the topologies of the 'tiles' forming the infinitely extended plane should be closed subsets of the topology of the plane, which should itself be the topological *sum* of the topologies of the tiles. Given this, the plane's dimension should equal the maximum dimension of the tiles (The Sum Theorem for Dimension n, Hurewicz and Wallman, 1946, Theorem III 2, p. 30), hence if the tiles are 2-dimensional the entire plane should also be 2-dimensional. More generally, any countable composition of 2-dimensional surfaces can only be 2-dimensional, hence a dense but countable collection of unit squares 'stacked' one on top of the other, as in the diagram below

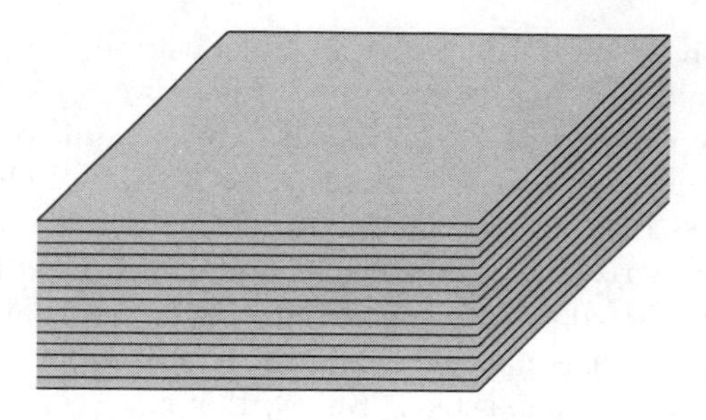

should be 2-dimensional. *Ergo*, as we cannot arrive at 1-dimensionality *via* countable collections of 0-dimensional points, and we cannot arrive at 2-dimensionality *via* countable collections of lines, so we cannot arrive are 3-dimensionality—solidity—*via* countable collections of planes or other surfaces.

Where does this leave us? In order to characterize solidity topologically, it seems that we must go beyond surfaces and superposition—beyond the superficial to the spatial. This will be returned to from one perspective in Chapter 14, and from another perspective in Chapter 15.

12.7 On Orientability

Although metaphysically inclined philosophers, beginning with Kant, have been concerned with orientability largely because of its alleged implications for the existence of a '*Space*' that transcends the empirical, which we shall return to later, we shall confine ourselves in this section to drawing attention to connections between orientability and superposition. This comes out most clearly in an intuitive characterization due to Firby and Gardiner, 1982, p. 24:

> If, by following some closed loop on a surface, we can change clockwise rotations ↻ into anticlockwise rotations ↺ , we say that the surface is *non-orientable*. Otherwise we say that the surface is *orientable*.[16]

Both intuitive orientability and intuitive non-orientability may be illustrated in the figure below, which represents a sheet of paper with two circles drawn on its left and right halves, each of which bears a clockwise pointing direction arrow:

---

[16]Exact definitions, e.g., in Gramain, 1971, p.82 or in Bröcker and Jänich, 1973, p.35, define orientability in terms of tangent vector fields on differentiable surfaces, but these characterizations lie far above the simple operational level. Therefore, whether and how these characterizations, or in fact *any* differential topological characterization, might be related to empirical phenomena remains to be determined.

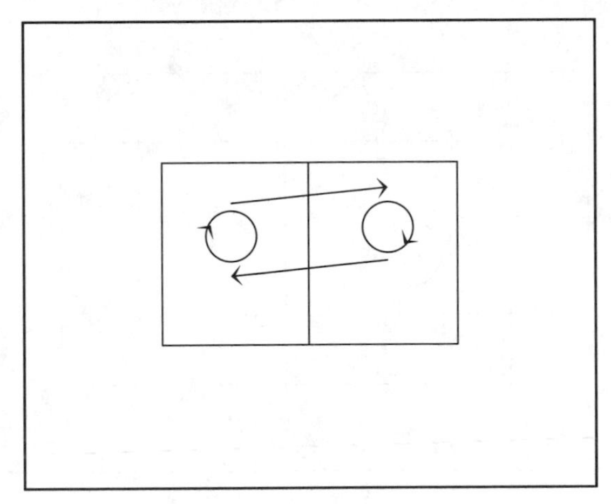

Figure 10

If the left-hand circle is slid along any path to the right, such as along the upper right-pointing arrow, until it comes into coincidence with the right hand circle, it does not change the clockwise direction of its arrow, and sliding it back along the lower left-pointing arrow to its original position also leaves the direction unaltered. This would follow a closed loop without changing a 'clockwise rotation into an anticlockwise' one, which suggests that the surface is orientable.

On the other hand, if the paper is folded onto itself along its midline, with the left half being folded up and over onto the right one, if the paper is translucent then when it is viewed with a strong light behind it, the circles should appear to be roughly superimposed, but one arrow would be clockwise and the other would be counterclockwise. This would suggest that the folded surface was non-orientable.[17]

The difference between the orientable, unfolded surface and the folded non-orientable one is significant for our purposes. The unfolded surface does not involve superposition and the folded surface does—albeit of a kind that we have excluded from consideration so far. We may generalize this. *Component surfaces, so long as they do not have superimposed parts like folds, are generally orientable, while composite surfaces are 'generally' non-orientable.* What we mean by 'generally' here is that composite surfaces involve face-to-face superpositions of 'simple' surfaces, such would be the case if two identical cubes were brought together, thus:

[17] The 'route' that the left-hand circle takes 'moving' onto the right-hand one and then back onto itself after the paper is folded can be pictured simply as shifting from the up-facing right half of the paper directly up onto the now down-facing part of the paper, which was originally its left half.

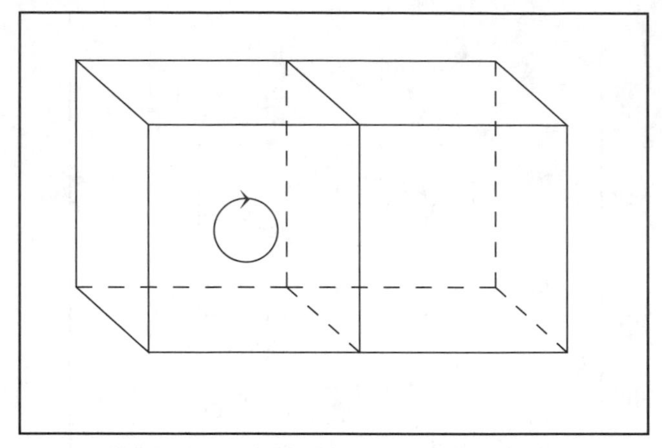

Figure 11

In this configuration a clockwise direction arrow on the front face of the left cube can be slid around and come back onto itself with the arrow pointing counterclockwise, if it is first slid around the right edge of that cube and onto its interface with the right-hand cube, then 'transferred' to the left face of the right-hand cube, then slid forward and onto the front face of *that* cube, and finally slid back across the division between the cubes onto the left-hand one. The composite surface formed from two cubical component surfaces is therefore non-orientable, thus illustrating our generalization about the relation between composition and orientability.

The foregoing improves in some respects on the claim on p. 24 of Firby and Gardiner, 1982, cited in Footnote 8: "Thus the surface of every solid object we see about us is a bounded orientable surface..." Combined with points made in section 12.5, this explains difficulties that often arise in understanding intuitive discussions of orientability and related phenomena. It is not strictly the case that every solid object that we see has an orientable surface. If all that is meant by 'solid' is that a thing is impenetrable, the folded sheet of paper in Figure 10 would be solid but non-orientable.[18] What is really required is that the thing's surface not have parts that are superimposed face-to-face.

[18] A perhaps more persuasive example can be imagined in which a solid torus is 'sliced' right around its circumference with a slice that rotates through 180° as it goes around. The resulting object would still not only be connected but be solid in the sense that it couldn't be deformed. Note that the faces that are superimposed in the slice would constitute a Möbius strip, which illustrates the well known fact that any non-orientable surface has a 'part' that is homeomorphic to such a strip.

The difficulties alluded to above, that arise in understanding many intuitive discussions of orientability, have largely to do with pictorial representations and descriptions of non-orientable surfaces like Klein Bottles and Möbius bands. Typical 'pictures' of Klein Bottles, e.g., Figure 2.4 of p. 67 of Arnold, 1962, show a gourdlike 'bottle' whose neck bends around and passes through the bottle's side into its interior. But the very fact that this represents a physical impossibility,[19] makes it difficult to imagine 'following a closed loop' on the surface in a way that changes a clockwise rotation into a counterclockwise one.

As to a Möbius band, on p. 259 of Courant and Robbins, 1951, the authors say that "A bug, crawling along this surface keeping always to the middle of the strip, will return to its original position upside down." But that ignores the fact that on a *physical* Möbius strip, created by twisting a strip of paper over by half a turn and gluing its ends (e.g., Figure 139, p. 260 of Courant and Robbins, *op. cit.*), the position returned to by the bug would actually be separated from the bug's original position by the thickness of the paper. In fact, granted that the paper has some thickness, the physical strip would be homeomorphic to a torus, which is an orientable solid.

A final example of the kind of confusion that is typical of ones that arise in giving intuitive meaning to ideas that are connected with orientability leads into the subject of the following chapter. The two congruent right triangles like the ones depicted below are said to have *opposite orientations*:

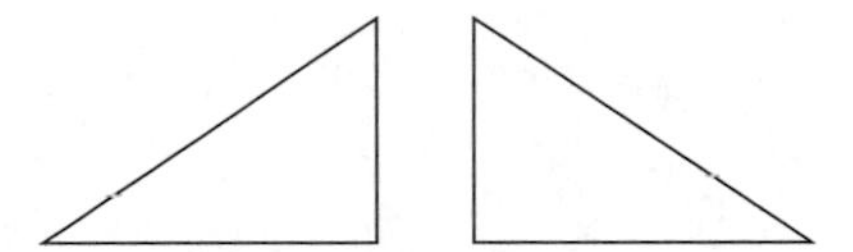

This is reflected in the fact that they cannot be displaced or slid around in the plane in such a way as to be brought into exact coincidence with one another. And, this is connected with the fact that the plane is 2-dimensional while 'Space' is 3-dimensional, because if the left-hand triangle is lifted off of the plane, turned over, and placed down onto it again, then it can be brought into exact coincidence with the right-hand triangle.[20]

Now, what is possible in three dimensions but not in two is often generalized to what may be possible in four dimensions but not in three,

[19] In fact, it illustrates the claim that "the surface of every solid object we see about us is . . . orientable."

[20] Another 'phenomenon of dimensionality' is the number of bodies that can share a common face. This will be commented on in Section 13.3.

which includes changes in orientation. But unlike three-dimensional phenomena, which are difficult but not impossible to represent in two-dimensional pictures, as above, we have no experience of the fourth dimension to aid us in understanding so called 'representations' of it, either in two or in three dimensions. But that does not prevent popularizers from trying to represent it, and Figure 1.6.4 on p. 20 of Firby and Gardiner, 1982, provides an example of the confusions that can result. This depicts a series of five successive stages by which a knotted string that cannot be unknotted in three-dimensional space might be unknotted if it is 'lifted' into the fourth dimension, which are seemingly akin to the stages by which triangles that cannot be exactly superimposed in the plane can be taken out of it, turned over, and superimposed. The problem is that the second stage of Firby and Gardiner's picture of the process of unknotting the string shows one segment of it being made to 'pass through' another without cutting it, seemingly by magic, which leaves the reader in the dark as to what is supposed to be taking place.[21]

The way in which these examples lead into the subject of the following chapter is that they all concern in one way or another how things can *possibly* be superimposed.

[21] Perhaps this only reflects the author's lack of imagination, which is equally baffled by many other attempts to 'explain' the fourth dimension. Many such attempts are brought together in an amusing early collection "The Fourth Dimension Simply Explained," by Henry Manning, 1910, in which the knotted string appears in various places. Here we also find the 'hypercube,' also known as 'tesseract', dear to the hearts of science fiction aficionados, equally confusingly represented, for instance, typically, in Figure 2 on p. 48. Also found are typical metaphysical dimensionality arguments, e.g., on p. 230 that the possibility of a particular chemical transformation would prove that Space was four-dimensional. In fairness, it should be added that this collection includes a little essay "Difficulties in Imagining the Fourth Dimension" by one "N," that says on p. 126: "...to picture in the imagination a world of four dimensions, requires us to add to the three dimensions already known, other parts of which we know nothing whatever."

13

Possible Superpositions: Dissections and Solid Topologies

13.1 Introduction

The previous chapter focussed on the relation between composite surfaces and their components, and here we will focus on the composite surfaces that *might* be constructed from given components, and their relations to one another. The questions that we will be concerned with may most readily be grasped using the example of counters that are placed in various ways on a plane surface. Five conceivable and one inconceivable placement of rectangular counters A, B, and C are depicted in the figures below:

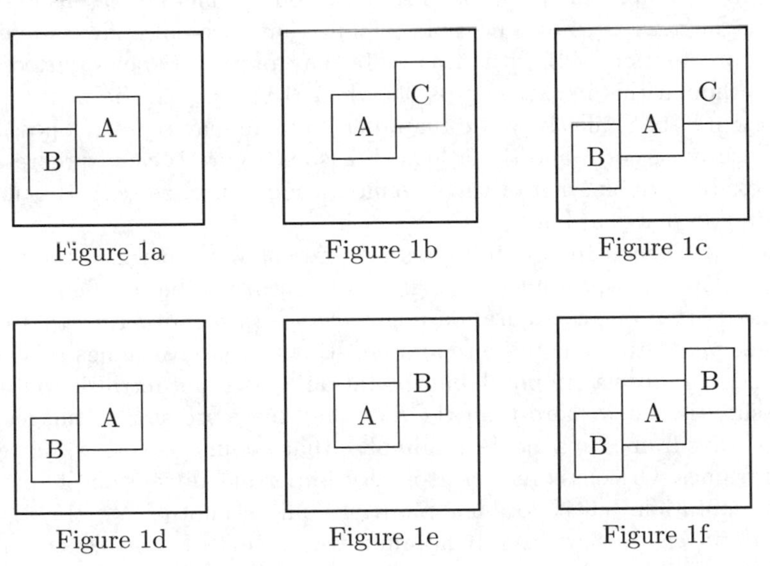

Figure 1a Figure 1b Figure 1c
Figure 1d Figure 1e Figure 1f

The 'placements' above define composite surfaces, Σ^{1a}–Σ^{1e}, the components of which are the surfaces Σ^{plane}, Σ^{A}, Σ^{B}, and Σ^{C}. Figure 1f does not represent a real possibility, since it seems to show rectangle B in two places at once. However, putting aside the fact that the plane and the counters have some thickness, which we will ignore for now, the composite surfaces are confined to the plane, and all of them involve superposition. Moreover, as Figures 1d and 1e illustrate, the same components, rectangles A and B, can be 'composed' or fitted together in different ways, and to allow for this we introduce a new topo-geometrical primitive, namely that of *superpositional possibility*, which might be expected to play a role in the theory of superposition akin to the one that features that might be constructed played in the theory of individual surface spaces that was developed in Part II.

One thing that must be stressed at the outset is that the idealization of our theory of surface feature constructions, which leaves time and change out of account, will carry over to our account of possible superpositions. It follows that in supposing that rectangles A, B, and C might be placed in the plane either as shown in Figure 1a or as shown in Figure 1b, we ignore the possibility of one placement being possible at one time but not at another.

On the other hand we do allow that possible placements might be dependent on circumstances, as illustrated in Figures 1a–1f. Figure 1a might represent a possible configuration of rectangles A and B and Figure 1b a possible configuration of A and C, but for factual reasons Figure 1c might not represent a possible combination of these configurations—the combination of B with A as in 1a may prevent C being 'fitted' to A, although this might be possible when B was not in the picture, as in Figure 1b. Similarly, while Figures 1d and 1e may represent possible configurations of A and B, for logical reasons Figure 1f cannot represent a possible combination of these configurations since, as said, B cannot be in two places at once.

That Figures 1d and 1e represent different ways in which the same things, rectangles A and B, might enter into possible configurations, suggests that one configuration might *change into* another, or that one might *move* from one place to another. But to conceive things this way would introduce a temporal 'interpretation' that is not intrinsic to these possibilities—any more than the fact that there are similar images in successive frames in a movie film implies that the images 'move' between the frames. Of course motion is a very important dimension of spatial representation, but it does not enter into our account of it.

However, even without temporality the theory of possible superpositions, while similar in some ways to that of possible surface feature constructions, may be expected to be more complicated than the latter simply because while surfaces are 2-dimensional, the bodies that we are concerned with are 3-dimensional and they can be fitted together 3-dimensionally. Here we will only make unsystematic comments on some special topics connected to superposability, beginning with a digression on superpositionality assumptions in *The Elements*.

13.2 Speculative Remarks on Superpositionality Assumptions in *The Elements*

Euclid's most obvious uses of superposability assumptions are in the notorious proofs of Propositions 4 and 8 of Book I of *The Elements*, and we will comment on Proposition 4:

> If two triangles have the two sides equal to two sides, respectively, and have the angles contained by the two equal straight lines equal, they will also have the basc equal to the base, the triangle will be equal to the triangle, and the remaining angles will be equal to the remaining angles respectively, namely those which the equal sides subtend. (Heath, Vol. I, p. 247).

More briefly, triangles with two sides and the included angle equal are congruent. The proof pertains to two triangles ABC and DEF pictured on p. 247 roughly as:

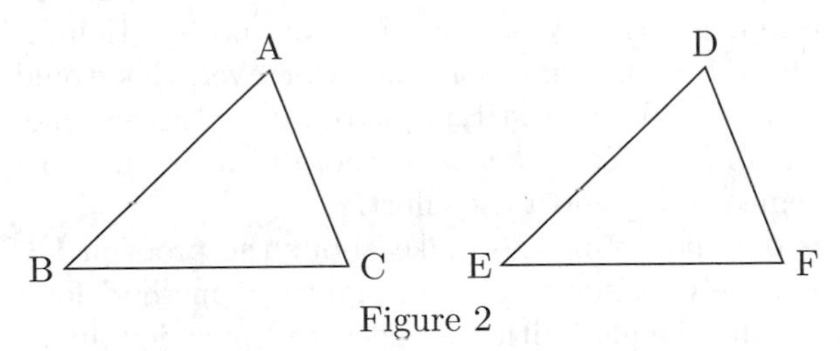

Figure 2

The steps in the proof that we will focus on are:

> . . . if the triangle ABC be applied to triangle DEF, and if the point A be placed on the point D and the straight line AB on DE, then the point B will also coincide with E, because AB is equal to DE.
> Again, AB coinciding with DE, the straight line AC will coincide with DF, because the angle BAC is equal to the angle EDF; hence the point C will also coincide with the point F because AC is again equal to DF.

> But B also coincided with E; hence the base BC will coincide with the base DF.[1]

The first sentence above presupposes it is possible to 'apply' one triangle to another, but a striking thing about the entire argument is that it never makes a direct appeal to sensory observation—for instance, in the way that the argument that $\pi = 22/7$ cited in section 11.1 from the 12th century manuscript *Artis Cuislibet Consummatio* does. Euclid's argument is no more literally about the figures that accompany it than a geology textbook is literally about the pictures of geological formations that appear in *it.* The figures in the proof and the pictures represent in the textbook *other* things, which are the things that the proof and the textbook are *about.* Therefore we cannot agree with might be thought to be implicit in Heath's suggestion (Vol. I, p. 249), that superposition "... may furnish an empirical basis on which to found a postulate," *viz* that Euclid's superposition argument makes a direct appeal to sense experience.

What Euclid really did was appeal to *principles of superposition* that are every bit as abstract and idealized as his postulates, although they are not explicitly stated as such. Thus, the proof of Proposition I.4 assumed that given two triangles, it is possible to superimpose them in such a way that a vertex and a side of one come into coincidence with a vertex and side of another. But no argument is given for this,[2] and if Euclid had attempted to give one he would have been faced with the task of developing a theory of superposition akin to Helmholtz's—and that might have been a formidable one. Moreover, this would only have been done for the sake of justifying two out of the enormous body of theorems set forth in *The Elements* (more than 600), and that might not have seemed to be worth the effort.

There is one more point to make about the proof of I.4. It can be interpreted naively as suggesting an empirical method for testing the proposition that triangles with two sides and included angles equal are congruent. That would simply be to 'cut out' the triangular figures ABC and DEF, place one on top of the other, and *observe* that when side AB 'fits' over side DE, angle BAC fits over angle EDF, and side AC fits over side DF, then side BC fits over side DF. This would fit in very well with the practical-constructive and technological approach to Geometry

[1] Heath's translation, pp. 247–8. This adds parenthetically the necessary step that the argument that BC coincides *as a whole* with DF depends on the assumption that two straight lines cannot enclose a space.

[2] This is evident in the fact that no definition, axiom, postulate, or previous theorem is cited, accompanying this step in the argument, although such justifications accompany the steps in the proofs of Euclidean propositions.

in *The Elements*, and the fact that he included such an argument hints at the practical origins of the science. This is related to some speculations on another hint relating to superposition that is found at the very beginning of *The Elements*.

Consider Definition 7 of Book I:

> A plane surface is a surface that lies evenly with the straight lines on itself. (Heath, 1956, 171).

Heath speculates that this was meant to suggest the same thing about planes as was previously suggested in Definition 4 about straight lines, namely "A straight line is a line that lies evenly with the points on itself." But contrary Heath's suggestion and to something else that he suggests, namely that the definition "... was intended simply express the same idea without any implied appeal to vision" (p. 171, below),[3] it seems plausible to the author to interpret the definition as, without explicitly referring to a drawing instrument, describing a plane surface as one to which such an instrument could be applied 'evenly' between any two points on it—where 'evenly' means that there are no 'gaps' between the edge and the surface to which it is applied, or superimposed. That would fit in with the idea that a straight line is *drawn*, and drawing it requires the use of an instrument, a straightedge, that fits 'evenly' against the surface on which it used to draw things. [4]

Are there any other hints concerning superposition and superposability to be found in *The Elements*? Curiously, no instrument for drawing circles is hinted at in the primitive propositions that pertain to circles, either in Postulate 3, which Heath translates as "To describe a circle with any centre and any distance," (p. 199), or in Definition 15:

> A circle is a plane figure contained by one line such that all straight lines falling upon it from one point among those

[3]Such as is explicit in a Platonic definition of straightness: "straight is whatever has its middle in front of (i.e., so placed to obstruct the view of) both of its ends" (from the *Parmenides*).

[4]Note that the use of the straightedge in drawing straight lines avoids the criticism of superposition methods that was set forth in section 12.5, that portions of surfaces that are superimposed face-to-face are inaccessible to observation. In this case all that needs to be seen are the portion of plane on which the line is drawn, and the edge of the straightedge used to draw it. It follows that if everything could be reduced to straight line constructions and comparisons, as Descartes could be interpreted as affirming in the first sentence of *La Géometrie*, "All the problems of geometry can easily be reduced to such terms that thereafter we need to know only the length of certain lines in order to construct them" (Olscamp translation, 1965, p. 177), then all objections to superposition methods and arguments could be met. This is partly vindicated by Hilbert's 'reduction' of the problem of determining the congruence of plane triangles, essentially to determining congruences of straight line segments.

lying within the figure are equal to one another (Heath, 1956, p. 183).

What the definition and postulate seem to suggest is that although the circle-drawing instrument, the compass, was known and widely used long before Euclid's time (cf. Knorr, 1986, p. 16), it was only described by reference to the properties of the figures, the circles, produced by its use—unlike the straightedge, which according to our speculation was described by the way it 'lies on the plane'.

One might have expected to find hints at other drawing instruments and superposition methods in the stereometric books of *The Elements*, namely Books XI, XII, and XIII, which treat of solids among other stereometric topics. That is because the theory of solids, which Euclid brings to a triumphant conclusion in Book XIII by constructing the five Platonic solids and then proving that they are the only regular ones, might have been expected to rest on comparisons of solids by superposition. However, no new stereometric constructive methods are introduced in Books XI–XIII.[5] In fact, superpositions of *solids* hardly enter at all, since the 'solid figures' of which these books treat are not really solid.[6]

Our final and somewhat tangential speculation is that it is because the 'stereometric' books treat of geometrical relations *in Space* that ideas of superposition and superposability play a correspondingly less promi-

[5]It might seem that the sphere is an exception, since it is described in Definition 14 (Heath, *op. cit.*, p. 261) as a (solid) figure that is generated by rotating a semicircle, essentially around the diameter between its ends. But this is not a practical method for producing a sphere, and, as Heath points out (*op. cit.* p. 269), the 'definition' is less a definition than a theoretically useful *description* of the sphere, which Aristotle and others had defined more 'naturally', if no more practically.

[6]'Stereometry', which has passed into English as 'solid geometry', is the wrong term for Books XI–XIII (Legendre's "Géométrie dans l'Espace," 1848, p. 290, would be preferable). In the first place many of these books' propositions treat of spatial relations among non-solids. E.g., Proposition I of Book XI (Heath, Vol. III, p. 272) simply states that a straight line cannot lie partly in one plane and partly in another plane that is 'more elevated'. In the second place there is nothing solid about the 'solid figures' which these books do treat of, since they allow that one solid figure can be 'comprehended' in another. Thus, Proposition 15 of Book XIII, (Heath, Vol. III, p. 478) describes how to comprehend a cube in a 'given sphere', which would itself be hard to comprehend if it were interpreted as purporting to tell the reader how to construct a *solid* cube inside a *solid* sphere. In the third place 'constructing' these solid figures requires even less than constructing the plane figures of which the earlier books treat. For example, the proof of Proposition 15 of Book XIII declares that the cube has been constructed when only its eight corners have been produced, which is less than what the proof of Proposition I of Book I (Heath, Vol. I, p. 241) 'delivers' in constructing an equilateral triangle, since the demonstration describes how to construct not only the triangle's vertices but also the lines that connect them.

nent role in them than they in the planimetric books. For instance, the author believes that it is not coincidental that Books XI–XIII do not include any propositions describing the construction of planes. To be sure, there are propositions implying their *existence*, e.g., Proposition 2 of Book XI, "If two straight lines cut one another, they are in a plane." (Heath, *op.cit.*, p. 274),[7] but nothing is said about how to construct it, and so far as the author is aware, 'plane-producing instruments' akin to line-producing straightedges have never figured in geometrical theory. Lines are drawn *on* planes, but on what are planes produced?[8]

Whether Euclid could have developed 'pure stereometry' without reference to Space will be considered later, but we may note that even pure stereometry could not be developed intrinsically, without reference to things other than bodies. Something related to this was commented on in section 11.1, where it was noted that Helmholtz's 'kinematic stereometry', in which superposition does play a central role,[9] was dualistic, because the 'things' that it requires to be superimposed on one another are bodies and the regions of space that they occupy. Now we see that any stereometry based on a superposition relation must involve non-corporeal elements, because bodies cannot be superimposed on one another in such a way as to fill the same space. And, we just saw that

[7]A modern formulation would be more explicitly existential, e.g., "If two straight lines cut one another, there exists a plane in which they both lie." Compare, e.g., Axiom I, p. 21 of Borsuk and Smielew, 1960, which affirms that for any three points there exists a plane that contains them.

[8]Significantly, the third-dimensional, spatial element of the stereometric books makes it necessary to apply propositions of Book I in a non-constructive way in the proofs of propositions in the later books. Thus, Proposition 11 of Book XI (Heath, *op. cit.*, p. 292) describes how to draw a perpendicular to a plane from point not in it, and its proof applies Proposition 12 of Book I to draw a straight line from the given point to point on a line in the plane. But this requires interpreting Postulate 1 as asserting that a straight line can be drawn between *any* two points, even ones not already supposed to lie in a plane.

[9]That Helmholtz had a kinematic theory is evident in his presupposing the existence of '*mobile* but rigid bodies' ("On the Facts Underlying Geometry," in Cohen and Elkana, 1977, p. 42; author's italics). This was in the tradition of Helmholtz's time, but it may be considered whether *motion* was essential to the theory, as against timeless possibilities.

Perhaps it requires no explanation as to why Helmholtz should have fallen back on an absolute Space to supply what *corpora* alone cannot supply in stereometric theory, since in his day absolutist traditions had been entrenched from time immemorial – even if anti-absolutist views were shortly to be taken seriously in science (*viz.* Poincaré's writings). It remains to be seen whether *our* anti-absolutist view, in which *corpora* are supplemented by non-corporeal surface features, can serve as a basis even of a classical stereometry. Of course, even if that should be the case it would be far from proving that it would serve the needs of modern science, in which Space, or spaces, play an increasingly active role.

this was borne out in Books XI–XIII, where Euclid also followed the body+space approach. Our approach, of course, has been to supplement *corpora* with the ontologically distinct category of incorporeal surface features, and in section 4 of this chapter we will comment on whether *this* might be sufficient for pure 'topological stereometry'. But first we turn to a purely topological law of superposability that is important.

13.3 A Special Law of Superposability

A fact that is peculiar to 2-dimensionality is that at most two figures can have a common boundary.[10] For instance, while four squares in a plane can meet in a *common point*, only two of them can have a *common boundary* in the plane, as below:

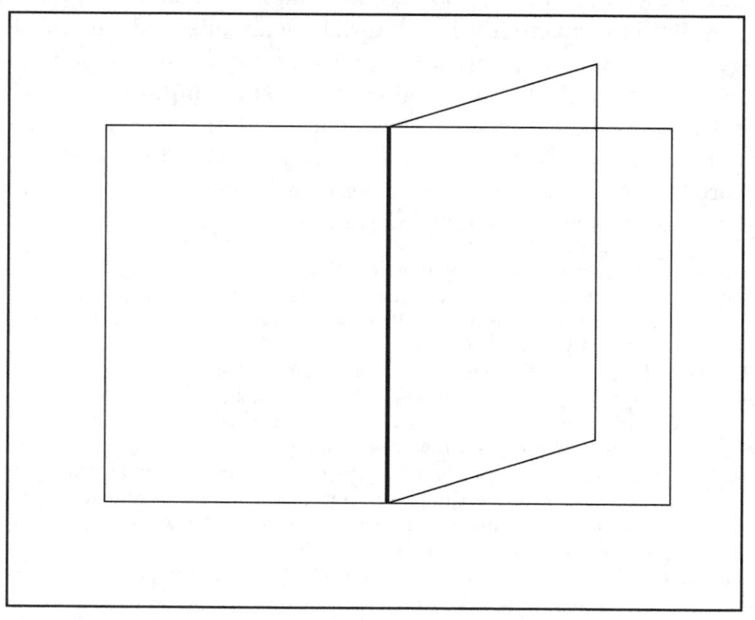

Figure 3

On the other hand if they are embedded in a third dimension then 'leaves' can be added, which have a common boundaries with two of them, as shown. Now, while we have not as yet considered 'Space' in a systematic way, much less argued that it is 3-dimensional, there is an analogue to the common-boundary law that applies to surfaces, and how many of them can be superimposed 'face-to-face'. Specifically, it is a fact

[10]This is closely related to the Lebesgue 'tiling' criterion of dimension referred to section 9.3.

of 3-dimensionality that while any number of solid bodies can meet in a common point in a 3-dimensional space, only two of them can have a common face in the space – although if they were embedded in a higher dimensional space, then more solid 'leaves' could have a single common face in that space.[11] Thus, assuming that the bodies whose surfaces we have been concerned with are in a 3-dimensional space, a very general restriction on the ways in which their surfaces can be superimposed is placed on them by the dimensionality of that space. But this is not quite as straightforward as it might seem.

If the bodies we were concerned with had common *parts* then any number of them could share a face, in the same way that more than two figures in the plane could have a common boundary if they had common parts, as pictured below:[12]

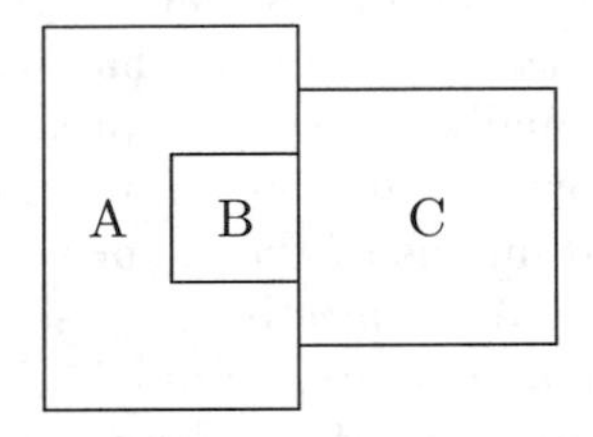

Therefore, in saying that three bodies cannot have a common face we presuppose that they have no common parts. But this raises the question of what it is for two bodies not to have parts in common, even if they have faces in common, which could seem to be related to logical problems of superposition that were raised in section 11.2.

If bodies have faces in common don't they have parts in common? Well, the term 'part' is vague, and some mereologists might even accept this,[13] but the resolution of these problems that was sketched in section 11.3 only showed how one surface could be superimposed on another without being fused with the other and 'losing its identity' in the process. This has nothing to do with *parts*, and therefore it does not resolve the problem of distinguishing between having parts in common and having faces in common. Hence we still have not clarified the 'law' that only two bodies without common parts can be superimposed face-to-face. And the reason is clear. So far the bodies we are concerned with could be conceived to be like the cube described in Proposition 13

[11]This is implicit in the idea of a *complex* and its *simplicial decomposition* in Euclidean 3-space (cf. Pontryagin, 1952, pp. 10–13).

[12]Here simplexes A and B are not *properly situated* (Pontryagin, op. cit., p. 12).

[13]Stroll, 1988, sometimes holds that bodies' surfaces are parts of them; viz., "... a surface is part of a ball bearing" p. 185.

of Book XIII of *The Elements* (Heath, op. cit., Vol. III, p. 478), which is a 'solid in name only' because it is fully 'constructed' when only its eight corners are constructed. Of course, we have not followed Euclid, but the topologies that we have described so far have only been topologies of bodies' surfaces, and these do not provide us with a means of characterizing their interiors.[14] Now we will sketch a way of doing that.

13.4 Decompositions and Their Spaces

Look again at the composite formed of the rectangular counters A and B in Figure 1a. The 'union' of counters A and B could itself be conceived to be a 'basic body', and A and B could be conceived to be pieces into which this body was 'dissected' or 'decomposed'. In that case, if pieces A and B were 'recomposed' by being fitted back together in the way they were originally, the composite surface topology that would result from this would be the same as the one that would have been formed if A and B had not originally been pieces of the union. Moreover, this topology would have points that were in a sense 'interior', not to the individual pieces, but to unions of them, such as the points interior to the region of the composite surface on the face that is common to A and B. Very roughly, the points interior to a body might be conceived of as points that could be 'uncovered' on the surfaces of pieces into which the body might be cut. This would have the advantage of explaining the inaccessibility of these points to direct superficial inspection, since they would be inaccessible in the sense of section 12.1. But a lot would be still have to be explained.

Bodies, as against their surfaces, are intuitively 3-dimensional, but how do we know that the 'spaces' of points that could be uncovered by dissecting them is 3-dimensional? We don't have a clear idea of this space, much less of the topology in terms of which its dimension might be defined. In fact, if only countably many cuts could be made in dissecting a body then the composite space formed by reconstituting it would be a countable sum of surface spaces, and we saw in section 9.2 that the dimension of this sum would be the maximum of the dimensions of the surfaces composing it—which by Theorem 9.3.8 would be plausibly 2-dimensional. What follows will briefly sketch a hypothetical approach

[14]The theory of simplicial decomposition distinguishes having parts and having faces in common by reference to the difference between boundary and interior points of convex bodies, which have parts in common if they have interior points in common. But convexity is a metrical, non-topological concept, which falls outside our purview. Moreover, so far our theory deals only with the surfaces of bodies, and we have not yet even assumed that they have interior points—though the decomposition-space theory sketched in the following section will assume that they have interior *parts*.

to these problems, that doesn't assume that bodies can be dissected or broken in pieces in uncountably many ways, or that it is immersed in a 3-dimensional space. This recapitulates ideas outlined in section 4 of Adams, 1988.

Picture a 2-dimensional rectangular 'body' with cracks or fissures running through it in an irregular pattern, looking something like a jigsaw puzzle whose pieces have been fitted together:

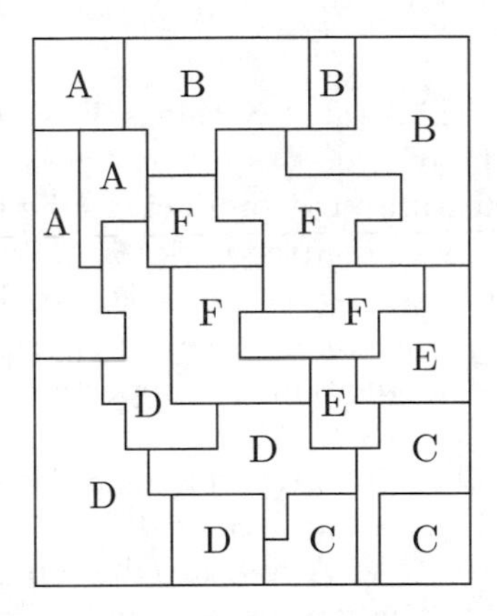

Now imagine 'disassembling' the puzzle by breaking off pieces bit by bit, beginning with the group **A** of pieces labelled 'A' in the upper left-hand corner, then breaking off the group **B** with pieces labelled 'B' in the upper right, and so on until there are 6 groups **A**–**F**, each of which is of an irregular shape like **A** and **B**:

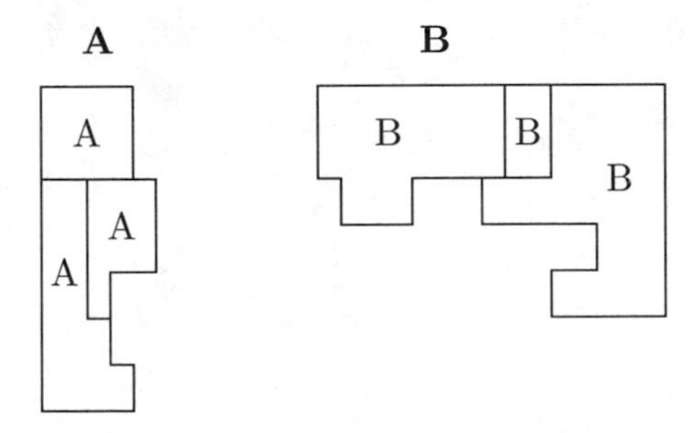

The breaking down process' may then be carried further, for instance breaking **A** and **B** into their component pieces:

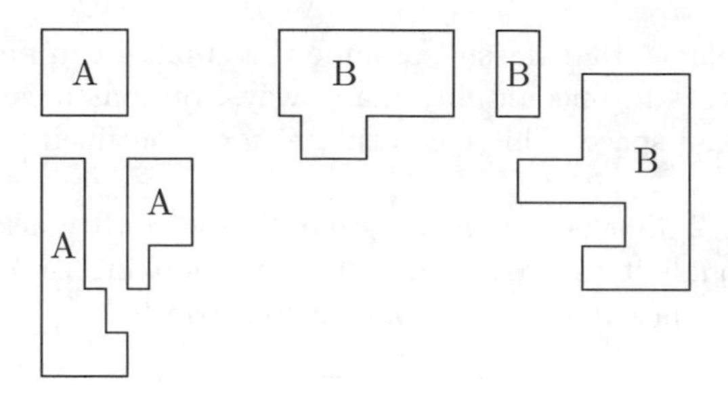

The examples above illustrate two things. First, breaking down, though destructive, can still be regarded as a 'constructive process', analogous to producing surface features, but producing solid pieces with their own surfaces, which, when they are fitted back together, form their own composite surfaces. Second, the pieces formed at each stage are pieces *of* something else, perhaps of larger pieces, perhaps of the original rectangle. And, the 'piece of' relation has important structural properties (Figure 4).

The three pieces labelled 'A' are pieces of 'group' **A**, which before it was broken up itself formed a piece of the whole rectangle. Calling A an *immediate piece* of **A**, and **A** an immediate piece of the whole rectangle, we can call A a *mediate* piece, or simply a *piece* of the rectangle. Given this, it is clear that the graph of the piece-of relation forms a *tree* (Ore, 1963, Chapter 3) which can be graphed as below (here the A's are subscripted in order to be able to refer to them individually):

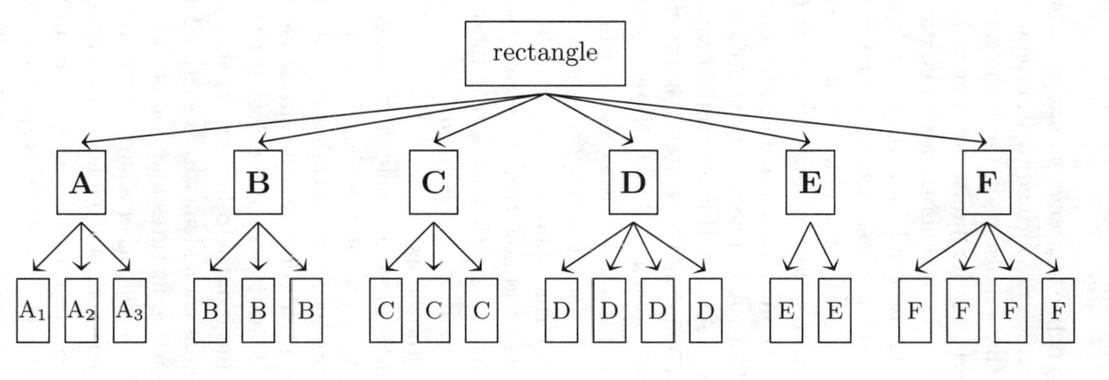

Figure 4—Graph of the ‘piece’ relation

It is tempting to define '*material points*' or '*atoms*' in a dissection as the smallest pieces which are formed by it, or equivalently, as the lowest points in the graph that represents it. Or, if the graph is infinite, material points can be taken to be the 'limits' approached as dissection 'goes on to infinity', which can be associated with maximal ideals or 'chains of maximum length' that descend from the graph's apex (the *root* of the tree, Ore, op. cit., p. 34). For instance, here the left-most 'material point' is A_1, which corresponds to the chain: rectangle $\rightarrow\mathbf{A}\rightarrow A_1$. And because of the tree structure a topology can be defined on these material points by taking a neighborhood of such a point to be any member of the chain that represents it. Thus, **A** is a neighborhood of the point corresponding to the chain: rectangle $\mathbf{A}\rightarrow A_1$. But the trouble with this topology is that it is necessarily 0-dimensional, because it is totally disconnected—i.e., all point-sets are both open and closed (Hurewicz and Wallman, op. cit., p. 10).[15] This can be regarded at least as strong evidence that a body's topology is not entirely determined by its material composition. Beyond that, how the body is formed and how its pieces fit together—which ones touch which other ones and how their surfaces are superimposed—must be taken into account. Doing this leads to a more complicated 'space', that we will call a *formation-superposition space*, or fs-space, which can be informally characterized as follows.

The *elements* of the fs-space yielded by dissecting a body, perhaps to infinity, may be taken to be the class including the body and all of the pieces that are produced in dissecting it (call these elements *solids*), together with all of their constructible surface features. Two such elements can then be defined to be coincident if either: (1) they are both solids and one is a piece of the other, or (2) they are both solids and a surface feature of one is coincident with a surface feature of the other, or (3) they are both surface features and one is coincident with the other, or (4) one is a solid and the other a surface feature of it. In effect, so far as coincidence is concerned, surface features are treated as though they were themselves pieces of the solids on whose surfaces they lie. Nevertheless, in spite of this the theory of fs-spaces is ontologically dualistic, among other things because surface features are arguably immaterial and 2-dimensional while solids are material and 3-dimensional. But of course it must be shown that when coincidences are defined in the way we have just done, they generate spaces somewhat in the way that surface features alone generate surface spaces. To do this the Separation Postulate must be generalized and an assumption must be added which,

[15]This assumes that a thing can be divided into only finitely many immediate pieces.

unfortunately, implies the infinite divisibility of the solids to which the theory applies. This is one of the respects in which the theory of solids is more 'hypothetical' than the theory of their surfaces.

The new assumption and the generalization of the Separation Postulate are designed to deal with situations like the one depicted below:

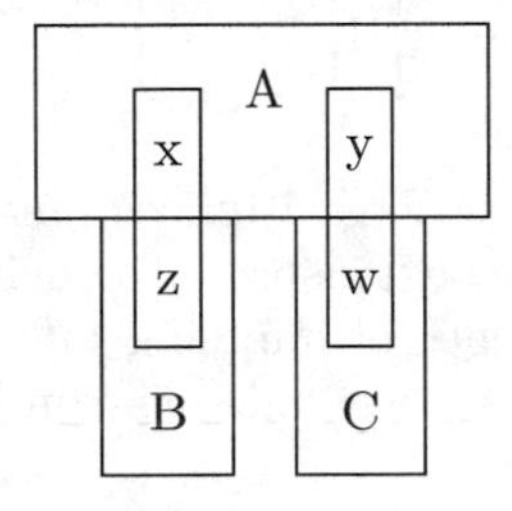

Here A, B, and C are all supposed to be solids, B and C are separate, and A is coincident with both of them in the sense that it has a surface feature, x, that is coincident with a surface feature, z, of B, and another surface feature, y, that is coincident with a surface feature, w, of C. The Separation Postulate therefore demands that it should be possible to cover A with a finite number of 'elements', which can now be either solids or surface features, no one of which is coincident with both B and C. Now, the set of features {x,y} does not have any members that touch both B and C, but we do not want to say that it 'covers' A – that would imply that the solid, A, was coextensive with its surface.[16] But if A is 'more than its surface' then any system of solids and surface features that covered A would also have to contain a solid or solids different from A.

Perhaps the simplest way to deal with the problem described above is first to generalize the Separation Postulate to say that if a solid, A, is coincident with two separate solids, say B and C, then A can be broken into a finite number of pieces, say A_1 and A_2, that jointly cover it, no one of which is coincident with both B and C, and second, to postulate that A has a piece, say D, that is *internal* to it in the sense that is separate from its external surface, as shown:

[16]But we should be careful here, since the doctrine of Plato's *Timaeus* has it that matter is composed of atoms that are no more than space-enclosing shells (cf. Chapter 3 of Vlastos, 1973).

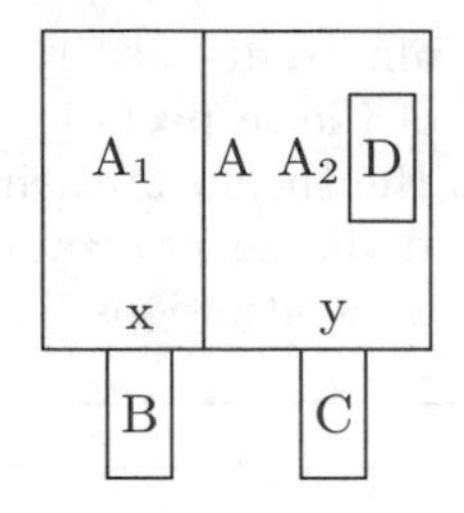

Of course, the postulation of the existence of internal pieces, which guarantees that solids are not coextensive with their surfaces, has, as noted, the questionable consequence of implying infinite dissectabiliy. But we can only acknowledge this here, and pass on to its less objectionable consequences.

Granted, then, that the dualistic system of bodies and their surface features satisfies the Separation Postulate, and assuming the hypothesis formulated in 6.6.6, it follows that it has the other properties set forth in Chapter 6, and in particular *fs-points* can be characterized as having the property that any two fs-elements are coincident if and only if they have an fs-point in common, and a finite set of them (an fs-U-system) covers another such system if and only if all fs-points of the latter are fs-points of the former (Theorems 6.5.4). Moreover, it is plausible that the postulates of Chapters 7 and 8 also hold for fs-spaces, hence they have topologies generated by open sets that are complements of the fs-point sets of finite sets of fs-U-systems. And jumping to conclusions, we may hazard that the dimensionality criteria cited earlier apply to them in ways that are analogous to those in which they applied to surfaces in Chapter 9. Specifically, following and slightly changing the notation of Theorem 9.2.2, a body (more exactly, its fs-topology) is at least 3-dimensional if its fs-space contains pairs of separate elements or fs-U-systems (A_i,B_i) i = 1,2,3, that cannot be 'separated' by 'separate separators', C_i, i = 1,2,3, roughly with the property that C_1, C_2, and C_3 do not meet in a common point, and for each i = 1,2,3, C_i 'divides' the fs-space outside of itself into separated subspaces in such a way that A_i lies on one side of the division and B_i lies on the other.

How do we know that a solid's fs-space satisfies the above condition? The following is a plausible argument that is does. Imagine a cube formed from 27 subcubes in 3 square layers, each composed of 9 smaller cubes:

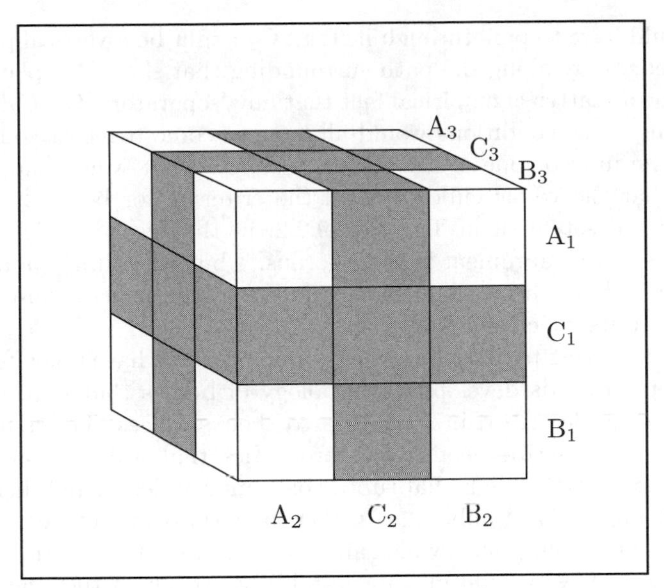

Figure 5

Now let the 9 cubes in the top layer the fs-U-system form A_1 and those in the bottom layer form the fs-U-system B_1. Similarly, let the 9 cubes in the left front-to-back vertical 'slice' form A_2 and those composing the right front-to-back slicc form B_2, while those in the front crosswise vertical slice form A_3 and those in the back slice form B_3. Given this, layers A_1 and B_1 are separate, the front-to-back slices A_2 and B_2 are separate, and the crosswise slices A_3 and B_3 are separate, but cach layer intersects each slice, and the crosswise slices intersect the front-to-back ones.

Finally, anything, C_1, shown shaded in, that 'separated' the top and bottom layers would have to extend right across the middle layer in such a way that a 'path' from the top to the bottom layer would pass through it. E.g., C_1 might be a horizontal plane with edges on the line surrounding the middle layer. Similarly, any thing, C_2, also shaded in, that separated the left and right front-to-back vertical slices would have to extend all the way through the middle front-to-back slice in such a way that a path from the left hand slice to the right hand one would have to pass through it. E.g., C_2 might be a vertical plane lay whose edges lay along the path surrounding the middle front-to-back slice. And, any thing, C_3, also shaded, that separated the front and back crosswise vertical slices would have to extend all the way through the middle crosswise slice in such a way that a path from the front slice to the back one would have to pass through it. E.g., C_2 might be a vertical plane lay whose edges lay along the path surrounding that slice. But plausibly it would be a matter of empirical fact that any 'separators' C_1, C_2, and C_3 satisfying these conditions would all have to come together somewhere inside the interior one of the 27 small cubes. That would support the thesis that the whole cube satisfied the criterion for being at least n-dimensional set forth in Theorem 9.2.2, in the case in which $n = 3$. Admittedly this argument is not rigorous,[17] but we will not pursue this in greater depth here, and instead pass to some general observations relating to its conclusion.

In attempting to describe fs-space topologies, we have taken tentative first steps towards developing a topology of bodies that is to them as the topology described in Part II is to their surfaces. The fs-topology attempts to describe geometrical properties that are invariant under deformations of the kinds that rubber balls may undergo, and therefore it can be thought of as 'rubber ball' rather than 'rubber sheet' geometry. Its bodies are not solid solids, which are often supposed to be undeformable, but they do have a kind of impenetrability. Although they can be cut in pieces (we have supposed), the pieces cannot penetrate one another, and it is presupposed that separated pieces can be put back together in the way they were before they were separated. Thus, they are not like fluids that can be mixed so perfectly that originally separate quantities of them can come to occupy exactly the same place.

There is also an important technical point to make about dimensionality. While we assume that a body is infinitely dissectable, it does not follow that all points of its fs-space lie on the surfaces that are revealed when it is dissected. Dissection may involve only countably many cuts, and it may produce only countably many pieces, in which case the composite surface of its dissection space is at most 2-dimensional. But the body's fs-space may be topologically 3-dimensional, as we have just argued. *But*, if every piece that is produced when it is dissected has *cavities* then the fs-space has the structure of a *fractal foam* (Mandelbrot, 1983, p. 133) and both the dissection and the fs-space can be 2-dimensional. This leads to another important technical point.

The idea of a *cavity*—of an empty space—is *spatial*, and Adams, 1988, gives a very simple argument that, assuming that a body's topology is the same as that of the space that it occupies, if it has cavities in

[17] One respect in which the present argument is only intuitively plausible is in assuming that for any $i = 1, 2, 3$, a continuous (connected) path from A_i to B_i would have to pass through C_i. Formally, this should be derived from empirical hypotheses about C_1's relation to A_1 and B_1, together with the definition of a continuous path, as applied to the fs-space topology.

every piece then it is at most 2-dimensional. But while our description of the fs-space of a body may make it seem to be topologically equivalent to an empty space, we haven't shown that fs-space topologies can be 'embedded' in the topologies of a background space or spaces, which we have yet to characterize.[18] Concepts of Space or of spaces, possibly devoid of matter, will be taken up in the following chapter, although we will not be able to settle the question of how they relate to the fs-spaces of the bodies in them, but the point for now is the following. Fs-spaces are described intrinsically and they do not presuppose concepts of space. To this extent we can give a positive answer to a question raised at the end of Section 13.2. Given that stereometrical theory must involve more than bodies' surfaces, we are at least able to develop one independently of concepts of Space, which Euclid's stereometry did not do.

But now let us turn to more specifically geo*metrical* concepts.

[18]A comment by Weyl suggests something like this, even if he held that the identity of points is 'unrecognizable': "... in a completely homogeneous substance without any quality, the recognition of the same place is as impossible as that of the same point in homogeneous space" (Weyl, 1949, p. 165).

14

Rigidity

14.1 Aspects of Rigidity

An object is said to be rigid if it cannot be enlarged, compressed, or otherwise deformed. However, deformation is a process of change, of objects having one form at one time and another form at another, and because classical geometrical theory had no temporal component, deformation, deformability, and its opposite, *rigidity*, played no part in this theory.[1] Nonetheless, rigidity has atemporal aspects, and while these are not explicit in the classical theory they are importantly presupposed by it. These and their importance to geometry are what we will concentrate on primarily here, but it should be stressed that in doing this we will side-step the problem with which readers of recent writings on the philosophy of space and time are most likely to be familiar,[2] namely that of guaranteeing the invariance over time of the properties of geometrical measuring instruments, which has already been commented on in section 4.7.

Two atemporal aspects or indices of rigidity are the following: (1) rigid bodies cannot be placed in coincidence with or fitted up against other bodies as easily as can non-rigid bodies, and (2) if two or more rigid bodies are brought into 'rigid coincidence', then as long as they remain in coincidence in that way they form a composite 'rigid frame'. Point (1) can be illustrated by the counters depicted in Figures 1a–1e of Chapter 13. But now compare Figure 1a of Chapter 13, which is reproduced below, with Figure 1g:

[1]Unlike Helmholtz's theory, which postulated the 'free mobility of fixed bodies' (Cohen and Elkana, 1977, p. 43). It is noteworthy that Schlick, commenting on Helmholtz's 'fixed bodies', called them *rigid corporeal models*, cf. Note 11 on p. 28 of Cohen and Elkana, op. cit.

[2]Cf. especially Reichenbach and Grunbaum.

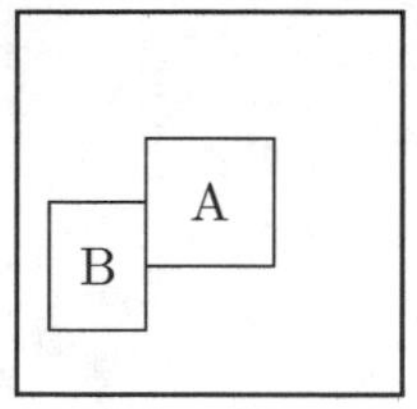

Figure 1a

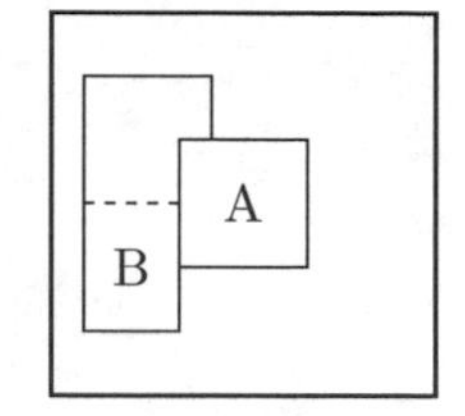

Figure 1g

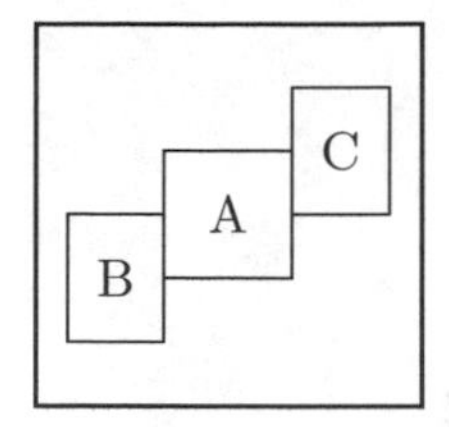

Figure 1c

After rectangle B was originally fitted to rectangle A in Figure 1a, it appears to be have been extended up and bent part way around rectangle A in Figure 1g, although the two composites cannot both be possible, assuming that A and B are rigid. That is, the surfaces Σ^{1a} and Σ^{1b} are represented as being formed from the same components, Σ^{A} and Σ^{B},[3] but composites with the same components cannot differ in the way they do in Figures 1a and 1g if the components are rigid. Intuitively, B cannot be fitted 'flatly' against A, as in Figure 1a, and also 'crookedly' against it as in Figure 1b. This requires clarification, and a step in that direction is to say that, assuming that A and B are rigid, a 'sufficiently large part' of a composite surface formed from them, like Σ^{1a} or Σ^{1b}, determines all the rest. In this case one part of Σ^{1a} that determines all the rest is the region in which they coincide.[4] Of course this is not the whole story,[5] and another important aspect of it will be returned in the following section. But let us first consider point (2).

Now consider rectangles A, B, and C in Figures 1a and 1c. Assuming that these 'bodies' are fitted together as shown in the figures, then, *so long as they remain fitted together in these ways*, the composite bodies that they form will be rigid. Thus, the composite formed from rectangular counters A and B in Figure 1a will be rigid if A and B are, and the same is true of A, B, and C in Figure 1c, whose composite forms the start of a 'rigid frame'.[6]

[3] Assuming that the counters have some thickness, they should meet in common faces when they were fitted together. However, for the present we will idealize and pretend that the counters are no more than regions of 2-dimensional surfaces.

[4] A much smaller part that determined all the rest would be any three non-collinear points of coincidence, but it may be easier to observe that the bodies coincide over extended regions.

[5] For instance, this leaves out of account aspects of rigidity that are characterized *metrically*, e.g., that if rectangles A and B in Figure 1a are rigid then B cannot stretch uniformly relative to A. To the extent that this kind of stretching must be defined metrically, it presupposes that the instruments used to determine it are rigid.

[6] But note that the converse, that if a composite surface is rigid then its components must be too, does not follow. Thus, each of A, B, and C, as well as the 'frame' formed from the three of them in Figure 1c might be rigid, but the composite of B

It is important that it is a factual matter whether or not a given body or composite of bodies or their surfaces is rigid. This depends both on the bodies and on how they are fitted together, and, conceivably, on the 'forces' that act between them when they are fitted together.

Of equal importance is the conceptual fact that rigidity, or at least the indices of it noted above, is *relational*; it is a matter of how bodies or surfaces can be fitted *together*. Furthermore, to the extent that possible 'fittings together' define rigidity, and frames and places in them are defined in terms of rigidity, a classical relativism of place and position is supported, at least in a frame. In fact, this kind of relativity is even more extreme. It is consistent with everything said so far that bodies of one kind could be rigid relative to other bodies of the same kind, while bodies of a different kind could be rigid relative bodies of *that* kind, but bodies of either kind might not rigid relative to bodies of the other (in this case we might be inclined to say that the bodies were formed of different kinds of matter). So far as the author can see, however, this possibility does not arise in Nature. If this should be so it would be a natural law; it would be a matter of fact that there should be only one kind of rigidity in Nature.

But now let us turn to the geometrical significance of one atemporal aspect of rigidity.

14.2 An Atemporal Rigidity Presupposition of *The Elements*: Constructive Reference and Abstraction

Perhaps the most important atemporal rigidity presupposition of *The Elements* is implicit in Postulate 1 of Book I, concerning which Heath says "There is still something more to be inferred from the Postulate combined with the definition of a straight line, namely... that the straight line joining two points is *unique*," (Heath, Vol. I, p. 195). The significant point for us is that if straight lines are unique the instruments that draw them must be rigid. For example, it should not be possible to fit a straightedge in essentially different ways between corners like ' > ' and ' < '. If that were possible then it would be possible to draw *different* straight lines between the corners, thus: >═<.[7] This is also very important to geometrical theory.[8]

---

and C alone obviously would not be.

[7]Clearly the surface on which the lines are drawn must itself be rigid.

[8]Related points might be made concerning circles and the compasses used to draw them. But this is complicated by the fact that circles can be drawn by 'swinging arcs' at the end of stretched strings, and these are obviously not rigid. Even the metal compasses used in schoolrooms are only temporarily rigid, during the intervals

When we speak of "the" unique straight line between two points we are using 'straight line' in an abstract sense. We do not mean that the line cannot be *drawn* twice and thereby produce two concrete features, but rather that what is drawn instantiates one abstract 'thing', a geometrical line,[9] as was discussed at length in Chapter 4. Moreover, most of this abstract thing lies between the things it connects, hence much of it may lie in a blank space on the surface. For the same reason, a point or place in a blank space may be referred to 'constructively', e.g., as "the point where the diagonals of the rectangle below intersect":

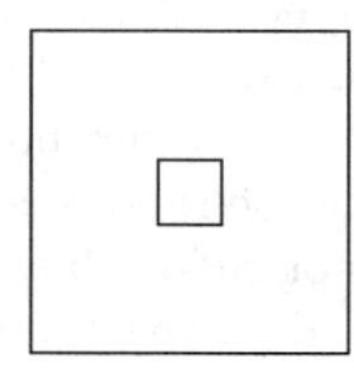

That it is possible to refer to abstract loci like straight lines, diagonals, and their intersections means that the 'vocabulary' of our spatial language has been extended from the *topological* to the *geometrical.* The language of topology, as applied in the way outlined here, only allows us to refer uniquely to loci that are bounded by actually existing things—e.g., to boundaries, intersections, unions, exteriors, and interiors of concrete features and systems of such features. In contrast, geometrical language allows us to refer to lines other than boundaries, like diagonals, their points of intersection, and a great deal else that transcends the topological. And, verifying claims that use the extended vocabulary, such as "There is a small hollow square surrounding the point where the diagonals of the rectangle above meet," requires the use of rigid instruments, in this case straightedges, to draw the diagonals and 'see' that they meet in the square. There would be no way to verify such a claim in 'rubber sheet geometry', and it might be said to be meaningless in that context.[10]

when they are used to draw individual circles. Though 'circumstantial rigidity' can be accommodated in our theory, doing so complicates it too much to be entered into here.

[9]This is obviously rough. Concrete features produced by the use of straightedges should be visible, hence they must have breadth, and therefore they cannot instantiate ideal lines, straight or otherwise. In fact, different features produced connecting two corners may not be exactly coincident, and therefore they may not instantiate the same abstract point set. What is required of straightedges is that certain parts of the boundaries of features produced by their use—the parts that are in contact with the edges said to be straight—should be coincident.

[10]This overlooks the fact that rubber objects have forms to which they return

It is equally important that extending the language to the geometrical makes it possible to extend it to 3-dimensional spaces that can be associated with rigid frames. For example, not only can a point be referred to as being 'where the diagonals of a square meet in the plane', but also as 'where a line of length equal to the side of a square would end', if it were erected perpendicular to the square from one of its corners. In fact this allows us to describe complete spaces of points in rigid frames, and inquire into their dimensionality and relations, both to the spaces of other rigid frames and to the 'material' fs-spaces of the bodies in them, as described in section 13.4. These matters will be discussed in the following chapter, but let us now turn to rigid motion.

14.3 Rigid Motion

Recall the proof of Proposition 4 of Book I, of *The Elements*, that triangles ABC and DEF (pictured below) are congruent, which says "For, if triangle ABC be applied to triangle DEF, and if the point A be placed on the point D and the straight line AB on DE, then B will coincide with E, because AB is equal to DE" (Heath, Vol. I, p. 247).

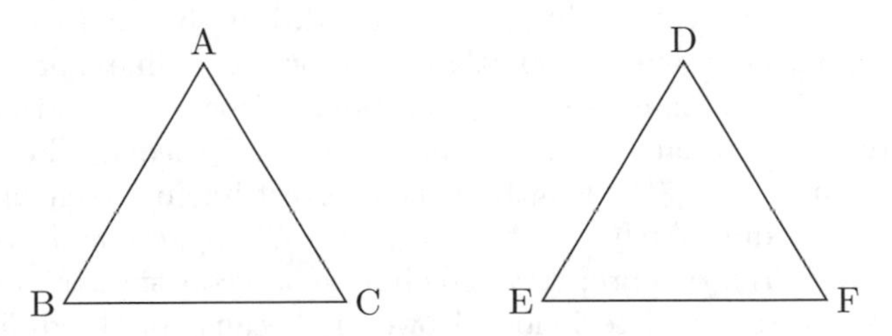

Clearly this presupposes rigid motion, because line AB is assumed to retain its length and straightness when it is moved from its original position to its new position, where is superimposed on DE. This is invariance over time, and that transcends the atemporality characteristic of the 'Platonic' Euclid. How does this relate to the atemporal aspects of rigidity commented on in the previous section?

If we think of the different positions occupied by triangle ABC as different ways in which it is superimposed on a plane surface, then each of the places to which it can be moved or displaced corresponds to a possible superposition of the triangle on the plane. An entire rigid motion then corresponds to a succession of possible superpositions, and if their temporal order is left out of account we are led to something like William of Ockham's conception of motion: "For motion is nothing more than

naturally, after they have been stretched or otherwise distorted. But to this extent 'rubber sheet geometry' is not quite a suitable subject matter of topological theory.

this; the movable body coexists with different parts of space, so that it does not coexist with any single one while contrary statements are made true" (Boehner, 1957, p. 156). Such a conception has the advantage of not construing motion as a 'state',[11] which leads to paradoxes like Zeno's Arrow, but it leaves a great deal out of account. Order in time is 'real', at least as it is now conceived in Science. Whether or not it is reversible, it cannot be chaotically rearranged, and the laws of motion describe possible rates at which motion can take place.

Given the foregoing, clearly the atemporal aspects of rigidity commented on earlier do not exhaust its 'physical content'. Whether they shed any light on the more current methodological questions alluded to in section 14.1 will be commented on again in section 16.3, and we will end this chapter with comments on the concepts of *length*, and of *distance* between points on a surface, and their relations to rigidity and congruence.

14.4 Length, Distance, and Rigidity, and Their Relation to Congruence

Lengths and distances are obviously related, but they mustn't be confused with one another since lengths are properties of lines and distances are measures of relations, e.g., between points. However, length and distance are closely related, at least in geometrical theory. The distance between two points is the length of the shortest line between them, and for many including Archimedes (p. 3 of *On the Sphere and the Cylinder*, Heath, 1897), this property is definitive of 'geodesic straightness',[12] which will be returned to below. However, because of the difficulty of measuring the lengths of curved lines, classical geometrical theory found great difficulty in dealing with them, and Euclid in particular restricted his linear metrology to straight segments. Interestingly, this was not the case with *area*, since *The Elements* includes theorems pertaining to the areas of circles (e.g., Proposition 2 of Book XII, which says in effect that the ratio of the areas of circles is equal to the square of the ratio of their diameters, Heath, Vol. III, p. 371). Why area should have been easier to deal with than length is worth commenting on, since this throws light

[11]Of course this conception is non-relativistic, since according to Special Relativity a moving body undergoes a Lorentz contraction in its direction of motion. In a sense, according to modern physics, being in motion is a state.

[12]In his fascinating commentary on Book I of *The Elements*, Proclus says that Archimedes *defined* the straight line the shortest of all lines having the same extremities (cf. Morrow, 1970, p. 89), but, more exactly, Archimedes stated this as an Assumption, "Of all lines which have the same extremities, the straight line is the least."

on 'the very idea' of length, and its relation to *congruence.*

Figures with different shapes can be compared in area because one can be represented as part of another, e.g., as is possible with the areas of the outer and inner squares below, S and s, and the circle, C, which is inscribed in one and which circumscribes the other:

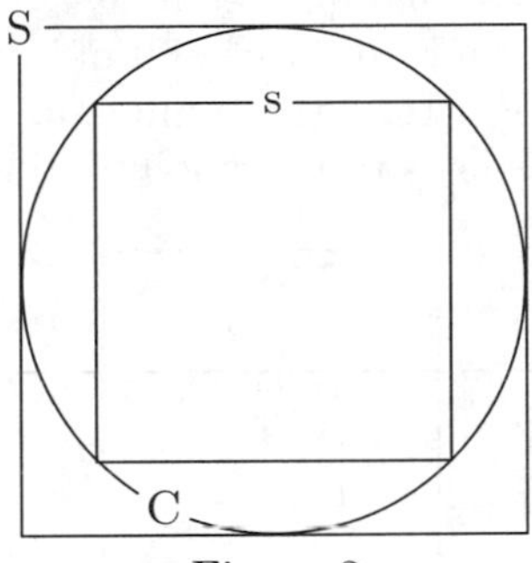

Figure 2

Because s is included in C and C is included in S, the area of C lies between that of s and S ("The whole is greater than the part", Common Notion 5, p. 232 of Heath, Vol. I).[13]

But compare the problem of measuring the areas of circles with that of measuring their perimeters. We might assume that the perimeter of C must be less than that of S and greater than that of s. But how? Though s as a whole is part of C as a whole, and the latter is a part of S as a whole, the *boundary* of S is not a part of the boundary of C, and the latter is not a part of the boundary of S. In fact, not even the smallest arc of the circumference of C can be directly compared with any part of the perimeters of the squares. Now, this is posited in Archimedes' proof that the value of π lies in the open interval $(3\frac{10}{71}, 3\frac{1}{7})$ (Heath, *The Works of Archimedes*, p. 98), but that the perimeter of a circle is less than that of any regular polygon in which the circle might be circumscribed was never proved. Of course one supposes that Euclid knew this, but that he was unable to prove it,[14] which suggests the following.

[13]In fact, Eudoxus' method of exhaustion, involving circles inscribed in and circumscribing regular polygons of arbitrary many sides, is used in the proof of Proposition 2 of Book XII, referred to above.

[14]Recent short papers of Eenigenburg, 1997 and others have returned to this topic, and it is worth quoting an abstract of Eenigenburg's paper in particular:

> The assumption that the ratio of arc length to chordal length tends to 1 as the ends of the arc approach one another is not always true. The presupposition only remains true if the function describing the arc is confined in some way. The assumption is valid when the function

Euclid's primary method of proving the equality or inequality of geometrical magnitudes—lengths, areas, angles, etc.—was by combinations of demonstrations of congruence, possibly requiring displacements as in the proof of I.4, and by comparisons between parts and wholes, as in the proof of XII.2. But these methods cannot be used to demonstrate what Archimedes assumed, that an arc of a circle is less than the sum of the tangents drawn from its ends to their point of intersection, e.g., that the quarter circle below is less than the sum of the horizontal and vertical tangents drawn from its ends to their point of intersection, as shown:

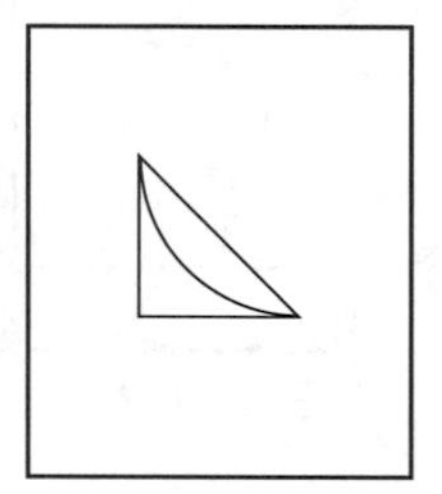

Figure 3

It is perhaps more obvious that the quarter circle should be longer than the chord connecting its end points, since that follows from the assumption that the straight line is the shortest line between two points. But it is no more possible to demonstrate this by congruence plus part-whole methods, which are putatively the primary methods of magnitude comparison employed in *The Elements*, than it is to prove that the quarter circle is less than the sum of the tangents from its ends to their point of intersection, and Euclid defined straightness differently. Moreover, it is implausible that he could have presupposed that straight lines are geodesics, for if he had the proof that the sum of any two sides of a triangle is greater than the remaining one (the "Triangle Inequality," Proposition 20 of Book I, Heath, Vol. I, p. 286), would have been trivial.

So we may ask: If lengths were not to be compared by congruences possibly involving rigid displacements, together with part–whole relations, how were they supposed to be measured? Well, pending but possibly presupposed by modern methods of integration, these may well have been very archaic methods indeed. The Babylonians knew that π

has a continuous derivative (it is 'smooth'), but not when it has a discontinuous derivative.

Note that the assumption that the quarter circle in the figure below is less than the sum of the tangents from its ends to their point of intersection is trivially false in spherical non-Euclidean geometry, in which indeed the ratio of a circle's circumference to its diameter is not constant.

was approximately equal to 3 (Van der Waerden, 1963, p. 81). And presumably the Egyptians knew this too, of whom Democritus wrote "No one surpasses me in the construction of lines with proofs, including the so called rope stretchers among the Egyptians" who Van der Waerden says "were probably surveyors, *whose principal measuring instrument is everywhere the stretched cord*," (quoted in Van der Waerden, op. cit, p. 15, author's italics). Thus, stretched ropes or cords probably were the (or a) primitive practical method of determining straightness.[15]

The important point for us is that ropes or cords may be used to measure things when they are not stretched, as is suggested in the medieval 'practical proof' that $\pi = 22/7$ cited in section 11.1. The idea was that if a straight line whose length is 22/7 of the diameter of a circle could be curved around and superimposed on its circumference it would fit exactly over it. This suggests in turn that a rope or cord could be stretched along a straight line equal to 22/7 of a circle's diameter, then curved around and superimposed on its circumference, and this would provide practical proof that the circumference was approximately equal to 22/7 of the diameter.

The foregoing suggests that the 'curved cord' was a practical method of length comparison not restricted to straight lines, which could conceivably have been used by the ancients to compare the perimeters of circles with their diameters. However, this method is not as well suited to geometrical theory as congruence determination involving rigid displacement, because it involves the use of ropes or cords which do not retain their geometrical forms when they are displaced. What is invariant about a rope is not its form but its length, the very thing it is used to measure.[16] The rope is conspicuously *non-rigid*, but for this very reason it and what it is used to measure transcend the topic of this chapter, rigidity, and, at least for Euclid, it transcends Geometry itself. But another point is worth making about the method.

Two advantages that idealized ropes have over rigid rods as length measuring devices are as follows: (1) Because they are non-rigid their use is not restricted to spaces of constant curvature, which are the only ones in which free rigid motions are possible, and (2) The key 'invariant property' of stretched ropes is that they satisfy the *triangle inequality*,

[15]This may not have died out entirely in Greek times. For instance, Heath cites "A line stretched to the utmost" as one of Heron of Alexandria's definitions of straightness (Heath, Vol. I, p. 168; Heron was born approximately 150 BC).

Something related to this is suggested by the etymology of the word "straight," which is given on p. 2033 of the Shorter OED as originating from the past participle of the Middle English verb *strechen*, to stretch.

[16]Or, idealizing we might say that all ropes had the same form, namely that of the Platonic interval $\mathbf{I} = [0,1]$ described in Chapter 10.

that if three of them can be connected end-to-end to form a triangle then any two of them taken together are longer than the third, which makes them suitable for defining metrics in arbitrary topological spaces. Both of these properties are important in connection with distance measurement. Modern physical theory teaches us that Space, which has not yet been considered in this work, does not have constant curvature, and therefore if distances are to be measured at all it must be by means that don't require rigid instruments.[17] Connected with this is the fact that while ropes can be superimposed or 'laid out' on surfaces,[18] they can also be stretched across spaces, and, ideally, determine distances 'as the crow flies' by this means. The measuring *tape* might be the better metaphor for distance measuring in the spaces of modern Physics than is the measuring *rod*. There is a final point to make.

Given that measuring tapes can be used to measure lengths and distances, it follows that although they are not rigid, their invariance, as it were, goes along with that of the rigid bodies and frames to which they are applied and in which they are employed. Thus, given a rigid body represented by the rectangle below,

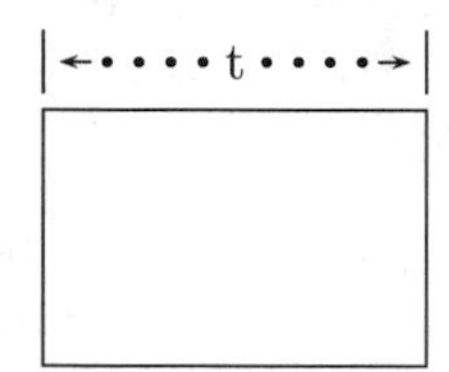

a tape t stretched at one time along the dotted line between the upper left and the upper right hand corner of the body will remain that way until it is removed, and it will be tight again if it is stretched between these corners a second time. This will assist us greatly in describing the 'intrinsic spaces' of rigid frames, and in describing the interconnections both between frames and between them and the fs-spaces of bodies in them, to which we now turn.[19]

[17] Admittedly standard physical theories regard space (or space-time) as 'locally Euclidean', and the means that are used to build locally Euclidean 'neighborhoods' into non-Euclidean spaces are far more sophisticated than crude ropes. However, at least it is interesting to consider how one might start *ab initio* from methods that don't presuppose even local rigidity.

[18] Though this kind of superposition is not so clearly of the kind that Part III of this work is concerned with, since idealized ropes don't have surfaces, and 'real' ropes are less like bodies of the kind we have been considering than they are like fiber bundles—albeit these may have little connection with the 'fiber bundles' of topological theory.

[19] The subject of 'right' or 'appropriate' distance measures will be returned to in section 15.5.

15

Rigid Frames and Their Spaces

15.1 Introduction

This chapter is concerned with the possibility noted in section 14.2, that rigid construction instruments make it possible to describe abstract loci in blank spaces on surfaces, and in the process moves us from the topological to the geometrical. Not only can loci in empty spaces on surfaces be described, but also loci off the surfaces of bodies can be described by these means, which moves us from surfaces to spaces of frames associated with rigid bodies. To develop this theme systematically would require us to retrace the steps gone through in Part II, first to characterize the points in the 'space of a rigid frame', then to characterize its topology, and then to examine its geometry and its relation to other spaces. This would be a very tall order, far beyond anything that we are able to undertake here, and we will restrict ourselves to speculating on certain things that one would expect to find in such an investigation.

As long as we are restricted to rigid frames and constructions, we are restricted to spaces in which free rigid motions are possible, and these are spaces of constant curvature; and, as long as we are restricted to spaces of constant curvature we might as well consider Euclidean spaces, where we can make free use of the principles of Euclidean geometry. We will confine ourselves principally to those, and in the following section we will comment on aspects of plane geometry, now built upon the surface feature topologies and their possible superpositions described in Part II and in the previous chapters of Part III. Following that we will comment very briefly on the extension of the constructive space of the plane to the 3-dimensional space of points *off* the plane, but 'in its space'. And, finally, we will comment on the possibility of weakening the rigidity requirement, and replacing it with a kind of 'measuring tape-geometry' that doesn't presuppose constant curvature.

15.2 Euclidean Plane Geometry

Section 10.6 left open the problem of giving intrinsic necessary and sufficient conditions that a surface topology must satisfy if it is to be homeomorphic to the Euclidean plane or closed region of it. However, let us assume that that problem is solved, and that a given surface, possibly composite, is 'topologically flat' and satisfies these conditions. We will also assume that for each surface feature or figure of the surface, not only is the region or class of points it occupies defined, but so is the class of regions that could possibly be occupied by it if it were moved rigidly about on the surface. And, finally, let us assume that we have rigid instruments—straightedges and compasses—e.g., for producing another feature whose edge is a straight line between any two features,[1] and for producing figures, points on whose edges always lie a fixed distance from a given point. Of course given this much, the theory of this surface must be Euclidean Geometry pure and simple, and our approach might seem to add nothing to it.

But one thing is added, namely a partial account of the *application* of Euclidean geometry to concrete things, and even, conceivably, a way of testing the theory's postulates. The theorem that the medians of a triangle are coincident is a familiar example,[2] and although it is too complicated to discuss in detail here, some points concerning it may be noted.

Assume that our surface is flat and three 'streaks', α, β, and γ, are drawn, whose straight edges meet in corners D, E, and F, and which form triangle DEF, as illustrated in Figure 1:

---

[1] E.g., these must be ones that satisfy the requirement of applying 'evenly' to the plane, which section 13.2 suggested might explain Euclid's definition of a straight line as "a line that lies evenly with the points on itself" (Definition 4 of Book I, Heath, V. I, p. 163), and which might also be regarded as partly definitive of the plane itself.

[2] I have been unable to find this in Euclid's *The Elements*, but, as one would expect, it is implicit in a theorem in Archimedes' *On the Equilibrium of Planes; of the Centers of Gravity of Planes*. This is Proposition 13 of Book I, "In any triangle the center of gravity lies on the straight line joining any angle to the middle point of the opposite side," cf. p.198 of the Dover edition of Heath, 1897.

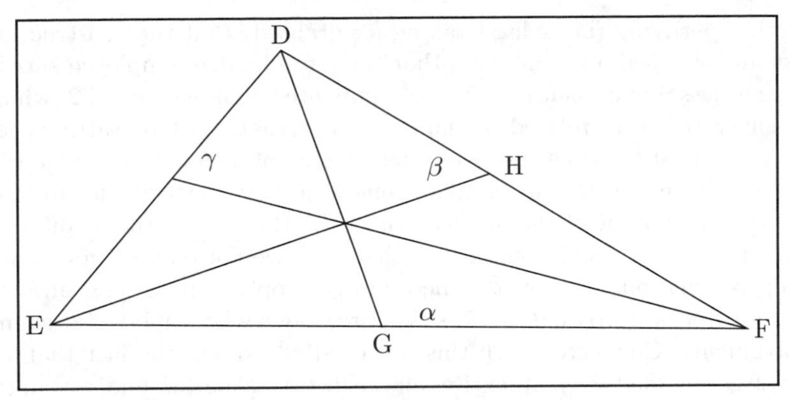

Figure 1

Now let two median straight lines DG and EH be drawn, connecting corners D and E with the midpoints, G and H, of the sides opposite to them.[3] Then if the third median is drawn connecting corner F with the midpoint of the side opposite, the theorem of the coincidence of the medians predicts that it will pass through the point of intersection of DG and EH. No doubt if the reader studied geometry before the era of the New Math, she or he will have carried out just such a practical test of the theory, and, hopefully, will have been impressed with the accuracy with which the result confirms it.[4]

Now we can describe in a bit of detail how the operationalist foundation of topology and the theory of superposition developed in Part II and the earlier chapters of Part III applies to the theorem of the coincidence of the medians. The page on which Figure 1 lies 'supports' a topological point-space, as described in Chapters 6 and 7, and streaks α, β, and γ define sets of abstract points in this space, as described in Chapter 6. The edges of α, β, and γ are their boundaries, as characterized in Chapter 7, and their intersections constitute I-systems that, if they are carefully drawn, define PI-systems that individuate abstract points in the space, D, E and F. These are the topological aspects of the application, which does not yet involve superposition.

Moving to the geometrical level brings in superposition, but it also

[3]Of course locating the midpoints of EF and DF requires the use of a compass of fixed radius to draw circles with centers at D, E, and F, finding their points of intersection, and drawing straight lines between them, which is a matter of some complexity.

[4]The author, having been educated well before the New Math made its appearance, had this pleasure, and he regrets that among the subjects covered in modern 'Geometry' textbooks, geometrical construction finds hardly any place. Section 10 of Adams, 1994, comments on modern geometrical pedagogy.

requires verifying (or at least taking for granted) that the construction instruments used are rigid, e.g., that the straightedges employed satisfy straightness requirements like those commented on in section 14.2 (which we suggested were related to the conditions that must be satisfied for the page itself to be geometrically flat, which of course it is not quite). Our comments on the latter conditions were very cursory, and to that extent our account of the application of the theorem of the medians is incomplete. It is also incomplete in that we have not offered an account of *approximation*, i.e., how the theory might apply to figures on surfaces that are not perfectly flat, or ones that are drawn with imperfect drawing instruments. Connected with this, we need to explain the fact that we are often satisfied to treat 'entire' marks like streaks α, β, and γ as lines, rather than their edges. But we hope that the reader will accept what has been sketched so far as a start towards accounting for these things; moreover, one which is not merely a barren claim to the effect that the concrete things we have been speaking of are mere representations of things in some suprasensible realm.[5]

But let us now turn to more difficult problems that arise in accounting for the application of theorems of solid geometry to objects in 3-dimensional spaces, which we will assume are defined by rigid frames.

15.3 Rigid Frames and the Application of Geometry to Objects in Them[6]

We may begin by noting that the diagrams that accompany theorems of Euclidean solid geometry have a much more remote relation to the 'things' of which the theorems treat than those of plane geometry. Take the first 'constructive' proposition of Book XI, namely Proposition 11, "From a given elevated point to draw a straight line perpendicular to a given plane" (Heath, Vol. III, p. 292). The figure below accompanies the proof of this proposition:

[5] Adams, 1999, discusses similar questions that arise when theories formulated in first-order logic are applied to phenomena that do not perfectly satisfy the theories' axioms. This paper notes alternative deductive 'routes' that can be taken from premises to conclusions, depending on whether or not one 'idealizes' by assuming that the 'world of the senses', in which the theory's axioms are only approximately true, 'resembles' in a well defined sense an 'ideal world' in which they are exactly true. However, the final section of the paper points out difficulties that arise in extending this account to geometrical theory.

[6] Many of the points in what follows are gone into in more detail in section 5 of Adams, 1986.

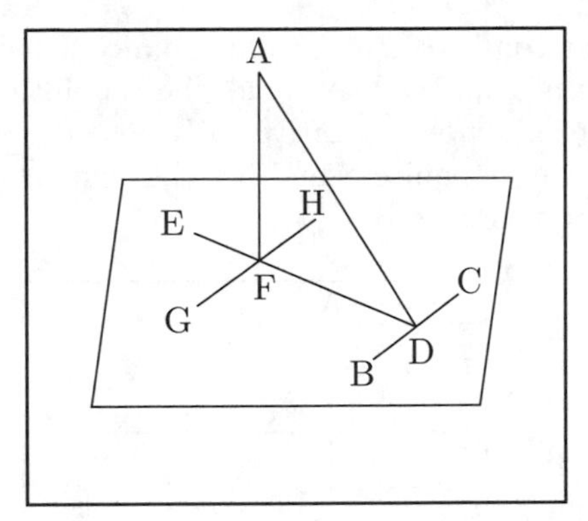

Figure 2

The key steps of the proof are as follows:

> Let A be the given elevated point, and the plane of reference be the given plane; thus it is required to draw from the point A a straight line perpendicular to the plane of reference.
> Let any straight line BC be drawn at random in the plane of reference, and let AD be drawn from point A perpendicular to BC (Proposition I.11 describes how this is done).
> If the AD is perpendicular to the plane of reference, that which has been enjoined will have been done.
> But, if not, let DE be drawn from the point D at right angles to BC in the plane of reference, let AF be drawn perpendicular to DE (Proposition I.12 describes how these things are s done)...
> AF is at right angles to the plane of reference. q.e.f.

Obviously what Figure 2 pictures is not itself the kind of thing that Proposition XI.11 putatively pertains to. If nothing else, the figure is 2-dimensional but because point A is not in 'the plane of reference' to which a perpendicular is to be drawn, what is depicted is, presumably, 3-dimensional. Figure 2 depicts *something*—it is an idealized perspective drawing of a something, but the relation between it and this something is very complicated. In any case, how it informs the viewer of the way the procedure described in the proof applies to 'things in the world' is significantly different from the way that Figure 1 does this, which is related to the fact that the latter is an approximation and not a picture *of* anything. Let us consider this in greater detail.

First, and probably most fundamentally, how do we know that point A is 'in the same space' as R? Presumably it is a point of something, and if we are to relate the theorem to the concrete, A ought to be a point of something concrete. Let us jump to a conclusion about this. So far as application is concerned, A should be a point of the surface of a body, say N, that is immobile relative to R, and the way to insure this immobility would be to require N and R to be elements of the same rigid frame. Let us depict that:

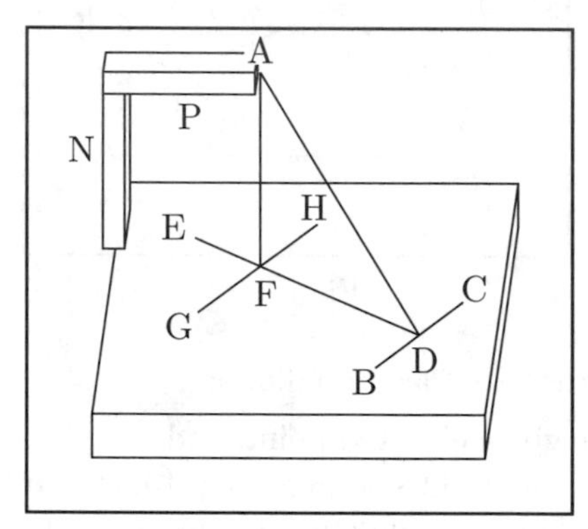

Figure 3

Here we have A on a corner of the end of a 'crane' formed of 'bars', N and P, that are rigidly attached to each other and to R, thus guaranteeing that A and R are in the same space.

The next step is to draw line BC on the surface of R, and assuming that the surface is flat, this poses no new problems. But what about drawing the line AD from A to BC and perpendicular to BC? Having to draw *any* line from A seemingly 'through space' down to plane R seems doubtful. On what is it to be drawn? It is true that Postulate 1 says "Let the following be postulated: to draw a straight line from any point to any point" (Heath, Vol. I, p. 195), but that was stated in the context of plane geometrical theory, and it is plausible in that context because the plane is always there to draw on. But it is much less plausible to postulate that figures can be drawn in empty space. Perhaps it is taken for granted that a *plane* can be constructed, containing point A and intersecting the plane of reference, and a line from one to the other can be drawn on that, but it is curious that although there are numerous propositions in the later books of *The Elements* affirming the existence of planes with various properties,[7] no postulates are formulated that affirm their *constructability* in the way that Postulate 1 affirms the constructability of straight lines.

[7]E.g., Proposition 2 of Book XI, "If two straight lines cut one another, they are in one plane," Heath, Vol. III, p. 274.

It is even worse with the construction of a straight line from A *perpendicular* to the line BC. The possibility of this is said to be guaranteed by Proposition 12 of Book I, "To a given infinite straight line, from a given point which is not on it, to draw a perpendicular straight line" (Heath, Vol, I, p. 270). But proving this requires drawing a circle with center A intersecting BC at two points, which is said to be guaranteed by Postulate 3, "To describe a circle with any center and any distance" (Heath, Vol. I, p. 199). However, constructing such a circle would require it actually to pass through the plane of reference, and that only seems possible 'in thought'. This is suggestive.

The proof of XI.11 does not so much give a practical 'recipe' for dropping a perpendicular from a point to a plane—rather, it gives a '*gedanken* procedure' for this. One does not imagine a 'geometrical engineer' following the directions laid out in the proof of our theorem *drawing* a line from a point outside the plane down to and perpendicular to it. This and many other propositions like it do ultimately lead to practical conclusions, e.g., the concluding theorems of Book XIII that describe the five Platonic solids and prove that they are the only regular solids. But many of the intervening steps are impractical. The intention of the proof of XI.11 is not really to tell us how to draw a concrete straight line from any point in space perpendicular to any given plane, but rather to convince us that such an abstract line *exists*. This is also suggestive.

That an argument is given for the existence of an abstract object for which no practical means of construction is described, such as for producing a perpendicular from a point to a plane, suggests that new principals of abstraction are invoked, beyond the principles adumbrated in 6.6 that guarantee the existence abstract entities that are individuated by constructible surface features, and secondary features such as their boundaries. What guarantees the existence of the perpendicular is a *gedanken* procedure, allegedly for constructing it, when the procedure cannot be carried out in practice? This section concludes with speculative comments relating to this question.

The first point is that it seems that in the geometry of Space, abstract existence propositions that are supported at best by *gendanken* constructions that take the place of 'practical' constructability propositions. The *gedanken* constructive arguments are obviously persuasive, since they persuade us, but one can only speculate on the nature and epistemological status of the new principles of abstraction that are implicit in them. One principle seems evident, as follows.

We seem to assume that constructions that are physically possible in the plane are only 'abstractly possible' in space. For example, we can draw a straight line from any point to any point in a plane, and in Space we seem to assume, not that straight lines can be drawn between points, but rather that there exist abstract straight lines connecting them. More generally, we seem to conceive of Space as though it were filled with abstract planes on which abstract figures satisfying the principles of plane geometry exist. But concomitant with their abstractness, we, along with Helmholtz, treat these planes and the figures on them as though they were interpenetrable—'tactually transparent'.[8]

Of course this doesn't explain the persuasiveness of the general principle of abstraction that is implicit in conceiving Space as though it were filled with abstract planes. [9] It could even seem, were it not that it would leave too much unexplained, that the only requirement of abstract existence in the Geometry of Space is absence of contradiction, as Poincaré once suggested ("Non-Euclidean Geometries" in *Science and Hypothesis*, p. 44). But this wouldn't explain why we should regard Space as 3-dimensional, given that it is logically consistent to suppose that it has a higher dimensionality. We cannot answer this question, but cursory comments on the topologies of spaces bear on it.

15.4 Remarks on the Topologies of Spaces of Rigid-Frames

To simplify, let us limit attention to a room with closed doors, which forms a rigid frame and defines a limited space of 'points in the room', whose topology we are concerned to describe. Call this the *topology of the room.* The details would have to be worked out carefully, but points of the fs-spaces of objects and combinations of objects rigidly connected to the 'frame of the room', like tables standing on its floor or light fixtures attached to its walls or ceiling, also define places and points in this frame.[10] And, the open sets of the fs-topologies of these objects can be supposed to correspond to open sets of the topology of the space of the room itself.[11] If the room is finite, as seems plausible, its topology

[8] Stereometric diagrams like Figure 2 reflect this in representing bodies as though they were hollow shells.

[9] Are there analogous principles in other abstract realms, e.g., those of color and time?

[10] Among other things, the 'coincidence postulates' that rigidly attachable objects and their features satisfy must be considered, and in particular whether they satisfy the separation and finiteness postulates, which would allow point-individuating I-systems to define abstraction points of the space of the room. An essential preliminary to this would be to examine the conditions that have to be satisfied for the coincidence of features of distinct possible rigid attachments.

[11] It is also plausible that basic open sets of the room's topology could be equivalently defined as sets of points of the room that are not points of the given object or combination of objects.

should be Hausdorff, compact, and metrizable, which would permit its dimensionality to be determined. In particular, given that the fs-spaces in question are subspaces of the space of the room, and they are 3-dimensional, as in Section 13.4, and supposing that the room itself is a 'finite sum' of these spaces, it would follow as a basic theorem of Dimension Theory that the space of the room was also 3-dimensional. There are two significant points to make in this connection.

Although rigidity is fundamental to the defining the room's topology and dimension, motion, and *a fortiori* free motion, has not entered the picture. Therefore, so long as it is 3-dimensional, the topology of our room can be fairly arbitrary. For example, free *rigid* motion is not presupposed, and therefore the space of the room need not have constant curvature. On the other hand, since the room and the spaces within it (e.g., a closed cupboard) 'wall in' and keep objects inside them from moving outside of them, dimensionality is closely related to motion. Walls, which are 2-dimensional, restrict the motion of objects, and much of the practical import of a space's being 3-dimensional is that it has room for containers—i.e., it is a 'container space'. Of course being *in* the space is not the same as being *contained* in it, which means being prevented from moving out of it. Incorporeal, 'ghostly' things could be in the room but not contained in it, since nothing would prevent their passing out through its walls.[12]

But now let us comment very briefly on relations between spaces, including spaces generated by frames that are in motion with respect to one another.

15.5 Relations Between Spaces

Even if the 'room of reference' above were the entire universe, there might be other rooms and spaces within them, just as a rigid container might be contained in a finite room. And even if the container were fi-

[12]But the 'container conception' is closely related to speculations of Poincaré's about the 3-dimensionality of Space. Thus, in *Last Essays* he says:

> Let us now consider the case of space. It cannot be divided into several parts either by forbidding the passage through certain points or by forbidding the crossing of certain lines; these obstacles could always been circumvented. It will be necessary to *forbid the crossing of certain surfaces*. And that is why we say that space has three dimensions ("Why Space has Three Dimensions," Bolduc translation, Dover ed., p. 29).

This 'cut' conception is a considerable evolution from earlier views of Poincaré's concerning space, in which motion nevertheless plays an important part. Thus, in "Space and Geometry" he wrote "...sight and touch alone could not have given us the idea of space... Not only could this concept not be derived from a single sensation, or even from a *series of sensations*; but a *motionless* being could not have acquired it..." (W.J.G translation, Dover ed., p. 59).

nite, the 'space' of its possible rigid extensions might include the entire universe. If the room and the containers were stationary with respect to one another they could be regarded as parts of the same frame,[13] and the only problem would be that of correlating points originally described with respect to one frame with those originally described with respect to the other. So long as the frames had Euclidean or even more general metrics that could be 'coordinatized', so that the points of each frame could be described, e.g., by triples of real numbers, correlating points described by one system of coordinates with those described by another would become a standard problem of geometric transformation theory, and the only new problem for us would arise if the points in the frames had no such coordinate representations. All that we can say about this here is that the correlation would be expected to 'retreat to the concrete', wherein abstract points of the frames involved are described by reference to the systems of constructible concreta that individuate them. Actually, this is probably the most fundamental way of correlating geometrical representations, which would suggest that the problem of describing correlations between them, e.g., of moving from one system of Cartesian coordinates to another, is essentially a linguistic one, not unlike that of transforming from Roman to Arabic numerals.

No further problems arise in describing correlations between frames that are in motion with respect to one another, so long as these motions are conceived 'classically', as successions of possible static positions. It is only when time plays a 'dynamic' rôle, as in the theory of Relativity with its Lorentz contractions, that new problems arise. Our only contribution to this problem, if it really is one, is to suggest again that one may expect to go behind abstract representational 'surfaces', and consider the concrete phenomena that they represent. Of course the phenomena themselves may have to be reexamined, since, as they were described in Chapter 2 they involve timeless relations of coincidence and separation among timelessly existing 'things'—*viz*, features of bodies' surfaces. But this is an idealization, and while it may be useful to point out its centrality to problems of relative position and motion, it transcends the modest aims of the present investigation to deal with these matters systematically.

[13]But determining that the two were stationary with respect to each other might present difficulties, and if they were far apart it is conceivable that visual means would have to be employed. Recall that so far our account of spatial concepts has been based solely on relations of coincidence and separation among surface features of bodies, and except as these features were *essentially* visual objects like colored spots, visual observation has not been essential to determining relations of coincidence and separation among them.

One last point is relevant in this regard. While the present remarks concern relations between the spaces of all rigid frames, nevertheless all of these the spaces are relative, if not to an actual frame, at least to a possible one. It follows that we have not adopted either a relativist or an absolutist position as far as Space is concerned, nor have we 'tilted' towards a preferred class of invariances, Galilean, Lorentz, or whatever. For its own reasons, Physics may concentrate on a particular invariance group, but these transcend our present considerations—although some points of contact between our present concerns and Physics and its philosophy will be noted in Chapter 16. But to end this chapter we will return briefly to 'stretched rope' or 'measuring-tape' geometry.

15.6 Comments on Measuring-Tape Geometry

Let us idealize by assuming that our measuring tape is infinitely thin, that it is quite but not necessarily perfectly flexible, and that it is calibrated, say in meters. Let us also retreat to the superficial and focus on the surface of just one body—say the earth, including its mountains and valleys, but leaving out of consideration liquid bodies like oceans, lakes, and rivers, which chapter 16 will comment on separately. The earth's surface will be supposed to support a topology as described in Part II, and if the measuring tape is constrained to lie at or near it, it will yield a metric, as roughly characterized at the end of section 15.1. Here are some advantages might accrue from conceptualizing Geometry in this way.

First, the measuring-tape approach allows us to measure curved lines, as already pointed out in section 15.1.

Second, all that is required of measuring tapes is rigidity in the 'linear dimension', which avoids the difficulty of requiring 'total rigidity' or undeformability. The problem of assuring that measuring tapes have constant length is one that is shared by all standards such as colors, weights, etc. (cf. section 4.7).

Third, the approach does not presuppose the laws of geometrical theory, Euclidean or otherwise. In fact, distances on the earth's surface obviously do not exactly conform to the laws of Euclidean plane geometry. Thus, geodesics can be characterized as the loci where tapes might be stretched on the earth's surface, and figures like triangles can be defined in terms of these loci. But clearly the figures do not satisfy the postulates of Euclidean or any other 'regular' geometry. *Ergo*, they do not have the properties that follow from Euclidean and other geometries, such as having coincident medians,[14] or of having 'congruent replicas'

[14] We should be careful about this, since certain coincidence theorems hold in 'super

everywhere on the surface.

Fourth, the method does not assume that lines of sight are straight, hence that transits can be used for surveying. The properties of this 'measuring-tape space' are independent of those of light. Therefore, characterizing the geometry of the space does not give rise to problems in ascertaining the conditions under which light travels in straight lines—i.e., in which lines of sight are straight.

Finally, the method is *practical*. Length and distance measurements are practically useful, e.g., for the purpose of estimating times and difficulties of travel, and measuring tapes provide useful assistance in this.[15] Interestingly, this utility transcends superposition, since what moves over the earth's surface tends to maintain only limited contact with it. Therefore to be useful in 'measuring' the time and difficulty of this kind of travel, the measuring-tape must also be in 'loose contact' with the surface, and suitable care in using the measuring tape can arrange that.

This 'loose superpositional' use is precisely what is required to meet a difficulty raised by Mandelbrot in his discussion of the length of the coastline of Britain in Chapter 5 of his (1983). This length, he says, is not just vague, but increases without limit, the more detail of promontories, bays, and progressively smaller inlets are taken into account. And, in view of the arbitrariness of the level of detail used in the measurement, he suggests that length is an inadequate concept in terms of which to describe the coastal outline, and that fractal concepts of measure and dimension would be an improvement (Mandelbrot, op. cit., pp. 25a,b). Now, the author agrees with Mandelbrot's premises, but he emphatically disagrees with his conclusion, that methods of measurement, of length or anything else, should be straitjacketed by a mathematical formalism, of whatever degree of sophistication. In fact, Mandelbrot seems to have taken literally the idea that 'true' measures of length, say of outlines or coastlines, should be carried out with instruments that conform exactly to the outlines being measured. But this overlooks the fact that the results of measurement should be useful. Of what practical use would

absolute geometry', which applies to all metric spaces. One such is the theorem that the perpendicular bisectors of the sides of a triangle are coincident, which is used by Adams and Adams, 2000, in application to arbitrary metric spaces, including the so called 'city-block' metrics.

[15]They are also obviously useful for non-geometric purposes like estimating amounts of materials of different kinds needed to accomplish various tasks, e.g., the length of cloth needed to make a suit. The general view of measurement as a relatively objective but *useful index of a phenomenon*—in this case time and difficulty of travel—is developed at some length on Adams, 1966. A view of scientific concept formation in general is advanced in Adams and Adams, 1987.

a fractal representation of Britain's coastline be to a sailor aiming to circumnavigate it?[16]

But in pointing to the practical importance of *not* measuring distances by means requiring exact superposition, of measuring tapes or anything else, we are going beyond the aims of this book.

---

[16]The author would criticize certain other recent and not-so-recent measurement-theoretic formalisms for similar reasons. Euclid's own metrology, formalized in Book V of *The Elements*, served as a straightjacket in the representation of magnitudes of all kinds, including lengths, areas, volumes, weights, times, velocities, and momentum, up through Galileo and beyond. More recently there have been controversies involving abstract 'representational' measurement theory (cf. especially, Krantz, Luce, Suppes, and Tversky, 1971, also two succeeding volumes by the same authors appearing in 1989 and 1990), one of whose aims is to avoid the ontological reduplication of reporting the results of measurement in 'denominate numbers' (e.g., '5 grams', '5 meters', '5 seconds') with their peculiar algebras (e.g., you can divide 5 meters by 5 seconds to get 1 meter per second, but you can't subtract one from the other). Adams, 1979, comments on some of these trends, but, as said, criticizes formal views for neglecting the practical uses for which measurements are designed.

Part IV

Miscellaneous Topics

16

Connections with Physical Theory

16.1 Introduction

We have said repeatedly that the reflections put forth in this essay have no direct implications for physical theory. One would expect such indirect connections as exist to be those of the geometrical theories presupposed by physical theories, but two obvious difficulties even stand in the way of this. First, unlike at least classical geometrical theory, the geometries presupposed in the formulations of physical theories pertain to 'spaces' that are special to those theories. As said in section 1.1, problems of characterizing the geometrics of physical theories are peculiar to them, and very little is said about those problems here. Second, an ontology of 2-dimensional and non-material features of the surfaces of bodies is fundamental to the present theory, and these play no part in physical theory. Worse, not only do physical theories not concern themselves with these things, but they implicitly call in question many of the assumptions that we have made about them, perhaps the most fundamental of which is that bodies *have* well defined surfaces, and that relations of coincidence and separation between features on them are well defined. This will be returned to in section 16.3, but some comments on the first difficulty may be suggestive. It is not so much a difficulty as a limit to the reach of the present analysis.

16.2 The Rôle of Non-Geometrical Considerations in Defining Spatial Relations in Physical Applications of Geometry

Consider the most classical of 'modern' physical theories, namely Newtonian mechanics. It is explicitly formulated in terms of positions in an 'absolute' Space, which is assumed to be Euclidean. Thus it is an application of that theory, which, except for leaving out surface features, is very close to the applications of the topology of 3-dimensional Euclidean geometry commented on in section 15.4. In particular, points of the space are incident in bodies, and coincident bodies can even individuate single points.[1] The problem, of which Newton was acutely aware, was that of motion and rest—of maintaining that it is 'meaningful' to say that a body *moves*, or equivalently that different points are incident in it at different times. Newton assumed that this is meaningful, and he described the famous Bucket Experiment to support the assumption.[2] This is important for us and it reinforces a point made in the preceding chapter, namely that in common with 'pure' geometrical theory, the present analysis is 'purely static' and cannot deal with motion, with changing or remaining in place. By contrast, the Bucket Experiment invokes a non-geometrical, *causal* law, of the effect of rotational motion partially implicitly to define motion.[3] This illustrates the point that non-geometrical considerations, physical, astronomical, or other, *must* enter into the characterization of motion—and when they do they are less general than pure geometrical theory because they are specific to the physical, astronomical, or other domain whose laws are invoked.[4] But, as said, this only reinforces the point that our present considerations do not bear on the geometrical problems that are special to physical theories, Newtonian or Relativistic. However, there are other bearings to be considered.

16.3 Marks in the Application of Physical Theory[5]

Applications of physical theories generally involve applications of Geometry as well, and those involve not only the kinematic principles that may be special to the theory but also static principles of pure Geome-

[1]A sphere on a flat surface for example, as in Routh, 1905, pp. 165–168. We have already commented on the principles of incidence of bodies in space that are formulated in Chapter 1 of Truesdell's Continuum Mechanics, 1977. Example 3 on p. 13 even seems to suggest that the author envisages possible bodies as occupying—even *being*—all regular, closed sets in a Euclidean space, which also suggests a way of bringing modal, 'constructability' ideas into the picture.

[2]But of course he admitted in Corollary 5 that this method would not give a complete implicit definition, since experiment cannot distinguish between systems of bodies at rest and ones in uniform rectilinear motion. Nor, until the advent of Relativity, and as Physics accustomed itself to working with hypothetical entities like forces, quantities of momentum, energy, etc., did the status of physical geometry become any clearer.

[3]Thus, it reminds one of Poincaré's claim that "The axioms of geometry are only definitions in disguise" (p. 50 of "Non-Euclidean Geometries")

[4]Thus, all of the Greek *physiologoi* took for granted 'absolute' rest and motion—as well as direction in Aristotle's case in physics and astronomy.

[5]This section amplifies ideas put forth in section 8 of Adams, 1994. However, it also deals with issues in the philosophy of Physics on which the author, who is not expert in these matters, can only comment with diffidence.

try. The considerations of this essay bear on the latter, and these in turn involve bodies' surfaces and the features on them—the very things concerning which the theory being applied is silent. For example, consider an application of the Law of Falling Bodies, according to which bodies falling from rest in a vacuum descend distances that are proportional to the squares of their times of descent. Even though this is extremely elementary, what is true of its applications is arguably true of applications of more advanced and sophisticated theories and laws.

Applying or testing the Law of Falling Bodies requires measurement, and this is the point at which observation and geometry enter the picture. Typical elementary laboratory apparatus involves a 'system' like the one pictured below (viz. Figure 9, p. 56, of Briggs and Bryan, 1903):

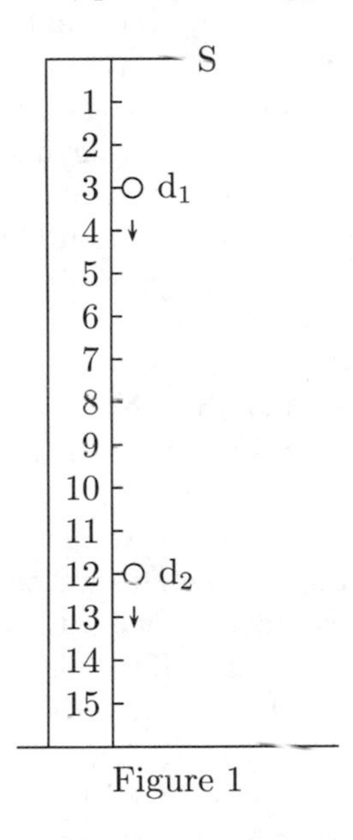

Figure 1

Here ball B is dropped from the upper 'suspension bar', S, and the distances that it has fallen at unit time intervals are measured. If the ball has fallen distance $d_1 = 1^2 \times 3 = 3$ 'units' in 1 second, it should fall a distance $d_2 = 2^2 \times 3 = 12$ units in 2 seconds, and so forth. These distances are supposed to be observed, and geometry enters both in ascertaining that the measuring rod is straight and vertical to the plane, and that its calibration marks are equal distances apart. For our purposes it is especially important that among the things observed are the calibration marks on the measuring rod, and the ball's near-coincidences with them at the different times. *The concrete application of the Law of Falling Bodies involves observing coincidences of marks on measuring instruments, the calibration of which involves geometrical operations.* But note that the law itself refers neither to the marks that are observed in this particular instance, nor to 'marks in general'. It is abstract, and it is silent about things like marks that must be observed in order to apply it. This suggests a general speculation.

It is typical of scientific theories to take for granted and not deal explicitly with the principles and procedures of measurement that are involved in applying them. In fact Science rarely pays explicit attention to these procedures, and when it does this it tends to be at times of 'crisis' in Kuhn's sense—for instance just prior to the advent of Relativity, when a few very gifted people like Einstein and Poincaré did this.[6] But then, scientific *theories* are only the 'tip of the iceberg' in Science—although like tips of icebergs, they are the most visible parts of Science, and they attract the most notice. But then too, is it really clear that marks like the calibrations on measuring instruments are ignored by scientific theory? This leads to our most philosophical comments.

Suppose that the calibration marks on the vertical measuring rod pictured in Figure 1 are colored black—they are short black stripes ending at the rod's edge. Now, Physics *is* concerned with colors, with wavelengths of light reflected from places on surfaces. Hence the laws of Physics do apply to the black calibration marks—it is only that the Law of Falling Bodies does not apply to them. But we must ask whether the reflected light rays that Physics deals with are the *same* as the calibration marks on the measuring rod. We don't want to say that the rays are like the marks in being *on* the surface of rod, or that they are 2-dimensional, or that they are coincident with other rays in the way that the marks may coincide with marks. There is a 'conceptual gap' between a calibration mark and the light reflected from it, and, we would suggest, it is only by mistaking one for the other that we are led to question the existence of the mark as a 'thing'. On the other hand the calibration mark has three strikes against it as a thing: (1) It is not self-subsistent, since its existence depends on that of the measuring rod it is on. (2) It is not itself subject to laws like the Law of Falling bodies. And (3), it has no causal properties, and therefore it is of no concern in human affairs. But (3) can be questioned.

It is the *mark* that we see, and not the light rays that are reflected from where it is. A philosophical observation too often forgotten is that

[6]The author does not claim intellectual kinship with such giants as these, nor that we are presently confronted with a scientific crisis—excepting, of course, the state of 'permanent crisis' in the foundations of Quantum theory. But an onlooker may at least be excused for reflecting on Science's methods.

we do not see the retinas of our eyes (only persons looking through ophthalmoscopes can do that), much less the light rays that impinge on them. If we see anything, it is the body from which the light rays are reflected, and if we see that, why deny that we see the mark? Moreover, seeing the mark *is* of concern in human and scientific affairs. Assuming that laws like the Law of Falling Bodies are tested by observation, some things must be observed when they are tested, and marks must be among those things. If the observer couldn't observe them and their near coincidences with the ball at given instants she couldn't carry out the test. This is indicative of the marks' epistemological significance.

Marks label places, but relations involving marks are of concern for what they are signs of, and not for themselves. In the falling body experiment, the coincidence of the mark with the ball is a sign of how far the ball has fallen. Its significance is like that of the level of a barometer or a clinical thermometer, the first of which is a sign of weather to come, and the second of which is a sign of a patient's health. So, although meteorology and medicine use barometers and thermometers, these instruments are not objects of meteorological or medical study. And, although Physics tests its theories by observing visible marks, the marks are not objects of study in that science. But there is a science whose objects are, if not marks, at least things that are closer to being marks than physical objects are.

Consider the law that the medians of a plane triangle are coincident, which was discussed in section 15.2. Figure 1 of that section was a concrete 'thing', namely a triangle formed by the edges of three streaks drawn on a flat sheet of paper, and the theorem of the medians was applied to it. If this is a real application, and not, as Plato's followers would have it, a visible reminder of an idea that can only be grasped by reason, then the law *has* a concrete application. It is only difficult to recognize it as an application because it does not fit the simple 'applied formal system' model.[7] In fact the previous parts of this essay have attempted to clarify the relation, if not of Geometry, to application, at least of the topology presupposed by it to application. Lacking such clarification, Geometry seems to have an ambivalent scientific status.

Because the objects of geometrical theory, e.g., points, are abstract, and often these abstractions are not uniquely individuated in concrete particulars, Geometry may have the appearance of a purely mathematical science, whose postulates are principles of pure reason. But the difficulty of justifying the Parallels Postulate by pure reason makes it clear

[7]This is the model that is formalized in the logical *theory of models*, but the intuitive ideas underlying it go back to antiquity, as well as being explicit in 19th century work on the foundations of geometry.

that Geometry has an empirical aspect. However, if the foregoing reflections are right, these aspects relate not so much to *physical* objects, i.e., to the objects of physical theory, as to the signs and language in terms of which the objects of Physics are described.

Concluding this chapter we will comment briefly on certain non-geometrical objects that marks are signs of.

16.4 Liquids and Matter[8]

Heretofore we have been concerned with marks and other features of the surfaces of solid bodies, the reason being that these things tend to be fairly permanent and to remain in fairly permanent relations of coincidence and separation. But quantities of liquid can also be regarded as bodies (e.g., Lake Erie is a body of water), and their surfaces also exhibit 'features', e.g., waves. What marks the difference between solids and liquids is the fact that liquids' surfaces tend to exhibit very ephemeral features. For example, even when two waves continue in existence, if they ever 'encounter' one another they are likely to separate again quickly. Related to this is the fact that even though waves can properly be regarded as features, they do not mark *places* on the surfaces they are on. In fact, the lack of place-marking features explains why 'extra-thalasic signs' must be employed for navigation at sea. There is another sense in which waves and other marine *ephemera* are also not 'markers'.

As Heraclitus has it, one cannot step twice in the same river,[9] and in a sense this is true. Although one may step twice in *the Mississippi*, one doesn't step twice in the same *water*. The Mississippi *flows*, and while the river itself remains relatively fixed in place the water in it moves. At any one instant the river may be said to be *formed* from a certain quantity of water (the water that forms the river may be called its *material content*), but that changes from moment to moment. This is not like a solid body, e.g., a block of ice or a nugget of gold, neither of which flows in the way the Mississippi does. The material contents of the ice and the gold are likely to remain fairly constant over a significant period of time. Moreover, to the extent that forces and the laws of dynamics apply to *bodies*, they apply to solids. If forces do apply to Lake Erie, we do not say that its rate of acceleration is proportional to their resultant. In fact, when dynamical laws are applied to liquids, as in Hydrodynamics, they do this by way of quantities of liquid of fixed material content. And laws

[8]Ideas in the section to follow were first set forth in sections 4.3 and 4.4 of Adams, 1984.

[9]At least, this is the schoolboy version of Heraclitus' 'grand idea'. But Kirk, 1951, pp. 35–42 (reprinted in Mourelatos, 1974, pp. 189–196) claims that Plato, Aristotle, and many later thinkers seriously misunderstood this.

of change of state in Thermodynamics, say from ice to water, also apply to such quantities. All of this is obvious and much of it was known to the ancients, but there are special points that are relevant in the present context.

It is not wholly coincidental that bodies with relatively fixed material content like gold nuggets are also ones that bear relatively permanent surface markings standing in relatively constant relations of coincidence and separation. In fact, it might be said that the markings are signs not only of physical geometry, but their constancy is a sign of the material content of what it is applied to.[10] For example, the gold nugget's color, gold, is a sign that gold is the kind of matter of which it is formed.

The laws that associate the substrates of bodies, including quantities of liquid, with the observable surface features that are associated with them are perhaps quite complicated, and we do not aim to give a comprehensive account of them here. Our point is simply that these laws provide 'windows' through which these substrates can be viewed directly, beneath the surfaces of the bodies that are formed from them. And, these windows constitute perhaps as important a connection between surface features and physical theory as their connection with physical geometry.

The next chapter follows up on this point, with comments on the relation between surface features and sense data that, we hold, are often mistaken for them, but whose relation to the external world is more difficult to explain. For better or for worse, this will bring us into the murky realms of philosophical psychology.[11]

[10] Of course a smoothly flowing river may have no surface features, unless smoothness itself is counted as a feature, and this in turn may count as surface-feature constancy. But in this case the signs are deceptive—as natural signs can always be.

[11] And what psychology is not 'philosophical'?

17

Surface Feature, Sense Datum, and Psychology[1]

17.1 Introduction

Section 16.4 commented on how features of a body's surface can serve as signs of its material substrate, and now we will add a third thing to the picture, namely the 'appearance' of the body, which we are accustomed to conceive as a subjective 'sense datum'. Although section 17.4 will question this conception, for the moment we will leave it unquestioned and focus on its relations to the body and its surface features. These can be pictured as forming a triangle whose vertices are the body, the surface-feature, and the sense datum. For example, consider a square wooden block with a small circle and line through it stamped on one of its faces. It and its relations to its surface features and to sense data can be diagrammed as follows:

[1]This chapter expands on ideas commented on very briefly in section V of Adams, 1984. However, there has been such a torrent of writing on these topics in recent decades, especially in the psychological literature, that even here only a few high points can be touched on. Except as they bring surface features into the picture, the author makes no claim to originality in the comments that follow.

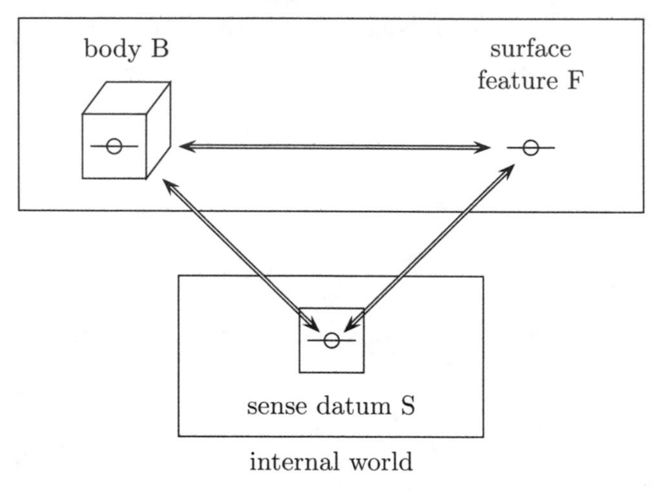

Figure 1

Now, traditional empiricism concerned itself with the relation between the sense datum and the body, and how the former can yield 'our knowledge of the external world'. The seeming unknowability of this relation has given rise to the well known skeptical doubts concerning the possibility of empirical knowledge that philosophers have wrestled with from time immemorial. But we have added a new element to the picture, namely the surface feature together with its relation to both the body and the sense datum. In one way this complicates the picture, but in another it could seem to solve some of the problems that arise when the only things dreamt of in empiricist philosophy are the mind and the body. We will comment very speculatively on a step towards a resolution.

For now let us take it for granted that the relation between the surface feature and the body is unproblematical, although it is difficult to characterize in detail, as we have seen. But if, somehow, we could account for the relation between the sense datum and the surface feature, and add to that the 'unproblematic' relation between the surface feature and the body, we might solve at least some of the traditional problems. What makes this plausible is the fact that the sense datum seems more closely to resemble the surface feature than it does the body, and if this 'seeming' could be substantiated it would yield a basis for a representative account of sense perception. Let us look into this, first by noting some ways in which the sense datum might be said to *resemble* the surface feature, and why it does this more than it resembles the body.

17.2 Similarities between Surface Features and Sense Data

Perhaps the most striking similarity between the surface features and the 'appearance' of the block is their dimensionality. Unlike the block itself, which is 3-dimensional, both the surface feature and the sense datum are 2-dimensional. Many epistemologists have noticed the dimensional dissimilarity between the solid and its appearances, and three quotations are worth citing. In paragraph 44 of the *New Theory of Vision*, Berkeley describes the 'visible moon' as a "round luminous plane, about 30 visible points in diameter," which, since it is plane, is clearly 2-dimensional.[2] On p. 95 of *The Structure of Appearance* Nelson Goodman says "A visual presentation of a baseball, for example, is spatially bidimensional, whereas the baseball, for example, is spatially tridimensional." And in section VII of "The Relation of Sense Data to Physics," Bertrand Russell speaks of a penny's appearance as being sometimes round and sometimes elliptical, which implies that the appearance is 2-dimensional since roundness and ellipticality are shapes of 2-dimensional figures.

The foregoing quotations also show that besides dimensionality, appearances are commonly described in terms like "round" or "black" that can also be applied to surface features. For example, the circle with the line through it on the face of the block pictured in Figure 1 is round and black, though it is a feature of the block's surface and not an appearance. In fact, some 'naive realists' have held that sense data are literally parts of bodies' surfaces, as the discussion in Chapter II of H. H. Price's *Perception* indicates,[3] though Price rejects this thesis. However, while one may well dispute the idea that sense data are *parts* of bodies' surfaces, or even that they resemble *features* of their surfaces (since they seem to have some of the attributes of the features), there can be no doubt that the same words are used to describe the sense data and the features. That is itself a datum that a logic of perception ought to take into account. Let us postulate a vague 'principle' of this logic that goes

[2]Recall Berkeley's dictum "An idea can like nothing but another idea" (*Principles*, par. 9), which, if likeness consists in sameness of dimension, is valid when only bodies and ideas (and *spirits* for Berkeley) are envisaged, but which becomes invalid when surface features enter the picture.

[3]"...Naive Realism holds that in the case of a visual or tactual sense-datum, belonging to means the same as *being a part of the surface of*: in the literal sense in which the surface of one side of this page is part of the whole surface of the page," *Perception*, p. 26. That a visual or tactual sense datum 'belongs to' a thing means here that it is a datum pertaining to the thing that is derived from looking at or touching it.

beyond the datum, and explains how sense data might be said to resemble features of bodies' surfaces more than they do the bodies themselves. The postulate, a special interpretation of which will be important, is as follows.

We use the same words to label the appearances or 'impressions' that accompany our most forceful perceptions of things as we use to label the things perceived.[4] So, for instance, seeing a circular mark we might say "the mark has a circular appearance," thus seemingly transferring our application of the term 'circular' from the mark to its appearance. Figuratively speaking, we seem to displace or transpose the external world into a fanciful world of the mind by 'gluing' labels for things in the former onto sense impressions or appearances in the latter.[5] This postulate also explains why we are more apt to transfer terms applicable to features of bodies' surfaces to appearances than we are to transfer terms that apply to whole bodies.

When we look at a body like the block of wood with the circle and line on one face, we are most forcefully aware of the features on its surface that are before our eyes. What is likely to 'jump into our minds' is the thought that we might express as "there is a circle with a line through it."[6] We are much less likely to jump to conclusions about attributes of the whole block, like being a cube, or attributes of parts of the block that are hidden from our view, either on the back or inside of the block.

Another thing that our postulate might explain is a connections between appearance and reality, as follows.

---

[4]The following quotation from paragraph 144 of Berkeley's *Principles* suggests something akin to our hypothesis:

> But when we find the same signs suggest the same things all over the world; when we know they are not of human institution, and cannot remember ever having learned their signification, but think that at first sight they would have suggested the same things to us they do now: all this persuades us they are of the same species as the things respectively represented by them, and it is by a natural resemblance they suggest them to our minds.

[5]But we shouldn't leap too quickly to the conclusion that the appearances are 'in the mind', since the paper of Russell's from which the quotation above is cited maintained that sense data are *physical.*

[6]Berkeley somewhere makes a similar remark about the relation between a sound, e.g., of a bell, and the bell that produces it. So, when we hear a bell ringing we may say "that is a bell," thereby displacing the name of the thing, the bell, to the sensation that accompanies our hearing it.

17.3 Appearance, Reality, Superposition, and Construction

Change "The mark has a circular appearance" to "The line through the circle has the appearance of being straight,"[7] and consider the difference between *looking straight* and *being straight*. Berkeley seems to argue that this difference is manifest *within appearances* when he writes "There is a *rerum natura*, and the distinction between realities and chimeras retains its full force" (*Principles of Human Knowledge*, par. 34). But this is doubtful.

The test for straightness that was discussed in section 14.2 involves superimposing a straightedge on the line whose straightness is to be determined. The important thing for us is that merely looking at the line doesn't apply this test: all that mere passive observation can do is lead us to think that if the test *were* applied to the line it would prove to be straight. We might even imagine or 'picture' carrying out the test, but that wouldn't actually apply it. In fact one of sense datum theory's most serious deficiencies is that it gives no place to tests, especially ones requiring superposition. 'Ideas' or appearances are not superimposed, and even Berkeley, who is the only 'ideaist' known to me who comments on superposition, is skeptical of its 'meaningfulness' (cf. the passage from the *Commonplace Book* quoted in section 12.1). That is why sense datum theory has so much trouble with spatial relations and 'exteriority'. Lines may or may not look straight and things may look near or far, but real straightness and distance are determined by tests that involve superposition, and those have no place in the realm of appearance.[8] This is also true of *pictures*, which are commonly mistaken for appearances because, one suspects, they suffer from limitations similar to those of appearances when it comes to 'revealing reality'. But pictures are *not* appearances, as will be noted in section 17.5, and first we will speculate on a positive account of the latter.

17.4 Towards a Positive Account of Appearances

Although appearances are often considered to be mental or at least subjective, so far our considerations have been consistent with a modest Ryleian view (Ryle, 1949), that there is no 'ghost in the machine', and nothing 'hidden in the mind'. The author is not a dyed-in-the-wool phys-

[7]More will be said in the following section about the transition from "The mark looks straight" to "The mark has the appearance (look) of being straight," which seems to reify the appearance.

[8]The same difficulty arises in accounting for visual illusions, like the Müller-Lyer illusion depicted below involving two line segments with inward and outwardly

icalist, but let us at least see where this leads.

Our 'postulate' that we use the same words to label the appearances or sense impressions that accompany our most forceful impressions of things as we use to label the things perceived is at least consistent with the assumption that all there are are things, their surface features, words, and sentences formed from them like "The mark has a circular appearance." In particular, even though "a circular appearance" seems to refer to something subjective (its logical form is that of a referring expression, so it seems to have to refer to something), the entire sentence can be used to express what would be more simply expressed as "The mark appears to be circular," which is a judgement about the shape of the mark, and not about its appearance. A seemingly infallible judgement about something 'immediately before the mind' is seen actually to be a fallible judgement about something outside of it—just as "The President's probable age is 52 years" really expresses a fallible judgement about the President's actual age and not an infallible judgment about his probable age. This is not to deny the existence of the ghost in the machine, but only to leave its nature somewhat more open than it is when it is conceived of as entertaining pictures or images or appearances that resemble their prototypes in the external world. Here are three things that might be accounted for on this more parsimonious conception.

First, the postulate that we use the same words to label the appearances or sense impressions that accompany our most forceful impressions of things as we use to label the things perceived now appears to be a special case of a simple grammatical fact. We apply the same word, 'circular', to the appearance of the mark that we apply to the mark simply because "the mark has a circular appearance" expresses the same thing as "The mark appears to be circular." If we add that this 'transposition' takes place in circumstances under which we have a forceful impression

---

pointing arrowheads at their ends:

What actually shows that the segments are equal is measurement, say by superimposing graduated rulers on the segments, and the problem is to explain why passive observation in the absence of rulers is so misleading here. In my belief, Berkeley was on the right track in his discussion of the Moon Illusion, which pointed out some cues that systematically mislead us in judging the size of the moon when looked at on the horizon, as against directly overhead. In the Müller-Lyer case we want to discover what cues mislead us in judging what the result of superimposing graduated rulers on the segments would be. Presumably they would have to do with our experience with rulers, and possibly with other things that have 'arrow points' like the ones in the Müller-Lyer figures.

of the mark, that is because those are the circumstances in which we are apt to form judgements as to the mark's attributes.

Second, a corollary of the above is that the alleged 'resemblance' between the appearance of the mark and the mark itself is not a mystery. The resemblance consists in our applying the same words to them, e.g., circular, and this is explained on purely grammatical grounds.

Third, there is still a distinction between appearance and reality, but it is 'within the external world'. "The mark has a circular appearance" and "The mark appears to be circular" correspond or fail to correspond to reality in the same way, according as the mark is or is not circular. And the *test* for this is applied to the mark, not to some ineluctable mental entity called an 'appearance'.

But let us return to a more full-blown mentalism, starting with comments on *pictures*. Their subjective surrogates, mental pictures, are often likened to appearances, but they are indubitably mental and they are brought in here because this leads ultimately to comments on visual 'geometries'.

17.5 Physical and Mental Pictures

Pictures are like appearances in some ways, but really they are a 'fourth quiddity', in addition to appearances, bodies, and surface features, and their connections to the latter complicate an already complicated picture. Let us comment on them briefly 'for their own sake', before turning to mental pictures and then to their alleged geometries.

Physical pictures often show how things and their surface features appear from particular points of view, and they may show some of the relations in which things in them stand. For example, the picture of block B with the feature F on one face of it in Figure 1 shows how they appear from the front, and it also shows the geometrical or at least the topological relation between them. In fact, F and the face it is on both correspond to surface features in the picture (although the face itself isn't one), and what shows the relation between F and the face it is on is the relation between the surface features to which they correspond in the picture.[9] Thus, the feature is near the center of what corresponds to the face, and therefore we conclude that the former is roughly in the center of the latter.

One respect in which the picture is like an appearance is that neither of them show *tests*, say for the straightness of the line through the circle.

---

[9]Shades of Wittgenstein's 'picture theory of meaning'! Thus *Tractatus*, 2.15: "That the elements of the picture are combined in a definite way, represents that the things are so combined with one another" (Ogden translation, paperback edition, 1981, p. 39).

Thus, the picture doesn't show a straightedge being superimposed on the line, and even if it did that wouldn't apply the test, which requires superimposing an actual straightedge on the line, not picturing one being superimposed on it. However, pictures and appearances can be very unlike.

If appearances *are* 'entities', they are not artificial. They are *of* things, and they can fail to correspond to these things in ways that can be determined by applying tests, which themselves fall outside the realm of appearance. On the other hand, a picture is an artificial thing, and its correspondence to a reality outside itself, if it exists at all, is an 'interpretation', either of the picture or of the intention of its maker. For example, a good picture of George Washington is a bad picture of Abraham Lincoln, but who the picture is considered to be of depends on how it is interpreted, which the picture's maker normally attempts to determine—perhaps by labelling it. More exactly, *qua thing*, the picture is not 'of' anything, and it is only relative to an interpretation that it makes sense to say that it corresponds or fails to correspond to this or that part of 'external reality'.[10]

On the other hand mental pictures, especially visual ones, though they are somewhat like physical pictures,[11] do not 'represent' anything and do not have to be interpreted. They are self-subsistent mental 'things', and for this reason alone they rate special comment, because more than other entities they seem most apt as subjects for the laws of visual 'geometries'.

[10]Given this, the claim that pictures show how things appear must be modified: Artists, specifically painters, *commonly intend* their pictures to be interpreted as showing how things appear from perspectives. Concomitantly, they commonly intend their pictures to be interpreted in a way in which their elements correspond to things that can be seen by a viewer from the pictures' perspectives. On the other hand, especially in modern art this is often not true, e.g., in Picasso's "Les Demoiselles d'Avignon," which simultaneously presents overlapping but distinct perspectives on certain of its figures (courtesy of Bill Adams). More generally, a painting, though it is made intentionally, may not only be non-representational, but it may be made without the artist intending to induce a specific interpretation in the beholder's mind. But then, the beholder is often puzzled to guess at the artist's intentions.

[11]Indeed, mental pictures, surface features, and appearances are all like physical pictures in that they are frequently pictured in the same way. Their *resemblance* to 'impressions' or ideas of other kinds, and the problem of distinguishing them from these other ideas, has been discussed by many sense data empiricists. In Chapter II of Part IV of the *Treatise of Human Nature*, Hume distinguished the impressions that we attribute 'reality' to from imaginings by reference to their constancy and coherence (Selby-Bigge, pp. 194, 195), and later on p. 629 of the Appendix to the *Treatise* he adds that the 'real' impression is distinguished from the 'fictitious' one by a different 'feeling', related to *belief* or judgement. Thus, appearances are about or 'of' things that are external to themselves, whereas imaginings simply 'are'.

17.6 Visual Geometry I: Two Philosophical Theories

We have already noted that sense datum theorists commonly apply terms for geometrical concepts to visual 'appearances'. Thus, Berkeley says the visible moon is *round*, Goodman says that the 'visual presentation' of a baseball is *bidimensional*, and Russell says that a penny's appearance is *sometimes round and sometimes elliptical.* It is only a step farther to suppose that such 'appearances' have a geometry of their own, *visual geometry*, and certain writers have attempted to formulate its principles and contrast them with those of standard Euclidean geometry. We will side-step questions as to whether these principles properly apply to appearances, which section 17.4 suggested might only be 'logical transformations' of things in the external world, and concentrate on mental images, which almost certainly are not. Consider first the Scottish 'philosopher of common sense', Thomas Reid, 1710–1796, chapter 6, section IX of whose book *An Inquiry into the Human Mind*, 1764, outlines a quasi-formal theory of a 'geometry of visibles'.

Reid's geometry was really a general theory of visual phenomena, and especially of visual perception, and, in common with Kepler, Descartes, Berkeley, Hume and many others, it fits squarely into the 'camera obscura model' described by Crary as being "... without question the most widely used model (from the early 1600s to the end of the 1700s) for explaining human vision, and for representing the relation of a perceiver and the position of a knowing subject in the external world" (*Techniques of the Observer*, 1990, p. 27, author's parenthetical insertion). The camera obscura or pinhole camera was known to the ancients. It functions by having light pass through a small 'pinhole' in one wall of a darkened chamber (the 'camera obscura') and fall on the opposite wall, thereby producing an inverted image of the scene from which the light emanated. However, it was Kepler's discovery, around 1604, that the eye functions essentially as such a camera in which the retina is the 'opposite wall', that led to the adoption of this as a model of human visual perception. And, Reid along with many other of its devotees devoted much space to problems raised by it—the inverted retinal image, binocular vision, visual illusions (e.g., the moon illusion), and so on. In fact, particularly because Reid's arguments often depend on ocular considerations, it may be a stretch to apply his geometry to mental pictures, which, even if 'visual', need not be perceptual or indeed have anything to do with the functioning of organs of sense. But let us make this application anyway, since at least some of the points to be raised in the following discussion are independent of whether the 'visual field' is perceptual or imaginary.

Two quotations epitomize Reid's views on the geometry of visibles:

> Supposing the eye placed in the center of a sphere, every great circle of the sphere will have the same appearance to the eye as if it were a straight line (p. 122).
>
> The properties therefore of visible right-lined (i.e., straight-sided) triangles, are not the same with the properties of plain triangles, but are the same with those of spherical triangles (pp. 123–4, author's parenthetical insertion).

Clearly this geometry is what we would now call *spherical non-Euclidian geometry*, and Reid formulates a number of its salient properties, such as that the sum of the angles of a triangle is greater than two right angles, that similar triangles must actually be congruent, and so on (pp. 124–5).[12] But let us concentrate on the thesis that every great circle of the sphere at whose center the eye is placed will have the same appearance to the eye as if it were a straight line. It will look like a straight line and it will *be* one in Reid's geometry. Why? So far as the author can see, Reid assumes that things that appear in retinal images, 'visible figures', look like the things from which the light that forms them emanates, and in the geometry of visibles they are endowed with the properties of these things. So, the geometry of visibles treats as straight any visible figure that is formed by light reflected into the eye from a straightedge.

Well! Assuming that this is Reid's argument for the basic thesis of his geometry, there are so many objections to it that it is hard to know where to start. The argument clearly doesn't apply to mental pictures—to 'images' that our minds may form even when our eyes are closed and there are no images on their retinas. Even when we are looking at things with our eyes open, we don't see our own retinal images. And, even if we could see our retinal images, it is not self-evident that ones that are formed by light emanating from straightedges would be perceived as straight. In fact, we 'perceive' as straight things that appear to *be* straight, and Reid's theory has no place for a *test* of straightness that applies directly to images, retinal or otherwise, which is analogous even to the test hinted at in Definition 4 of Book I of *The Elements*, which was commented on at greater length section 13.2. But the absence of tests, for straightness or any other geometrical property, is characteristic of many 'geometries' of vision, and it makes them more like uninterpreted formal systems than like *theories* of well-defined phenomena. Let us therefore turn to another theory, which has its own peculiarities although the objection just raised is less applicable to it.

---

[12]The author has even seen it said that Reid's theory should be recognized as an important precursor of non-Euclidean geometry.

Sections 9–13 of Reichenbach's *Philosophy of Space and Time* (1927, English translation 1956), present an interesting attempt to interpret Kant's doctrine that the laws of geometry are synthetic *a priori* in a way that involves detailed considerations of visualization, and which also avoids the standard criticism that Kantian arguments can only support Euclidean geometry. To begin with, Reichenbach distinguishes what he calls *image-producing* from *normative* functions of visualization, and some things he says about the former are subject to the criticism just levelled against Reid. Speaking on p. 39 of the visualization of an angle of a triangle, he says "If we want a more precise statement, we must try much harder; only then can we reproduce the triangle vividly enough in our imagination to estimate the size of the angle." Here a triangle is visualized so vividly that we can measure its angles—which is surely Reidian geometry with a vengeance! On the other hand, 'normative' uses of visualization are both less questionable and more interesting. Reichenbach uses one in considering how a geometrical proposition might be 'validated in a Kantian manner'.

Focussing on Pasch's Axiom he says on p. 39:

> I have a triangle and a straight line intersecting one of the sides of the triangle; if sufficiently prolonged will the line also intersect another side of the triangle? Visualization says "yes." It simply demands this answer and I can do nothing about it.[13]

And, on the same page the author also says that Kant's (view of) the "synthetic character of intuition" springs from this kind of visualization. But *this* visualization is an adjunct to a Gedanken experimental 'validation' of a geometrical proposition, of which image-production is only a part.

What is visualized in the example just cited is the *process* of prolonging a line that is actually seen (note the combination of perception and visualization!), and this could include visualizing the use of a straightedge to determine its straightness. Of course visualizing the use of a straightedge to determine the straightness of a visualized line is not actually determining the straightness of anything, but that no more makes the entire argument unconvincing than geometrical proofs that use diagrams are unconvincing because the lines in the diagrams are not perfectly straight. In fact, Reichenbach comments on similarities between diagrams and visualizations in geometrical arguments on p. 47,

[13]Around p. 284 of *The Bounds of Sense*, Strawson, discussing a 'phenomenal interpretation' of Kantian doctrine, makes similar comments relating to visualizations of intersecting circles.

and he also points out on p. 54 that the persuasiveness these 'devices' depends on experience. And, whether or not arguments that use these devices are rigorous, it is not an uninteresting interpretation of Kant's view of the postulates of geometry that they rest on such grounds.

In any case, while the 'normative' visualization commented on above relates to geometry, it no more has its own geometry than do the words used in geometrical demonstrations, and perhaps it should not be discussed 'in the same breath' as Reid's sort of visual geometry. Let us therefore turn to a recent psychological theory of visual perception, which could seem to be similar to Reid's, but which is infinitely more detailed. Although it is more limited in certain respects, it has interesting points of comparison with the theory of surface features and their topologies that is developed in Parts I and II of the present essay.[14]

17.7 Visual Geometry II: Marr's Theory

David Marr's renowned essay is titled *Vision, a Computational Investigation into the Human Representation and Processing of Visual Information* (1982), but in spite of this and in spite of such statements as

> ...the true heart of visual perception is the inference from the structure of an image about the structure of the real world outside. The theory of vision is exactly the theory of how to do this,... (p. 68),

what the book treats of is a computer model of some but by no means all aspects of human visual perception. The aspects treated can be illustrated by a hypothetical application to the visual inspection and 'processing' of a rectangular array of dots:

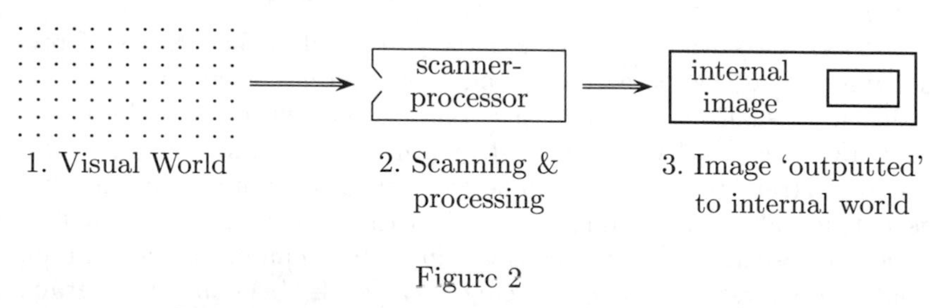

Figure 2

[14]The comments to follow are made with extreme diffidence, since the author has only a tenuous, outsider's grasp of Marr's theory. This theory is singled out for comment from almost a torrent of recent works on visual perception, because of the close relation between certain of its aspects and ideas discussed in the present essay, which justify and even require notice.

This pictures a three-stage process in which the array, which is part of the 'visual world' external to the observer, is 'viewed' at stage 1 by an electronic scanner, which is the computer analogue of the eye. At stage 2 the computer 'processes' the information retrieved by the scanner, which putatively models the way in which the brain processes the optic nerve impulses that are stimulated in the retina as a result of light emanating from the array and impinging on its receptors. At stage 3 an 'image' is 'outputted' by the computer into the internal world of the brain, or what corresponds to it in Marr's model. The image 'represents' the array, not in exactly resembling it, but rather for purposes of information-retrieval. Here, for example, the image is rather like what Marr calls a 'raw primal sketch' (section 2.2), which enables 'the mind's eye' to pick out the edge or outline of the array.

The foregoing example illustrates both what Marr's model subtracts from and what it may add to human visual processing. It certainly leaves out a stage corresponding to the viewer's making an 'inference from the structure of an image about the structure of the real world outside'. That is, Marr's process ends with the image, which is the analogue of something in the mind of the viewer, and it does not go back to the world outside by any 'inferential process'. And, this is connected to something that Marr's model adds, which is currently hotly debated.[15] That is the very existence of images in the minds of visual observers, analogous to the one produced at stage 3 above, which concomitantly intimates that there are 'homunculi' inside our heads that scan *these* images, and perhaps make the inferences to the world outside that Marr speaks of.[16] The present author, while recording his skepticism concerning the image and the observing homunculus,[17] will leave the controversy surrounding them at this point, and end the discussion of Marr's theory with comments on ways in which it may contribute importantly to the enterprise of surface topological analysis sketched in Parts I and II of this essay.

---

[15] At least it *was* hotly debated at the time that Marr's book appeared; see for instance the collection of essays edited by Block, 1982. Our own acquaintance with the controversy stops short at about this time.

[16] Adding the homunculus turns the picture into a modern version of Plato's Allegory of the Cave, from the opening of Book VII of *The Republic*.

[17] Recent theoretical investigations by Patrick Suppes and his associates, which are bolstered by new empirical studies of electrical activity in the brain, develop an account of sensory processing in which patterns of oscillations in the brain's electrical field take the place of mental images, and in which wave-comparison mechanisms take the place of the observing homunculus (from "A Physical Model of the Brain's Computation of Truth," *Eleventh Annual Alfred Tarski Lecture*, Berkeley, March 19, 1999).

The things scanned and analyzed according to Marr's theory include marks like '—ө—', which, while they are described as being parts of the 'visual world',[18] are also among the surface features that are dealt with and theorized about in Parts I and II of the present essay. Moreover, the processes that Marr describes yield images that 'manifest' topological properties of the marks and their relations to one another that closely parallel topological properties and relations of marks discussed in Part II of our essay. Thus, the properties that scanning and processing 'make manifest' include *spatial continuity* (esp. pp. 49–50), *edges* (esp. pp. 68–73), '*terminations*' (esp. pp. 76–79), and even aspects of *dimensionality*. It is impossible to go into detail concerning Marr's way of dealing with these things, but quick comments on continuity and edges are in order.

Some of the things that Marr says about continuity are reminiscent of things that our section 7.5 says about connectivity. For example, under the heading 'Spatial continuity' Marr speaks of "*Markings generated on a surface by a single process*," seeming to suggest that such markings tend to be continuous, which mirrors our Connectivity Hypothesis, 7.5.1. But linking continuity to mode of generation is really more characteristic of the approach taken in the present essay than it is of Marr's thinking. The most fundamental relations of our theory, coincidence and separation, are characterized in section 2.3 in terms of the *history* or *generation* of the things that are related. But the history of a thing is something that computer scanning cannot reveal, and what Marr focuses on instead are 'cues' that scanning can pick up, like 'smoothness of contour' (p. 49), much as scanning a person's appearance usually furnishes a reliable estimate of his or her age within a few years, even though 'real' age cannot be determined in this way. This difference, which could be described as 'the difference between concept and cue', is perhaps better illustrated in the detection of *edges*, as pictured in Figure 2 in the case of the rectangular array of dots.

The concept of a *boundary*, say of a mark, which corresponds roughly to an *edge* in Marr's theory, is operationalized in the present work in terms of a *boundary cover*, which is defined as a (finite) system of marks that are touched by every connected mark that is touched but not covered by the given mark (Theorem 8.2.2). And, since connectivity is defined genetically, this property cannot be detected by scanning. On the other hand, Marr's way of analyzing edges of things involves the detec-

[18]Thus, on p. 69 of Marr's book he says "...the visual world is made of contours, creases, scratches, marks, shadows, and shading, and these are spatially localized." Not all of these, and specifically, contours, shadows, and shading, are surface features (though they may be descried on surfaces), but creases, scratches, and marks certainly are.

tion of gradients in their illumination, which can be detected by scanning (pp. 68–75),[19] though it presumably results from gradients in the intensity of the 'unscannable light' that is reflected from the objects scanned.[20] *But*, here it seems that both the methods and the results differ markedly. According to Marr, the 'edge image' of the array of dots is a connected rectangular outline, while for us a boundary cover of it is a union of boundary covers of the dots that compose it, which could be highly disconnected (or empty, if the dots were single points). The question "Which, if either, is 'right'?" leads to our final series of comments.

17.8 Concluding Philosophical Reflections

The operational criteria developed in this work for determining connectivity, boundaries, and other topological properties and relations of surface features pose as being 'real', whereas it is suggested that Marr only analyzes certain cues that visual scanning can use to determine these things, which are in principle not accessible to this kind of determination. And, the example of the array of dots shows that the results of these analyses can differ quite radically. But are our own criteria 'really real'? Let us return to continuity and this essay's analogue of it, connectivity.

The very expression 'Connectivity Hypothesis' makes it clear that the connectivity of marks *is* a hypothesis or, more exactly, that their point sets are connected in the topologies of the surfaces that they are on. Consider, however, the streak pictured in section 7.5,

x y

▬▬ ,

the connectedness of which only close inspection would seem to be able to tell. But can it? It follows from Theorem 7.5.2(2) that the streak is connected only if it is not a union of separate marks, say x and y, but according to the discussion in section 2.2 there is no way of determining

[19]This is strikingly illustrated in Figure 2–23, p. 74, where a highly distorted, 'quantized' figure of Abraham Lincoln is progressively transformed into an 'outline sketch', in a way that is very roughly illustrated by the transformation from the array of dots in Figure 2 into a rectangular outline.

[20]As an aside, note that while varying degrees of illumination of a scanner's receptors are produced by varying degrees of intensity of the light that falls on them, this light cannot be seen by the scanner any more than it can by the human eye (recall Einstein's observation that light, which moves, cannot be 'followed' by an observer). The supposition that persons can see the light that falls their eyes' retinas, which is related to the supposition that they see the retinas themselves, is, in Berkeley's phrase, part of "the humour of seeing by geometry" (par. 53 of the *New Theory of Vision*).

by direct visual inspection that 'x' and 'y' really label separate marks, since that depends on how the marks in question were formed. If they were drawn separately then they are separate if the first one was not 'encountered' while the second one was being drawn. But inspection 'after the fact' cannot determine this, and certainly the seeming appearance of a faint white trace between the halves labelled 'x' and 'y' does not settle the matter.[21] And, whether or not Marr's scanning and processing would pick up this trace and make it manifest in the resulting *image*, this would no more prove that it was continuous than the fact that the image of the dot-array was a continuous rectangular outline would prove that the *it* was continuous. But should we be so hasty in saying that Marr's cues to connectivity are *only* cues that may be 'reasonably reliable', although they may be incorrect when they are applied to things like the streak above?

In fact, the only empirical criterion we have for the connectedness of a surface feature is its apparent continuity, either 'at present' or in its history. Marr necessarily concentrates on the present, but neither can be established with certainty in borderline cases like that discussed above. And even if there were no vagueness in the history of a feature, the fact that very often human beings can only subject it to immediate visual or tactile inspection gives the analysis of the cues that such inspection can pick up, like the cues to a person's age, great practical importance. In fact one may wonder, á la Berkeley, why one should not forget about questions like "What *is* continuity?" and just concentrate on the empirical cues for it? Here is the author's very tentative answer.

While Marr's analysis of cues to continuity and boundaries has the greatest practical utility, these sets of cues could seem to be unrelated. It is as though we had one cue for being hot, a burning sensation to the touch, and another for being cold, a freezing sensation, but there was no connection between them. What theory does is to provide these connections—to unify a subject under a simple set of general principles.

[21] Readers familiar with Aristotle will recognize that the real question here ought to be "What is the streak?" This is a matter of feature-identity, or feature-unity, which is also discussed in section 2.2. However, here we are dealing with surface features, which are not among The Master's categories, and therefore his 'criteria' of unity (cf. *Metaphysics* Δ, 6) do not apply to them. Thus, he says "Things are said to constitute a unit ... because of their continuity: a bundle is *unified by a binder, and pieces of wood by glue*" and "Things are continuous if they move together and cannot do otherwise" (Hope, 1960, p. 95, author's italics). In other words, in addition to continuity, Aristotle adds that units should be *unified*—they must be held together by a 'binder' like glue that insures that they will 'move as a unit'. But this clearly doesn't apply to surface features, which if they move at all only do so as features of bodies that move.

The theory of heat with its temperature measure does this in the case of temperature. And topological theory does this in the case of concepts like connectivity, boundaries, dimension, etc. This will be returned to in section 18.2, and we will end here just by noting that the present work aims to augment topological theory with empirical criteria for applying it to figures on the surfaces of bodies.

18

Objectives, Theses, and Objections

18.1 Summary of Aims and Claims of This Essay

The primary aim of this essay has been to make progress in the study of applied geometry, focussing especially on topological ideas that are arguably presupposed by that science. More specifically, we aim to apply concepts and results of mathematical topology, like those in the part of it called dimension theory, to concrete, observable 'things' like the surfaces of material bodies in a way that makes it possible to give precise answers to questions such as about their dimensionality. The most important steps are taken in Parts II and III, the first of which takes its cue from Euclid's geometry by systematically developing an applied topological theory of surfaces of material bodies—solids—which, when they are flat, are parts of Euclidean planes. Part III sketches informally the extension of this theory to 'frames' that can be constructed by superimposing the bodies whose surfaces studied in Part II. Part I discusses central concepts presupposed in the theories developed in Parts II and III, and Part IV comments unsystematically on peripheral issues.

Somewhat contentiously, we have taken publicly observable but arguably two-dimensional and nonmaterial *surface features* like the small triangle '△' to be the fundamental concrete observables whose properties and relations are to be described in terms of our surface topologies. This approach is inspired by an interpretation of many propositions in the planar books of Euclid's *Elements*, which can be interpreted as giving *recipes for drawing diagrams* that are themselves compounded of surface features.

One important aspect of this interpretation is its *constructive* character, i.e., the recipes instruct their 'appliers' in drawing or constructing diagrams with specified properties – say equilateral triangles, as in Proposition 1 of Book I. A particularly important aspect of this approach is that it allows *abstract properties* of diagrams produced according to its recipes to be characterized by reference to what can or what cannot be constructed relating to them. E.g., parallelism of straight lines is defined by the requirement that it is not possible to extend them in such a way as to intersect. A particularly important property characterized in this way is that of any lines, straight or not, being *coincident*, i.e., it concerns the difference between the lines that form the triangle above *not* being coincident, and those that form the star, '✳', *being* coincident, and meeting in a point.

Answering the above question leads in Chapter 6 to a theory of *abstract points* in the topological spaces of surfaces, conceived of as what *coincident surface features have in common.* This is a theory of a relation between the concrete—the surface feature—and the abstract—the point, and based on it we are able to demonstrate: (1) that the surface is a 'container space' in which abstract points are 'everywhere' and every surface feature contains at least one of them; (2) there may be 'empty places' in the space where there are no perceivable surface features; (3) two or more surface features are coincident in the sense of satisfying the constructive test for coincidence if and only if they have at least one abstract point in common; and (4), no surface feature 'individuates' one and only one abstract point—although the smaller the feature the fewer points it contains.

The most important step towards topology 'proper' is taken in Chapter 7, where a *basic open set* of abstract points in a surface space is characterized as a complement of a set of abstract points that are contained in some finite set of constructible surface features. The system of all these basic open sets is then shown to generate a topological space in the abstract mathematical sense, which, given certain testable conditions on the surface features that generate it, also has important abstract properties, the possession of which allows standard topological theorems to be applied to these features and the surfaces on which they lie. Perhaps the most intuitively interesting application is the formulation in Section 9.3 of an 'operational test' that may be carried out on surface features to determine the dimensionality of the surface they lie on. The special case of dimension 2 is considered in detail, where it is made plausible that both surface features and the surfaces they lie on are two-dimensional.

Other surface properties are also considered and open problems are stated in Part II, perhaps the most interesting being that of formulating operational tests for a surface to be topologically equivalent ('homeomorphic') to a Euclidean plane or a portion thereof.

Part III takes informal steps towards extending the theory of bodies' surfaces developed in Part II to the *spaces* that are defined by rigid frameworks that may be formed when the bodies are fitted together in various ways. As a preliminary, Chapters 11 and 12 analyze from the topological point of view the ideas of fitting together or superimposing the surfaces of bodies that may not be rigid. One result is to resolve certain classical perplexities concerning physical superposition, which seem to entail that bodies that are fitted together would have to 'fuse' into a single body because their surfaces would no longer be distinct.

Beyond the perplexities of superposition, Chapters 14 and 15 take some steps towards solid geometry 'proper', first by analyzing the idea of *rigidity* (Chapter 14), and then that of a *rigid frame* that is formed by fitting rigid bodies together. Rigid motions, length, and distance measurement are considered (Sections 14.3 and 14.4), ultimately leading to an outline of key ideas of Leibnizian *relative spaces* (Chapter 15). However this part of the discussion is sketchy and informal, and aims only to make it plausible that a rigorous theory of superposition, rigidity, and spatial mensuration could be based on the fundamental topological theory developed in Part II.

But now we will comment on six major objections that might be raised against these theories. They are probably obvious, but they deserve consideration for this very reason. Other objections will doubtless have occurred to the reader, but as the discussions of these six objections will be fairly extended, discussing further ones here would draw out to unbearable length a work that is already overly long.

18.2 Objections Formulated and Discussed

Three of the following objections have to do with the general character of the theory developed in this essay, especially in Parts II and III, and three have to do with special features of the theory of surfaces in Part II. Each objection will be followed immediately by a discussion, prior to considering later objections. None of the discussions will resolve the matters at issue decisively.

Objection 1. Modern geometers do not regard Geometry as a system of 'recipes' for producing concrete entities that are accessible to sense observation.

Plato vehemently rejected this, and the allegedly platonistic Euclid should have been careful to avoid the 'technological' appearance of his work. In any event, modern geometers like Hilbert, Tarski, and many others have followed Plato and rigorously excluded such an appearance.

We would reply to this by pointing to the distinction drawn in Chapter 4 between the 'concrete constructible' and the 'abstract existent', and the principles of abstraction, which are simultaneously principles of application, that lead from one to the other. It may be argued that modern axiomatic geometry is non-empirical because it is concerned exclusively with the abstract side of the subject, and totally ignores its applications to the physical world. This is analogous to abstract number theory, perhaps based on Peano's Axioms, which ignores the connection between numbers and counting, and which could give the impression that we know the properties of numbers through pure intuition.

Objection 2. The use of superposition methods to measure objects and their distances apart is limited to middle sized objects and distances, and does not extend to the astronomical and submicroscopic objects that are of greatest concern to modern science.[1]

This is obvious, and it is related to the fact that the objects that modern physics is concerned with do not have clearly defined surfaces. Thus, it would be absurd to try to determine the dimensions of a star or an electron by fitting a measuring rod up against its surface.

We would also point out that classical methods of measurement are commonly restricted to classically familiar objects, or objects observed in familiar circumstances, and they must be supplemented by unfamiliar methods in application to unfamiliar objects or circumstances. However, this does not mean that familiar methods such as involve superposition are not worthy of study.

Objection 3. If properties and relations among bodies' surface features are the foundation of geometrical concepts, and geometrical concepts are fundamental to classical physical theories, how is it that surface features are not objects of those theories? Otherwise put, how can these theories not apply to the very entities that are not only most directly accessible to observation, but are most fundamental to their geometrical frameworks?

Geometrical concepts are involved in classical physical theories like Newtonian Mechanics, but that is a theory of the motions of three-dimensional material bodies, and these are clearly not two-dimensional and non-material, as we have supposed surface features like our triangle to be. On the other hand, *applying* physical theories generally requires making measurements, and if measuring rods are used for this purpose then marks on them must be observed, although they are not among the entities ranged over by the variables of the theories.[2]

[1]This is closely related to the objection that our theory is not concerned with special problems relating to the spatial and spatio-temporal frameworks presupposed by modern physical theories, which was stressed from the start.

[2]This is easily confirmed by perusing physics textbooks, which are devoted almost exclusively to theory and typically mention measurement only cursorily and in the first four or five pages. Basic theory comes first even in the engineering curriculum, and application is 'mere engineering' even to philosophers.

Note that it is seldom the case that scientific theories refer explicitly to all of

As to why physical theories should not refer explicitly to the methods of measurement employed in applying them, we have already pointed out that classical methods of measurement are commonly of restricted applicability, and they must be supplemented by other methods in application to unfamiliar objects or circumstances. Hence no one method is *the* method of measurement that 'coordinates the variables of a theory to the world'.

Objection 4. If scientific theories typically do not have unique procedures of application that coordinate them to the world, how can we claim to give 'coordinating definitions' relating topological concepts to empirical observables?

The main point to make about this is that while we have described operational procedures for determining, say, the dimensionality of a surface by reference to properties and relations among observable features of it, we have not pretended that they are logically necessary and sufficient, or that they function in the way that definitions in the context of a deductive theory do. We have already noted fractal concepts, in particular of dimensionality (section 9.4), and Marr's 'visual topological' ideas that include that of dimension 1.5 (section 17.7), and we have not claimed that these are *wrong*—only that our theory offers a practical test that arguably yields results that agree with our pre-theoretical intuitions for determining the dimensionality of a surface. However, it may not be amiss to expand on this, and comment on the general scientific role of tests like those described here.

Adams, 1966, and Adams and Adams, 1989 and 1991 argue that scientific concepts are introduced, modified, and refined for particular *purposes*, and they criticize the idea that these concepts must have well defined 'criteria of application' to the world of sense experience.[3] Of

---

the entities that are involved in their application, and it should not be surprising that the variables of classical physical theories do not range over surface features. Microbiological theories seldom refer to microscopes and astronomical theories seldom refer to telescopes, although these instruments are typically employed in verifying them.

[3]Other important theses are: (1) So called 'definitions' really serve as *aides memoires* that help the applied scientist to remember the metrological and other observational techniques that are characteristic of her trade, and (2) she has learned these techniques through 'hands on training' under the guidance of experts. (3) The ultimate test of mastery is simply the ability to get results that agree with whose of the experts. Finally (4), most important conceptually, the 'results' obtained by trained observers are often reported in terms that would be entirely inaccessible to observation if they were taken literally. In the case of classifying pot shards that is commented on in detail by Adams and Adams, these include the property of a shard coming from a 'wheel made pot', which the observer only in possession of the shard could not determine definitively, since she could not observe the making of the pot.

these two 'theses', the one concerning the purposes of scientific concept formation is illustrated by the Mohs Scale of hardness, which was introduced by F. Mohs in 1825, specifically for the purpose of assisting in the identification of mineral samples.

The thesis that scientific concepts need not have well defined criteria of application may be illustrated by an example from geometry itself, namely attempts by classical Greek geometers to define *straightness*. Euclid defines a straight line "... a line which lies evenly with the points on itself," (Heath, 1925, Vol. 1, p. 153), and on p. 168 Heath lists five other ancient definitions, including Heron's as "a line stretched to its utmost," and Plato's as "that line, the middle of which covers its ends." This disparateness may explain modern geometers' calling straightness an *undefined concept*, but in the author's opinion what this illustrates is the fact that the 'empirical meaning' of straightness is not logically connected with particular, precisely defined physical procedures. Thus, Heron's definition relates straightness to things, presumably like ropes, that are 'stretched to the utmost' (cf. section 15.5), while Plato's definition relates it to lines of sight, and to the fact that when such a line is viewed from one end, the other end is 'covered' and hidden from view.

What 'rightly' characterizes the empirical meaning of straightness? In the author's opinion there is no such thing, nor is there any need for it. For more than 25 centuries ordinary men and women, engineers, and other applied scientists have gotten along perfectly well using a variety of practical methods for determining straightness, and no one of these can claim priority over the others.

The lesson to be drawn from the foregoing seems clear. Physicists and other scientists may use whatever methods suit their purposes to determine whether lines are straight, surfaces are flat, and so on, and no one has the right to say that they must conform to particular *a priori* 'criteria'.[4] And if there are no clear criteria of physical identity, we should

A parallel in the present work relates to the 'genetics of diagrams', e.g., where it is said that the mark ▣ has a hole in it. But in referring to 'the mark' and 'the hole' we are imagining a particular way in which the figure was formed, and in particular that it was not formed by placing a small white dot in the middle of a previously formed square, ■ (cf. section 4.2 of this work). This 'figure-ground' distinction became especially important in Chapter 7, where it served as a basis for the non-geometrical distinction between open and closed point sets in the basic surface topology described there.

[4]Worse from the point of view of the present essay, it appears that the tests described in it for determining topological and geometrical properties are simply inapplicable to the sorts of things and phenomena that most concern modern physicists: namely the behavior of subatomic particles and of objects of astronomical size and at astronomical distances. As noted, it would be absurd to attempt to measure the diameter of a star or an electron by placing a measuring rod against itself surface.

be cautious in affirming that surfaces cannot be physical bodies because they do not satisfy criteria of physical identity—as will be remarked on again below. But things are not so simple.

Obviously, precise definitions have important parts to play *within* scientific theories. For instance, Euclid's definition of parallelism plays a part in his deductive theory (cf. the proof of Proposition 27 of Book I), but it is not as though it was never above 'extrinsic' criticism, as Heath's discussion (Heath, 1925, vol. I, pp. 190–4) makes clear. The fault of the 'criterion theorists' is to assume that the so called 'undefined terms' of deductive theories must have extrinsic 'coordinating definitions' that play the same part in a 'universal science' that formal definitions play within deductive theories. However, the fact that precise criteria of application are not required does not mean that imprecise ones do not partly control the formation of scientific concepts.

It is unthinkable that the most advanced of modern scientists should have completely outgrown the habits of thought and speech that have been ingrained in them almost from the cradle. Moreover, these habits include 'protoscientific' conceptions, especially of place and time. 'Put it in the drawer', 'He will come tomorrow', etc., are among the earliest acquired units of thought, and the spatial and temporal ideas they involve are not wholly unrelated to spatio-temporal ideas that enter into modern scientific theories. Perhaps this is most evident in the places and times that enter into the spatio-temporal frameworks of these theories, whose immateriality reflects the 'unreality'—the 'non-thinghood'—of the wheres and whens ('in the drawer', 'tomorrow') that enter into pre-theoretical thought and speech. And, there is reason to think that certain 'proto-topological' correspondences may hold as well.

Quine speaks of "limit myths" (section 51 of *Word and Object*, 1960), and this in combination with Euclid's "An extremity of a solid is a surface" (Definition 2 of Book XI, Heath, 1925, Vol. 3, p. 260) seems to suggest that surfaces aren't ordinarily conceived to be 'real things'. This corresponds to the already noted facts that Physics treats surfaces as dependent things, as surfaces *of* bodies, and only mentions surface features as 'epistemic entities'—which illustrates our observation that physical *theories* have little concern with the techniques employed in gathering the data used to test them.[5]

As said, our theory aims to show how concepts of mathematical topology may be applied to everyday things, not those that are of central concern to modern physical science. This does not mean that it mischaracterizes everyday applications of topological concepts to everyday objects.

[5]And it suggests an explanation as to why philosophers of physics tend to ignore these techniques.

So even though surface features are not objects of physical theory, we have explained why they are important to its 'epistemology' although neither the man in street nor the physicist regards them or the surfaces they are on as material objects.

But, again, this must be examined critically.

Objection 5. How sure are we that surface features like the triangle, '⧍', are two-dimensional and nonmaterial?

In partial response to this, there are two reasons for denying that a surface feature is necessarily material: (1) it is not necessarily composed of matter (it might be in Braille or incised in stone), and (2) its 'genetic' criteria of unity and identity (cf. section 2.3) are not the same as those of material objects. Both of these will be returned to in Objection 6, when we focus specifically on *printed* features like the triangle, but even so it doesn't follow that features like these are two-dimensional. The following argument does not prove this, but it gives a reason why they might be *conceived* to be two-dimensional.

Some but only some aspects of diagrammatic representations actually *resemble* what they represent. Thus, Diagram 18.1 represents a cube whose front face is a square, and that is *represented*

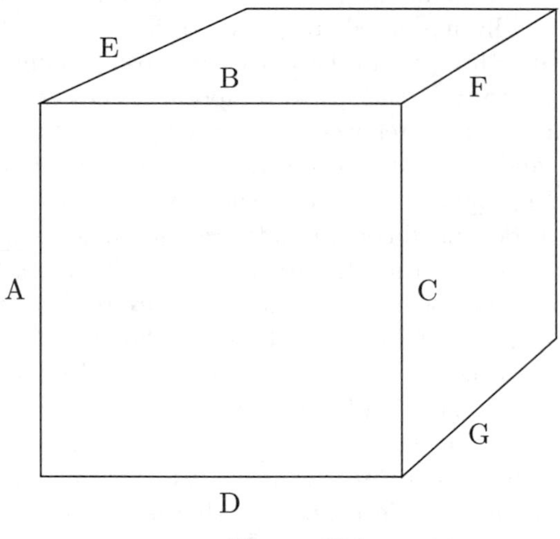

Figure 18.1

by square ABCD. On the other hand, while the lines E, F, and G, that slant backwards from the corners of the square, represent lines, the lines they represent do not slant, and the angle, say between lines B and E, does not resemble the angle between the lines that they represent. In short, the part of the diagram that represents a 'flat' two-dimensional aspect of the cube also *resembles* that part of the cube, but the rest of the diagram, while still representing aspects of the cube, does not resemble those aspects. This suggests that it is only as a two-dimensional object that the diagram resembles what it represents, and this might explain why it is conceived of *as* two-dimensional. Of course this allows that the *qua* 'thing', independent of what it might represent, the diagram might actually *be* three-dimensional.

This will be returned to in the discussion of our last, most serious, and possibly devastating objection, which reverts back to the discussion of connectivity in section 17.8.

Objection 6. We know a priori that the hypothesis that surface features are connected (the Connectivity hypothesis, section 7.5) is false.

Let us concentrate on visual surface features like the bar, '█████', commented on in section 7.5. This looks continuous, i.e., topologically connected, since one sees no 'gap' that would prove that it was discontinuous, such as appears in '██ ██'. But we know *a priori* that if the apparently continuous bar were examined under a microscope it would be revealed as discontinuous, e.g., a collection of ink particles. Even if microscopic examination did not reveal this, physical and physiological theories tell us that the appearance of continuity arises from the way that light is reflected from the surface of the page, and the effect of this light on the retinas of our eyes. The page itself and its surface are really only collections of submicroscopic particles, scattered at wide distances apart in space. Thus, the appearance of continuity not only fails to correspond to reality, it is almost totally at variance with it.

This is an extremely serious objection, given that the continuity of surface features is fundamental to practically all of the theory of surface spaces and topologies developed in Part II. For instance, the account of boundary relations developed section 8.2 is based on the idea of a 'boundary cover', and that depends crucially on continuity (connectivity), since a boundary cover of an object is defined as any connected surface feature that touches but is not covered by the object. Moreover, the theory of surface dimensionality developed in Chapter 9 depends crucially on the idea of a boundary, and everything falls to ground if the fundamental idea of connectivity comes in question.

At an even more fundamental level, the continuous appearance of surface features is presupposed in the conditions for their unity, identity, and incidence that are given in section 2.3, which depend on being able to refer to them ostensively, in a way that does not presuppose an analysis of their 'accidental' properties. Thus, we need to be able to say that while the figure '█████' is *one* surface feature, '█ ████' comprises *two*, without characterizing them geometrically, as longer or shorter.[6] And, labeling the bars ostensively by writing letters ‘a’ and ‘b’ over their ends, thus,

it should be clear that the letters label one thing in the first case and two in the second. Nor is this a distinction without a difference, since, as pointed out in section 9.4, if the gaps between the particles forming the figures were densely distributed, then they would be topologically at most one-dimensional. If, as Physics seems to imply, this should be true of all material things it would follow that although the connectivity hypothesis might be ‘true to the appearances’, it would be ‘totally false to the facts’.[7] Our final, highly speculative remarks bear on where this leaves us.

First, whatever one may say the bar ‘ ’ *really* is, it *appears* as one connected ‘thing’. Second, this appearance seems to be required for direct, ostensive reference to it to be possible. Third, this kind of reference is involved in the ‘Protokolsätze’ that report observational data, e.g. that two figures touch or are separate. Fourth, whether or not allegedly empirical science is *about* appearances, it depends on the gathering of data that necessarily *involve* appearances—how things appear to the naked eye, through a telescope, a microscope, or whatever. Fifth, and highly contentiously, our theories of surface topology and superposition are at least approximately ‘true of appearances to the naked eye’—though these are not all of the appearances. *Ipso facto*, these theories have empirical ‘content’, even if they are not what theorists of the ‘geometry is a branch of physics’ school might hold them to be. But much is left unexplained, and two points may be made in closing.

If Geometry partakes of the character of our theories, and it is only a theory of appearances of one or another kind, how can it form a ‘framework’ for physical theories that purport to describe reality? Is that reality itself partly one of appearance—just as it is a ‘fact’ that the bar ‘ ’

[6]Perhaps we may be aided in this by recalling the genetic characterization of feature identity set forth in section 2.3. We might say that the continuity in the appearance the bar, ‘ ’, consists in its looking as though it was drawn with a single continuous stroke. But this highly tendentious speculation will be put aside here.

[7]Mereology might be held to be the science that is ‘true to the facts’, since it is a theory of parts and particles are little parts. But the subatomic particles that are held by modern science to constitute matter hardly seem to be suitable to mereotopological theory, since they do not have well defined boundaries of the kind that that theory aims to characterize (cf. Chapter 5 of Casati and Varzi, 1999).

has a continuous appearance to the naked eye? Of course we have not answered these questions, which are as old as the hills, but it is hoped that this essay has brought some novel considerations to bear on them.

References

Adams, E.W. 1959. The axiomatic formulation of Rigid Body Mechanics. In *Symposium on the Axiomatic Method*, ed. L. Henkin, P. Suppes and A. Tarski, 250–265. Amsterdam: North-Holland Publishing Company.

———. 1961. The Empirical Foundations of Elementary Geometry. In *Current Issues in the Philosophy of Science*, ed. H. Feigl and G. Maxwell, 197–226. Holt, Rinehart and Winston.

———. 1966. On the nature and purpose of Measurement. *Synthese* 16: 225–269.

———. 1973. The Naive Conception of the Topology of the Surface of a Body. In *Space, Time and Geometry*, ed. P. Suppes, 402–424. Dordrecht, Holland: D. Reidel Publishing Company.

———. 1978. Two Aspects of Physical Identity. *Philosophical Studies* 34: 111–134.

———. 1984. On the Superficial. *Pacific Philosophical Quarterly* 65: 386–407.

———. 1986. On the Dimensionality of Surfaces, Solids and Spaces. *Erkenntnis* 24: 137–201.

———. 1988. A Note on Solidity. *Australasian Journal of Philosophy* Vol. 66, No. 4: 512–516.

———. 1993. Classical Physical Abstraction. *Erkenntnis* Vol. 18, No. 2: 145–167.

———. 1994. On the Method of Superposition. *British Journal for the Philosophy of Science* 45: 693–708.

———. 1996. Topology, Empiricism, and Operationalism, *The Monist* Vol. 79, No. 1: 1–20.

———. 1999. Idealization in applied first-order logic. *Synthese* 117: 331–354.

Adams, E.W., and W.Y. Adams. 1987. Purpose and scientific concept formation. *British Journal for the Philosophy of Science* 38: 419–440.

Adams, E.W., and I.F. Carlstrom. 1979. Representing Approximate Ordering and Equivalence Relations. *Journal of Mathematical Psychology*, Vol. 19 No. 2: 182–207.

Adams, J.F. and E.W. Adams. 2000. The geometry of voting cycles. *The Journal of Theoretical Politics* Vol. 12, No. 2, 131–153.

Adams, W.Y. and E.W. Adams. 1991. *Archaeological Typology and Practical Reality*. New York: Cambridge University Press.

Adams, W.Y., R.F. Fagot, and R.E. Robinson. 1970. On the Empirical Status of Axioms in Theories of Fundamental Measurement. *Journal of Mathematical Psychology* Vol. 7, No. 3:379–409.

Allison, H.E. 1983. *Kant's Transcendental Idealism*. New York: Yale University Press.

Arnold, B.H. 1962. *Intuitive Concepts of Elementary Topology*. Englewood Cliffs, New Jersey: Prentice-Hall.

Ayers, M.R. 1975. *Berkeley, Philosophical Works, Including the Works on Vision*. London: Dent and Sons, Ltd..

Barrow, I. 1916. *The Geometrical Lectures of Isaac Barrow*. Edition of J.M. Child. Chicago: The Open Court Publishing Company.

Bell, E.T. 1937. *Men of Mathematics*. New York: Simon and Schuster.

Block, N., ed. 1982. *Imagery*. Cambridge, Massachusetts: MIT Press.

Bourbaki, N. 1966. *General Topology*, Part I. Reading, Massachusetts: Addison-Wesley Publishing Company.

Borsuk, K., and W. Smielew. 1960. *Foundations of Geometry*, revised English translation. Amsterdam: North-Holland Publishing Company.

Bridgman, P.W. 1927. *The Logic of Modern Physics*. New York: The Macmillan Company.

Briggs, W., and G.H. Bryan. 1903. *Tutorial Dynamics*. London: University Tutorial Press, Ltd.

Bröcker, T., and K. Jänich. 1973. *Einführung in die Differentialtopologie*. Berlin, New York: Springer Verlag.

Cantor, G. 1915. *Contributions to the Founding of the Theory of Transfinite Numbers*, translated by P.E.B. Jourdain. Chicago: The Open Court Publishing Company.

Carmichael, C., ed. 1994. *Kent's Mechanical Engineer's Handbook*, 12th edition. New York: John Wiley & Sons.

Casati, R., and A.C. Varzi. 1995. *Holes*. Cambridge, Massachusetts: MIT Press.

Cohen, R.S. and Y. Elkana. 1977. *Hermann von Helmholtz, Epistemological Writings*. Dordrecht, Holland: D. Reidel Publishing Co.

Courant, R., and H. Robbins. 1941. *What is Mathematics?* London: Oxford University Press.

Crary, J. 1990. *Techniques of the Observer*. Cambridge, Massachusetts: MIT Press.

Descartes, R. 1637. *La Géometrie*, Olscamp translation, 1965. Indianapolis: Bobbs-Merrill Co. Inc.

Dickson, L.E. 1911. Constructions with ruler and compasses. In *Monographs on Topics of Modern Mathematics*, ed. J.W.A. Young, 353–388. Dover edition, 1955.

Dodgson, C. (Lewis Carroll). 1879. *Euclid and his Modern Rivals*. London: Macmillan and Co. Dover edition, 1973.

Donagan, A. 1970. Universals and metaphysical realism. In *Universals and Particulars*, ed. M.J. Loux, 129–158. Garden City, New York: Anchor Books, Doubleday & Co. Inc.

Eenigenburg, P. 1997. A Note on ratio of arc length to chordal length. *College Mathematics Journal*, Vol. 28, No. 5: 391–394.

Ellington, J.W. 1985. *Immanuel Kant Philosophy of Material Nature*. Indianapolis, Indiana: Hackett Publishing Company.

Firby, P.A. and C.F. Gardiner. 1992. *Surface Topology*. Chichester: Ellis Horwood, Ltd..

Franz, W. 1965. *Topologie*, Vol. I. Berlin: Walter De Gruyter & Co.

Frege, G. 1950. *The Foundations of Arithmetic*. English translation by J.L. Austin. Oxford: Basil Blackwell.

Goodman, N. 1951. *The Structure of Appearance*. Cambridge, Massachusetts: Harvard University Press.

Gramain, A. 1971. *Topologie des surfaces*. Paris: Presses Universitaires de France.

Graustein, W.G. 1930. *Higher Geometry*. New York: The Macmillan Company.

Grunbaum, A. 1963. *Philosophical Problems of Space and Time*. New York: Alfred A. Knopf.

———. 1968. *Geometry and Chronometry in Philosophical Perspective*. Minneapolis: University of Minnesota Press.

Hausdorff, F. 1914. *Grundzüge der Mengenlehre*, Chelsea edition, 1949. New York: Chelsea Publishing Co.

Heath, T.H. 1897. *The Works of Archimedes*. London: Cambridge University Press.

———. 1925. *The Thirteen Books of Euclid's Elements*. London: Cambridge University Press. Dover edition, 1956.

Helmholtz, H. v. 1868. The facts underlying geometry, Lowe translation. In *Hermann von Helmholtz, Epistemological Writings*, ed. R.S. Cohen and Y. Elkana, 1977, 39–58. Dordrecht, Holland: D. Reidel Publishing Co.

Hilbert, D. 1902. *The Foundations of Geometry*. Chicago: The Open Court Publishing Company.

———. 1971. *The Foundations of Geometry*, 7th edition, French translation.

Hirsch, E. 1992 *The Concept of Identity*. Oxford: Oxford University Press.

Hume, D. 1888. *A Treatise of Human Nature*. Silby-Bigge edition. Oxford: Clarendon Press.

———. *An Enquiry Concerning Human Understanding*, Open Court edition, 1952. Chicago: The Open Court Publishing Company.

Hurewicz, W., and H. Wallman. 1948. *Dimension Theory*, revised edition. Princeton: Princeton University Press.

Janiszewski . 1912. *Journal de l'Ecole Polytechnique*, 76–170.

Jowett, B. 1872. *The Dialogues of Plato*. New York: Charles Scribner and Company.

Kant, I. 1786. *Metaphysical Foundations of Natural Science*.

Kelley, J. 1955. *General Topology*. New York: Springer Verlag.

Kirk, G.S. 1951. Natural change in Heraclitus. *Mind* 60: 35–42.

Knorr, W.R. 1986. *The Ancient Tradition of Geometric Problems*. New York: Dover Publications, Inc..

Krantz, D., R.D. Luce, P. Suppes, and A. Tversky. 1971. *The Foundations of Measurement*, Vol. I. New York: Academic Press.

Kuhn, T. 1962. *The Structure of Scientific Revolutions*. Chicago: University of Chicago Press.

Lakatos, I. 1976. *Proofs and Refutations*, J. Worrall and E. Zahar, eds. Cambridge: Cambridge University Press.

Legendre, A.M. 1848. *Éléments de Géométrie*, 2nd. edition. Frères, Paris: Librairie de Fermin Didot.

Mandelbrot, B. 1977. *The Fractal Geometry of Nature*. New York: W.H. Freeman and Company.

Manning, H. P. 1910. *The Fourth Dimension Simply Explained*. London: Constable and Company, Ltd. Dover edition, 1960.

Marr, D 1982, *Vision: a Computational Investigation into the Human Representation and Processing of Visual Information*. San Francisco: W.H. Freeman.

Mates, B. 1996. *The Skeptic Way: Sextus Empiricus's Outlines of Pyrrhonism*. Oxford: Oxford University Press.

McKeon, R. 1941. *The Basic Works of Aristotle.* New York: Random House.

McKinsey, J.C.C., A.W. Sugar, and P. Suppes. 1953. The axiomatic foundations of classical particle mechanics. *Journal of Rational Mechanics*, 2: 253–272.

Morrow, G. 1970. *Proclus's Commentary on the First Book of Euclid's Elements*, translated and with introduction and notes by G.R. Morrow, Princeton: Princeton University Press.

Mourelatos, A.P.D. 1974. *The Pre-Socratics, a Collection of Critical Essays.* New York: Anchor Press, Doubleday.

Munkres, J.R. 1966. *Elementary Differential Topology*, revised edition. Princeton: Princeton University Press.

Newman, M.H.A. 1939. *Elements of the Topology of Plane Sets of Points.* Cambridge: Cambridge University Press. Dover edition (unaltered republication), 1992.

Ore, O. 1963. *Graphs and Their Uses.* New York: Random House.

Poincaré, H. 1905. Non-euclidean geometries. In *Science and Hypothesis*, 35–50. Dover edition, 1952.

———. 1905. Space and Geometry. In *Science and Hypothesis*, 51–71. Dover edition, 1952.

———. 1913. Why Space has Three Dimensions, J.W. Bolduc, translator. In *Last Essays*, 25–44. Dover edition, 1963.

Pontryagin, L.S. 1952. *Foundations of Combinatorial Topology*, translation from first Russian edition, 1947, by F. Bagemihl, H. Komm, and W. Seidel. Rochester, New York: Graylock Press.

Price, H.H. 1932. *Perception.* London: Methuen & Co., Ltd.

Quine, W.V. 1960. *Word and Object.* Cambridge, Massachusetts: MIT Press.

Reichenbach, H. 1927. *The Philosophy of Space and Time*, English translation by Maria Reichenbach, 1958. New York: Dover Publishing Company.

Reid, T. 1764, *An Inquiry into the Human Mind*, edited and with introduction by Timothy Duggan, 1970. Chicago: University of Chicago Press.

Routh, E.J. 1860. *Advanced Dynamics of a System of Rigid Bodies*, 6th edition. Dover edition 1955.

Russell, B.A.W. 1900. *A Critical Exposition of the Philosophy of Leibniz.* London: George Allen & Unwin.

———. 1902. *The Principles of Mathematics*, 2nd Edition, p. 405. New York: W.W. Norton & Company.

———. 1918. The relation of sense data to Physics. In *Mysticism and Logic.* London: Penguin Books.

———. 1919. *Introduction to Mathematical Philosophy.* London: George Allen & Unwin.

———. 1948. *Human Knowledge Its Scope and Limits.* New York: Simon and Schuster.

Steen, L.A. and J.A. Zeebach. 1970. *Counterexamples in Topology.* New York: Holt, Rinehart and Winston, Inc.

Strawson, P. 1959. *Individuals.* London: Methuen and Co. Ltd.

———. 1966. *The Bounds of Sense.* London: Methuen and Co. Ltd.

Stroll, A. 1988. *Surfaces.* Minneapolis: University of Minnesota Press.

Suppes, P., D. Krantz, R.D. Luce, and A. Tversky. 1989. *The Foundations of Measurement*, Vol. III. New York: Academic Press.

Suppes, P. 1999. A Physical Model of the Brain's Computation of Truth. *Eleventh Annual Alfred Tarski Lecture*, Berkeley (March 19, 1999).

Truesdell, C. 197., *A First Course in Rational Continuum Mechanics.* Vol. 1. New York: Academic Press Inc.

Twain, M. *Tom Sawyer Abroad.*

Van der Waerden, B.L. 1963. *Science Awakening*, English translation by A. Dresden, Science Editions. New York: John Wiley & Sons.

Victor, S.K. 1979. *Practical Geometry in the High Middle Ages.* Philadelphia: The American Philosophical Society.

Wedberg, A. 1955. *Plato's Philosophy of Mathematics.* Stockholm: Almquist and Wiksell.

Wiener, P.P. 1951. *Leibniz Selections.* New York: Charles Scribner's Sons.

Weyl, H. 1949. *Philosophy of Mathematics and Natural Science*, English translation by O. Helmer. Princeton: Princeton University Press.

Whitehead, A.N. 1929. *Process and Reality.* New York: The Humanities Press.

Wiggins, D. 1967. *Identity and Spatio-Temporal Continuity.* Oxford: Blackwell.

Wittgenstein, L. 1922. *Tractatus Logico-Philosophicus.* New York: The Humanities Press.

Zimmerman, D. 1996. Indivisible parts and extended objects; some philosophical episodes from topology's prehistory. *The Monist* 79: 148–180.

Index